Veterinary Epidemiology

Veterinary Epidemiology

Edited by:

Patricia Marques

Delve
PUBLISHING

www.delvepublishing.com

Veterinary Epidemiology

Patricia Marques

Delve Publishing
224 Shoreacres Road
Burlington, ON L7L 2H2
Canada
www.delvepublishing.com
Email: orders@arclereducation.com

ISBN: 978-1-77469-173-1 (Hardcover)

DECLARATION

ABOUT THE EDITOR

Patricia obtained her PhD from University College Dublin in 2010. Her interests are on Microbiology and Parasitology. She is currently working as a Postdoc at University of Maryland Baltimore, USA on Chlamydia infections.

TABLE OF CONTENTS

Section 2: Virus

Section 3: Rabies

Section 4: Bacteria

LIST OF CONTRIBUTORS

Mathilde Fre´rot
Department of Hospital Epidemiology and Infection Control, Dijon University Hospital, Dijon, France,

Annick Lefebvre
Department of Hospital Epidemiology and Infection Control, Dijon University Hospital, Dijon, France,
Department of Hospital Epidemiology and Infection Control, Reims University Hospital, Reims, France,

Simon Aho
Department of Medical Oncology, Lorraine Institute of Oncology, Nancy, France,

Patrick Callier
Department of human genetics, Dijon University Hospital, Dijon, France

Karine Astruc
Department of Hospital Epidemiology and Infection Control, Dijon University Hospital, Dijon, France,

Ludwig Serge Aho Gle´le´ ID
Department of Hospital Epidemiology and Infection Control, Dijon University Hospital, Dijon, France,

Edmund J Peeler
Centre for Environment, Fisheries and Aquaculture Science (Cefas), Barrack Road, Weymouth, Dorset, DT4 8UB, UK

Nicholas GH Taylor
Centre for Environment, Fisheries and Aquaculture Science (Cefas), Barrack Road, Weymouth, Dorset, DT4 8UB, UK

Lucie Collineau
CIRAD, AGIRs (Animal and Integrated Risks Management Unit, UPR22), Montpellier, France.

Raphaël Duboz
CIRAD, AGIRs (Animal and Integrated Risks Management Unit, UPR22), Montpellier, France.
Asian Institute of Technology, Computer Sciences and Information Management, Pathumthani, Thailand.

Mathilde Paul
UMR 1225 IHAP INRA-ENVT, National Veterinary School, Toulouse, France.

Marisa Peyre
CIRAD, AGIRs (Animal and Integrated Risks Management Unit, UPR22), Montpellier, France.

Flavie Goutard
CIRAD, AGIRs (Animal and Integrated Risks Management Unit, UPR22), Montpellier, France.
Faculty of Veterinary Medicine, Kasetsart University, Bangkok, Thailand.

Sinel Holl
National Veterinary Research Institute, Phnom Penh, Cambodia.

François Roger
CIRAD, AGIRs (Animal and Integrated Risks Management Unit, UPR22), Montpellier, France.
Faculty of Veterinary Medicine, Kasetsart University, Bangkok, Thailand.

Barend Mark de Clare Bronsvoort
The Roslin Institute and Royal (Dick) School of Veterinary Studies, University of Edinburgh, Roslin, UK.

Samuel Mwangi Thumbi
Centre for Immunology, Infection & Evolution, University of Edinburgh, Kings Buildings, Edinburgh, UK.

Elizabeth Jane Poole
ILRI, Nairobi 00100, Kenya.

Henry Kiara
ILRI, Nairobi 00100, Kenya.

Olga Tosas Auguet
Centre for Immunology, Infection & Evolution, University of Edinburgh, Kings Buildings, Edinburgh, UK.
Centre for Clinical Infection and Diagnostics Research, Department of Infectious Diseases, King's College London and Guy's and St Thomas' NHS Foundation Trust, London, UK.

Ian Graham Handel[1,]
[1]The Roslin Institute and Royal (Dick) School of Veterinary Studies, University of Edinburgh, Roslin, UK.

Amy Jennings
The Roslin Institute and Royal (Dick) School of Veterinary Studies, University of Edinburgh, Roslin, UK.

Ilana Conradie
Department of Veterinary Tropical Diseases, Faculty of Veterinary Science, University of Pretoria, Private bag X04, Onderstepoort, South Africa.

Mary Ndila Mbole-Kariuki
ILRI, Nairobi 00100, Kenya.
University of Nottingham, University Park, Nottingham, NG7 2RD, UK.

Philip G Toye
ILRI, Nairobi 00100, Kenya.

Olivier Hanotte
ILRI, Nairobi 00100, Kenya.
University of Nottingham, University Park, Nottingham, NG7 2RD, UK.

JAW Coetzer
Department of Veterinary Tropical Diseases, Faculty of Veterinary Science, University of Pretoria, Private bag X04, Onderstepoort, South Africa.

Mark EJ Woolhouse
Centre for Immunology, Infection & Evolution, University of Edinburgh, Kings Buildings, Edinburgh, UK.

Jose´ A. Barasona
SaBio IREC, National Wildlife Research Institute (CSIC-UCLM-JCCM), Ciudad Real, Spain,

Margarita Mulero-Pa´ zma´ny
Department of Evolutionary Ecology, Don˜ana Biological Station, CSIC, Seville, Spain,

Pelayo Acevedo1, Juan J. Negro
Department of Evolutionary Ecology, Don˜ana Biological Station, CSIC, Seville, Spain,

Marı´a J. Torres
Department of Microbiology, Universidad de Sevilla, Seville, Spain

Christian Gorta´ zar
SaBio IREC, National Wildlife Research Institute (CSIC-UCLM-JCCM), Ciudad Real, Spain,

Joaquı´n Vicente[1]
SaBio IREC, National Wildlife Research Institute (CSIC-UCLM-JCCM), Ciudad Real, Spain,

Michael Snow
Marine Scotland Science, 375 Victoria Road, Aberdeen, AB11 9DB, Scotland, UK

Ignacio García-Bocanegra
Departamento de Sanidad Animal. Facultad de Veterinaria, UCO, Campus Universitarios de Rabanales 14071, Córdoba, Spain.

Antonio Arenas-Montes
Departamento de Sanidad Animal. Facultad de Veterinaria, UCO, Campus Universitarios de Rabanales 14071, Córdoba, Spain.

Cristina Lorca-Oró
Centre de Recerca en Sanitat Animal (CReSA), UAB-IRTA, Campus de la Universitat Autònoma de Barcelona, 08193 Bellaterra, Barcelona, Spain.

Joan Pujols
Centre de Recerca en Sanitat Animal (CReSA), UAB-IRTA, Campus de la Universitat Autònoma de Barcelona, 08193
Bellaterra, Barcelona, Spain.

Miguel Ángel **González**
Laboratorio de Producción Animal. Consejería de Agricultura y Pesca, Sevilla, Spain.

Sebastián Napp
Centre de Recerca en Sanitat Animal (CReSA), UAB-IRTA, Campus de la Universitat Autònoma de Barcelona, 08193 Bellaterra, Barcelona, Spain.

Félix Gómez-Guillamón
Consejería de Medio Ambiente, Junta de Andalucía, 29010, Málaga, Spain.

Irene Zorrilla
Centro de Análisis y Diagnóstico de la Fauna Silvestre (CAD), Conserjería de Medio Ambiente (EGMASA), Junta de Andalucía, Málaga, Spain.

Elena San Miguel
Laboratorio Central de Veterinaria, Algete (Madrid), Spain.

Antonio Arenas
Departamento de Sanidad Animal. Facultad de Veterinaria, UCO, Campus Universitarios de Rabanales 14071, Córdoba, Spain.

Shelly Lachish
Edward Grey Institute, Department of Zoology, University of Oxford, Oxford, United Kingdom,

Becki Lawson
Institute of Zoology, Zoological Society of London, London, United Kingdom

Andrew A. Cunningham
Institute of Zoology, Zoological Society of London, London, United Kingdom

Ben C. Sheldon
Edward Grey Institute, Department of Zoology, University of Oxford, Oxford, United Kingdom,

João Rodrigo Mesquita
Departamento de Ciências Biológicas, Laboratório de Microbiologia, Universidade do Porto, Portugal, Rua Jorge Viterbo Ferreira, 228, Porto 4050-313, Portugal
Secção de Ciências Veterinárias, Instituto Politécnico de Viseu, Quinta da Alagoa - Estrada de Nelas, Ranhados, Viseu 3500-606, Portugal

Maria São José Nascimento
Departamento de Ciências Biológicas, Laboratório de Microbiologia, Universidade do Porto, Portugal, Rua Jorge Viterbo Ferreira, 228, Porto 4050-313, Portugal

Xiao Liu
Animal Biotechnology Center, College of Veterinary Medicine, Sichuan Agricultural University, Ya' an, China,

Shan Liao
Animal Biotechnology Center, College of Veterinary Medicine, Sichuan Agricultural University, Ya' an, China,

Ling Zhu
Animal Biotechnology Center, College of Veterinary Medicine, Sichuan Agricultural University, Ya' an, China,

Zhiwen Xu
Animal Biotechnology Center, College of Veterinary Medicine, Sichuan Agricultural University, Ya' an, China,
Key Laboratory of Animal Disease and Human Health, College of Veterinary Medicine, Sichuan Agricultural University, Ya' an, China

Yuancheng Zhou
Animal Biotechnology Center, College of Veterinary Medicine, Sichuan Agricultural University, Ya' an, China,
Key Laboratory of Animal Disease and Human Health, College of Veterinary Medicine, Sichuan Agricultural University, Ya' an, China

Terence P. Scott
Department of Microbiology and Plant Pathology, University of Pretoria, Pretoria, South Africa,

Melina Fischer
Friedrich-Loeffler-Institut, Greifswald-Insel Riems, Germany,

Siegfried Khaiseb
Central Veterinary Laboratories, Windhoek, Namibia

Conrad Freuling
Friedrich-Loeffler-Institut, Greifswald-Insel Riems, Germany,

Dirk Ho¨per
Friedrich-Loeffler-Institut, Greifswald-Insel Riems, Germany,

Bernd Hoffmann
Friedrich-Loeffler-Institut, Greifswald-Insel Riems, Germany,

Wanda Markotter
Department of Microbiology and Plant Pathology, University of Pretoria, Pretoria, South Africa,

Thomas Mu¨ ller
Friedrich-Loeffler-Institut, Greifswald-Insel Riems, Germany,

Louis H. Nel
Department of Microbiology and Plant Pathology, University of Pretoria, Pretoria, South Africa,

Evelyne Picard-Meyer
French Agency for Food, Environmental and Occupational Health & Safety (ANSES), Nancy Laboratory for Rabies and Wildlife, OIE Reference Laboratory for Rabies, European Union Reference Laboratory for Rabies, European Union Reference Laboratory for Rabies Serology, Malzeville, France,

Emmanuelle Robardet
French Agency for Food, Environmental and Occupational Health & Safety (ANSES), Nancy Laboratory for Rabies and Wildlife, OIE Reference Laboratory for Rabies, European Union Reference Laboratory for Rabies, European Union Reference Laboratory for Rabies Serology, Malzeville, France,

Laurent Arthur
Museum d'Histoire Naturelle de Bourges, Bourges, France,

Ge´ rald Larcher
SFEPM Chiroptera Group, Museum d'Histoire Naturelle de Bourges, Bourges, France,

Christine Harbusch
Nature and Biodiversity Conservation Union, Perl-Kesslingen, Germany

Alexandre Servat
French Agency for Food, Environmental and Occupational Health & Safety (ANSES), Nancy Laboratory for Rabies and Wildlife, OIE Reference Laboratory for Rabies, European Union Reference Laboratory for Rabies, European Union Reference Laboratory for Rabies Serology, Malzeville, France,

Florence Cliquet
French Agency for Food, Environmental and Occupational Health & Safety (ANSES), Nancy Laboratory for Rabies and Wildlife, OIE Reference Laboratory for Rabies, European Union Reference Laboratory for Rabies, European Union Reference Laboratory for Rabies Serology, Malzeville, France,

Luis E Escobar
Facultad de Ecología y Recursos Naturales, Universidad Andres Bello, Av. República 440, Santiago, Chile.
Center for Global Health and Translational Science, SUNY Upstate Medical University, Syracuse, New York, USA.

A Townsend Peterson
Biodiversity Institute, University of Kansas, Lawrence, Kansas, USA.

Monica Papeş
Department of Integrative Biology, Oklahoma State University, Stillwater, Oklahoma 74078, USA.

Myriam Favi
Sección Rabia, Instituto de Salud Publica de Chile, Av. Maraton 1000, Ñuñoa, Chile.

Veronica Yung
Sección Rabia, Instituto de Salud Publica de Chile, Av. Maraton 1000, Ñuñoa, Chile.

Olivier Restif
Disease Dynamics Unit, Department of Veterinary Medicine, University of Cambridge, Madingley Road, Cambridge CB3 0ES, UK.

Huijie Qiao
Key Laboratory of Animal Ecology and Conservation Biology, Institute of Zoology, Chinese Academy of Science, Beijing, China.

Gonzalo Medina-Vogel
Facultad de Ecología y Recursos Naturales, Universidad Andres Bello, Av. República 440, Santiago, Chile.

Tazumi A
Northern Ireland Public Health Laboratory, Department of Bacteriology, Belfast City Hospital, Belfast BT9 7AD, Northern Ireland
Laboratory of Molecular Biology, School of Environmental Health Sciences, Azabu University, Fuchinobe 1-17- 71, Sagamihara 229-8501, Japan

Maeda Y
Northern Ireland Public Health Laboratory, Department of Bacteriology, Belfast City Hospital, Belfast BT9 7AD, Northern Ireland
School of Biomedical Sciences, Centre for Molecular Biosciences, University of Ulster, Coleraine, BT52 1SA, Northern Ireland

Buckley T
Irish Equine Centre, Johnstown, Naas, Co. Kildare, Ireland
Department of Equine Studies, University of Limerick, Ireland

Millar BC
Northern Ireland Public Health Laboratory, Department of Bacteriology, Belfast City Hospital, Belfast BT9 7AD, Northern Ireland

Goldsmith CE
Northern Ireland Public Health Laboratory, Department of Bacteriology, Belfast City Hospital, Belfast BT9 7AD, Northern Ireland

Dooley JSG
School of Biomedical Sciences, Centre for Molecular Biosciences, University of Ulster, Coleraine, BT52 1SA, Northern Ireland

Elborn JS
Regional Adult Cystic Fibrosis Unit, Belfast City Hospital, Lisburn Road, Belfast, Northern Ireland, BT9 7AB
Department of Respiratory Medicine, The Queen's University of Belfast, Respiratory Medicine, Belfast City Hospital, Lisburn Road, Belfast, Northern Ireland, BT9 7AB

Matsuda M
Laboratory of Molecular Biology, School of Environmental Health Sciences, Azabu University, Fuchinobe 1-17- 71, Sagamihara 229-8501, Japan

Moore JE
Northern Ireland Public Health Laboratory, Department of Bacteriology, Belfast City Hospital, Belfast BT9 7AD, Northern Ireland
School of Biomedical Sciences, Centre for Molecular Biosciences, University of Ulster, Coleraine, BT52 1SA, Northern Ireland

Emer O'Mahony
UCD Centre for Food Safety, School of Public Health, Physiotherapy & Population Science, UCD Veterinary Sciences Centre, University College Dublin, Belfield, Dublin, Ireland,
Veterinary Food Safety Laboratory, Cork County Council, Inniscarra, County Cork, Ireland,

James F. Buckley
Veterinary Food Safety Laboratory, Cork County Council, Inniscarra, County Cork, Ireland,

Declan Bolton
Food Safety Department, Teagasc Food Research Centre, Ashtown, Dublin, Ireland

Paul Whyte
UCD Centre for Food Safety, School of Public Health, Physiotherapy & Population Science, UCD Veterinary Sciences Centre, University College Dublin, Belfield, Dublin, Ireland,

Se´amus Fanning
UCD Centre for Food Safety, School of Public Health, Physiotherapy & Population Science, UCD Veterinary Sciences Centre, University College Dublin, Belfield, Dublin, Ireland,

Gloria Chan
Department of Population Medicine, University of Guelph, Guelph, ON N1G 2W1, Canada.

Abdolvahab Farzan
Department of Population Medicine, University of Guelph, Guelph, ON N1G 2W1, Canada.

Glenn Soltes
Department of Pathobiology, University of Guelph, Guelph, ON N1G 2W1, Canada.

Vivian M Nicholson
Department of Pathobiology, University of Guelph, Guelph, ON N1G 2W1, Canada.

Yanlong Pei
Department of Pathobiology, University of Guelph, Guelph, ON N1G 2W1, Canada.

Robert Friendship
Department of Population Medicine, University of Guelph, Guelph, ON N1G 2W1, Canada.

John F Prescott
Department of Pathobiology, University of Guelph, Guelph, ON N1G 2W1, Canada.

Moses Y Otiende
Veterinary Services Department, Forensic and Genetics Laboratory Kenya Wildlife Service, Nairobi, Kenya.

Mary W Kivata
Department of Biochemistry and Biotechnology, Kenyatta University, Nairobi, Kenya.

Joseph N Makumi
Department of Biochemistry and Biotechnology, Kenyatta University, Nairobi, Kenya.

Mathew N Mutinda
Veterinary Services Department, Forensic and Genetics Laboratory Kenya Wildlife Service, Nairobi, Kenya.

Daniel Okun
Department of Biochemistry and Biotechnology, Kenyatta University, Nairobi, Kenya.

Linus Kariuki
Veterinary Services Department, Forensic and Genetics Laboratory Kenya Wildlife Service, Nairobi, Kenya.

Vincent Obanda
Veterinary Services Department, Forensic and Genetics Laboratory Kenya Wildlife Service, Nairobi, Kenya.

Francis Gakuya
Veterinary Services Department, Forensic and Genetics Laboratory Kenya Wildlife Service, Nairobi, Kenya.

Dominic Mijele
Veterinary Services Department, Forensic and Genetics Laboratory Kenya Wildlife Service, Nairobi, Kenya.

Ramón C Soriguer
Estación Biológica de Doñana, Consejo Superior de Investigaciones Científicas (CSIC), Avda. Américo Vespucio s/n 41092, Sevilla, Spain.

Samer Alasaad
Estación Biológica de Doñana, Consejo Superior de Investigaciones Científicas (CSIC), Avda. Américo Vespucio s/n 41092, Sevilla, Spain.
Institute of Evolutionary Biology and Environmental Studies (IEU), University of Zürich, Winterthurerstrasse 190, 8057 Zürich, Switzerland.

Daniel Nenene Qekwana
Section Veterinary Public Health, Department of Paraclinical Sciences, Faculty of Veterinary Sciences, University of Pretoria, Pretoria, South Africa,

James Wabwire Oguttu
Department of Agriculture and Animal Health, College of Agriculture and Environmental Sciences, University of South Africa, Johannesburg, South Africa,

Dries Venter
Gauteng Department of Agriculture and Rural Development, Veterinary Services Directory, Pretoria, South Africa,

Agricola Odoi
Section Veterinary Public Health, Department of Paraclinical Sciences, Faculty of Veterinary Sciences, University of Pretoria, Pretoria, South Africa,
Biomedical and Diagnostic Sciences, College of Veterinary Medicine, University of Tennessee, Knoxville, Tennessee, United States of America

Karla I. Moreno-Torres
The Ohio State University, Department of Veterinary Preventive Medicine, Columbus, Ohio, United States of America,

Laura W. Pomeroy
The Ohio State University, Department of Veterinary Preventive Medicine, Columbus, Ohio, United States of America,

Mark Moritz
The Ohio State University, Department of Anthropology, Columbus, Ohio, United States of America,

William Saville
The Ohio State University, Department of Veterinary Preventive Medicine, Columbus, Ohio, United States of America,

Barbara Wolfe
Morris Animal Foundation, Denver, Colorado, United States of America,

Rebecca Garabed
The Ohio State University, Department of Veterinary Preventive Medicine, Columbus, Ohio, United States of America,
The Ohio State University, Public Health Preparedness for Infectious Diseases Program, Columbus, Ohio, United States of America

Jennifer L Anderson
Marshfield Clinic Research Foundation, Marshfield WI 54449, USA.

Brian L Sloss
U.S. Geological Survey, Wisconsin Cooperative Fishery Research Unit, College of Natural Resources, University of Wisconsin-Stevens Point, Stevens Point WI 54481, USA.

Jennifer K Meece
Marshfield Clinic Research Foundation, Marshfield WI 54449, USA.

Hetron M Munang'andu
Department of Basic Sciences and Aquatic Medicine, Section of Aquatic Medicine and Nutrition, Norwegian School of Veterinary Sciences, Ullevålsveien 72, NO-0033 Oslo, Norway,
School of Veterinary Medicine, University of Zambia, , Lusaka, Zambia

Victor M Siamudaala
Zambia Wildlife Authority, Chilanga, Zambia

Wigganson Matandiko
Zambia Wildlife Authority, Chilanga, Zambia

Musso Munyeme
School of Veterinary Medicine, University of Zambia, , Lusaka, Zambia

Mwelwa Chembensofu
School of Veterinary Medicine, University of Zambia, , Lusaka, Zambia

Enala Mwase
School of Veterinary Medicine, University of Zambia, , Lusaka, Zambia

LIST OF ABBREVIATIONS

AEZ	Agro-ecological zones
AIC	Akaike Information Criterion
AHA	Animal heath assistants
BEAST	Bayesian Evolutionary Analysis by Sampling trees
BTV	Bluetongue virus
NoVs	Canine noroviruses
CC	Clonal complex
ENM	Ecological niche modeling
EFSA	European Food Safety Authority's
FAT	Fluorescent antibody test
FMD	Foot and mouth disease
GIS	Geographic information systems
HPR	Highly polymorphic region
HCMV	Human cytomegalovirus
IRA	Import Risk Analysis
QIC	Independence Model Criterion
IDEAL	Infectious Diseases of East African Livestock
IHNV	Infectious haematopoietic necrosis virus
IPNV	Infectious pancreatic necrosis virus
ISA	Infectious salmon anaemia
LBV	Lagos bat virus
LNNP	Lake Nakuru National Park
LRT	Likelihood ratio tests
MRS	Meru Rhino Sanctuary
MODIS	Moderate Resolution Imaging Spectroradiometer
MLST	Multi-locus sequence typing
NNP	Nairobi National Park
NgRS	Ngulia rhino Sanctuary

NHST	Null-hypothesis significance testing
PCR	Polymerase chain reaction
PCMV	Porcine cytomegalovirus
PRRSV	Porcine reproductive and respiratory syndrome virus
PFGE	Pulsed-field gel electrophoresis
SPD	Salmonid pancreas disease
ST	Sequence type
SNT	Serum neutralization test
SD	Sleeping disease
TTV	Tansfusion transmitted virus
TAGS	The absence of a gold standard
VAHWs	Village Animal Health Workers
VHS	Viral haemorrhagic septicaemia
VHSV	Viral haemorrhagic septicaemia virus
WSSV	White spot syndrome virus
WTO	World Trade Organization

PREFACE

Veterinary epidemiology is the study of the distribution and determinants of animal health-, welfare- and production- related states or events in specified populations and the application of this study to control of health problems.

Epidemiology is the study and analysis of the distribution (who, when, and where), patterns and determinants of health and disease conditions in defined populations. It is a cornerstone of public health, and shapes policy decisions and evidence-based practice by identifying risk factors for disease and targets for preventive healthcare. Epidemiologists help with study design, collection, and statistical analysis of data, amend interpretation and dissemination of results (including peer review and occasional systematic review). Epidemiology has helped develop methodology used in clinical research, public health studies, and, to a lesser extent, basic research in the biological sciences.

Epidemiology, literally meaning "the study of what is upon the people", is derived from Greek *epi* 'upon, among', *demos* 'people, district', and *logos* 'study, word, discourse', suggesting that it applies only to human populations. However, the term is widely used in studies of zoological populations (veterinary epidemiology), although the term "epizoology" is available, and it has also been applied to studies of plant populations (botanical or plant disease epidemiology).

The distinction between "epidemic" and "endemic" was first drawn by Hippocrates, to distinguish between diseases that are "visited upon" a population (epidemic) from those that "reside within" a population (endemic). The term "epidemiology" appears to have first been used to describe the study of epidemics in 1802 by the Spanish physician Villalba in *Epidemiología Española.* Epidemiologists also study the interaction of diseases in a population, a condition known as a syndemic.

This book is separated in 6 sections. The first section includes Chapters dedicated to general epidemiology and veterinary epidemiology including aquatic animals, livestock, ungulates, honey bees and genral animal health. The second section is dedicated to veterinary virus epidemiology studies. The third section is dedicated exclusively to Rabies, a zoonotic disease caused by a virus. The fourth section is dedicated to veterinary bacteria epidemiology, the fifth to Parasites and the last section, sixth to Fungus.

SECTION 1: INTRODUCTION

CHAPTER 1

What is Epidemiology ? Changing Definitions of Epidemiology 1978-2017

Mathilde Fre´rot[1], Annick Lefebvre[1,2], Simon Aho[3], Patrick Callier[4], Karine Astruc[1], Ludwig Serge Aho Gle´le´ ID[1]

[1]Department of Hospital Epidemiology and Infection Control, Dijon University Hospital, Dijon, France,

[2]Department of Hospital Epidemiology and Infection Control, Reims University Hospital, Reims, France,

[3]Department of Medical Oncology, Lorraine Institute of Oncology, Nancy, France,

[4]Department of human genetics, Dijon University Hospital, Dijon, France

ABSTRACT

Context

Epidemiology is a discipline which has evolved with the changes taking place in society and the emergence of new diseases and new discipline

Citation: Frérot M, Lefebvre A, Aho S, Callier P, Astruc K, Aho Glélé LS (2018) What is epidemiology ? Changing definitions of epidemiology 1978-2017. PLoS ONE 13(12): e0208442. https://doi.org/10.1371/journal.pone.0208442

related to epidemiology. With these evolutions, it is important to understand epidemiology and to analyse the evolution of content of definitions of epidemiology.

Objectives

The main objective of this paper was to identify new definitions of epidemiology available since 1978. Secondary objectives were to analyse the content of these definitions, to compare them with those used by Lilienfeld and to determine whether changes have taken place over the last forty years.

Methods

A review of grey literature and published literature was conducted to find the definitions of epidemiology written between 1978 and 2017.

Results

102 definitions of epidemiology were retained. They helped to highlight 20 terms and concepts related to epidemiology. Most of them were already used in the definitions used by Lilienfeld. Five terms were present in more than 50% of definitions from the period 1978 to 2017: "population", "study", "disease", "health" and "distribution". Several developments have occurred: strengthening of the terms "control" and "health" already used, the concept of "disease" was less frequently encountered whereas the concepts "infectious diseases", "mass phenomenon" are no longer used in definitions from 1978 to 2017.

Conclusion

This evolution of content of definition of epidemiology is absent from books on epidemiology. A thematic analysis of definitions of epidemiology could be conducted in order to improve our understanding of changes observed.

INTRODUCTION

Epidemiology is a recent discipline which has evolved with the changes taking place in society and the emergence of new diseases. This evolution has allowed epidemiology to remain a useful and relevant tool in bringing to light and understanding diseases and health events. Since its origins, more than a century ago, many definitions of epidemiology have been suggested.

In 1978, Lilienfeld published a seminal paper on definitions of epidemiology.

For him, there was no consensus among epidemiologists as to the definition of epidemiology. The objective of his work was to provide a single and understandable definition of epidemiology that was suitable for all types of diseases and populations. Lilienfeld based his work on 23 existing definitions of epidemiology and proposed the following definition [1]: "*Epidemiology is a method of reasoning about disease that deals with biological inference derived from observations of disease phenomena in population groups*".

The publication of this article resulted in many comments, including three letters to the Editor [2] each discussing the limits of this new definition. In his letter, Evans analyzed the content of the 23 definitions used reviewed by Lilienfeld. He listed the various terms and concepts used in these definitions and determined their frequency of use. After this work, he proposed a different definition of epidemiology [3]: "Epidemiology is the quantitative analysis of the circumstances under which disease processes, including trauma, occur in population groups, factors affecting their incidence, distribution, and the host response and use of this knowledge in prevention and control «. Since the work of Lilienfeld and Evans, new definitions of epidemiology have been proposed. During this time, many fields related to epidemiology (pharmacoepidemiology, molecular epidemiology, genetic epidemiology…) have expanded. Moreover, epidemiology is related to many disciplines such as ethics [4], philosophy [5] and epistemology [6]. These disciplines used existing definitions of epidemiology. Broadbend [5], in his work on philosophy and epidemiology, used the definition of epidemiology of Rothman and Last (p23) and insisted on the notion of comparisons of groups as did Morabia [7]. Given the evolution in fields related to epidemiology, it is important to understand epidemiology and to analyse the evolution of definitions of epidemiology.

The main objective of this paper was to identify new definitions of epidemiology available since 1978, including veterinary medicine and epidemiology subspecialties [7, 8]. Secondary objectives were to analyse the content of these definitions, to compare them with those used by Lilienfeld and to determine whether changes have taken place over the last forty years.

MATERIALS AND METHODS

Literature review

A review of grey literature and published literature was conducted to find definitions of epidemiology written between 1978 and 2017. It was conducted in English. Grey literature on the subject was retrieved using the search engine "Google scholar" with the following keywords: (*Definition* AND *Epidemiology*) OR (*definition of epidemiology*). Definitions of epidemiology included in books were retrieved via Google books (https://books.google.fr/) and Amazon (http://www.amazon.fr/), Library of Congress (https://catalog.loc.gov/), The British Library (http://explore.bl.uk/primo_library/libweb/action/search.do?vid=BLVU1#) and NLM Catalog (https://www.ncbi.nlm.nih.gov/nlmcatalog/). Among the epidemiology textbooks identified, the book "Epidemiology and the people's health: theory and context" by Nancy Krieger [9] contains a table with definitions of epidemiology from 1922 to 2007 retrieved from epidemiology textbooks. Works of Bhopal [10] and Zhang [11] were also considered as starting point. These definitions were selected as starting point for our work.

Definitions in the published literature were retrieved from "PubMed" and "ScienceDirect" using the following keywords: (*Definition* AND *Epidemiology)* OR (*Definition of epidemiology*) and the names of authors identified during the search of the grey literature.

Definitions for which the author could not be identified, those prior to 1978, those cited in Lilienfeld's article [1] or those that could be related to the definition proposed by Lilienfeld or another author were eliminated.

Definitions provided by online dictionary sites, encyclopaedias like Wikipedia or internet sites dedicated to students were also retained, using the strategy mentioned above (keywords: (Definition AND Epidemiology) OR (definition of epidemiology)).

Analysis of the Content of the New Definitions

The content of new definitions of epidemiology was analyzed to identify the terms and concepts present in these definitions. Once a term or concept was identified in at least two definitions, it was added to the list of terms.

Analysis of the evolution of definition content between 1978 and 2017

In 1979, Evans [3] analyzed the content of the definitions used by Lilienfeld to build his definition of epidemiology. In summary, he identified 23 different terms and concepts. For each term and concept, the frequency of occurrence in the definitions was calculated. Then, they were grouped into eight categories: status of the person (disease, infectious disease, physiologic conditions, injuries, health); group affected (populations, community, mass phenomena, outbreak); distribution of disease; spread (spread, propagation, dynamics); incidence, occurrence; etiology (causes, determinant factors, circumstances of occurrence, ecology); understanding disease (natural history or nature, understanding the process); prevention and control.

To study the evolution of the content of definitions of epidemiology over time, the content of the definitions used by Lilienfeld was compared with new content from the period 1978–2017.

For each term and concept identified by Evans, the frequency of appearance in the definitions from the period 1978–2014 was estimated.

For each term and concept identified using the definitions from 1978 to 2017, the frequency of appearance in the definitions used by Lilienfeld was estimated.

Then for each term and concept, whatever the period, the frequency of appearance in the definitions used by Lilienfeld and definitions from 1978 to 2017 were compared.

All statistical comparisons were performed using bayes factors [12] with R software [13] (package Bayes Factor). We have chosen bayes factor instead of p value, which is used in the context of null-hypothesis significance testing (NHST). Indeed, a non-significant p-value does not quantify evidence in favor of the null hypothesis. As stated by the American Statistical Association, "a relatively large p-value does not imply evidence in favor of the null hypothesis" [14].

RESULTS

Literature search

The search of published and grey literature revealed 102 definitions of epidemiology: 93 for human medicine, including subspecialties of epidemiology (n = 24) and 9 for veterinary medicine. We have selected 29 definitions of epidemiology from websites. The definitions of epidemiology were found in different media: articles, epidemiology online courses and excerpts from books. A total of 69 definitions from 1978 to 2017 were selected (Table 1).

Table 1. Definitions of epidemiology from the period 1978–2017.

AUTHOR (Reference)	DATE	DEFINITION
Frerichs RR et al. [33]	1978	Epidemiology is the study of the prevalence and dynamics of stages of health in populations
Barker DJP et al. [102]	1979	Epidemiology, the study of the distribution and determinants of disease in human populations, has always been an integral part of medical practice
Cole P [82]	1979	Epidemiology is the science dealing with the environmental causes of diseases of humans as inferred from observations of human beings
Rich H [38]	1979	Epidemiology is the science of the dynamics of health in populations
Evans AS [3]	1979	Epidemiology is the quantitative analysis of the circumstances under which disease processes, including trauma, occur in population groups, the factors affecting their incidence, distribution, and host responses, and the use of this knowledge in prevention and control
Lilienfeld A et al. [103]	1980	Epidemiology is concerned with the patterns of disease occurrence in human populations and the factors that influence these patterns
Kleinbaum DG et al. [27]	1982	Epidemiology is the study of health and illness in human populations
Miettinen OS [104]	1985	The discipline of epidemiology is "the aggregate of principles of studying the occurrence of illness and related states and events"
Weiss NS [105]	1986	Epidemiology per se is the study of variation in the occurrence of disease, and the reasons for that variation
Kelsey JL et al. [25]	1986	Epidemiology, the study of the occurrence and distribution of disease and other health -related conditions in populations, is used for many purposes
Shy CM [83]	1986	Epidemiology is a study of the occurrence and distribution of disease in populations and the factors that account for this distribution
Hennekens CH et al. [106]	1987	The study of the distribution and determinants of disease frequency' in human populations
Buck C [15]	1988	Besides its importance and usefulness in disease surveillance and prevention, epidemiology has an even more critical function to carry out- the gathering of knowledge for understanding the health-disease process. It can anticipate needs, identify risk conditions, and orient the definition of priorities and the use of available resources for planning and administering health systems. In short, by analyzing and evaluating health problems and health services, and their contexts, epidemiology can go beyond considering just specific health problems: it can help bring us closer to considering society as the source for explaining health problems and their solutions
Ahlbom A et al. [107]	1990	Epidemiology is the science of occurrence of diseases in human populations
Kuller LH [80]	1991	Epidemiology is the study of "epidemics" and their prevention
Vaughan P et al. [28]	1991	Epidemiology can be considered as the study of the distribution of problems related to health and disease and their determinants in human populations. The purpose of epidemiology is to collect, interpret and use information to promote health and reduce disease
Kelsey JL [108]	1996	Same definition as Kelsey, 1986
Berkman LF et al. [34]	1997	Epidemiology is the study of the distribution and determinants of states of health in populations
Savitz DA et al. [19]	1999	In defining epidemiology, some note that the principal applications of epidemiologic knowledge are to disease prevention and control, but all refer to some variant of "the study of disease", thus defining epidemiology as a science
Schoenbach V et al.[16]	1999	Well, some epidemiologists study the skin. But epidemiologists study all kinds of diseases and other aspects of health, also. Epidemiology is the study of health and disease in populations. It's a basic science of Public Heath
Schwartz S et al. [41]	1999	Epidemiology is a discipline dedicated to understanding the causes of health states in population
Vetter NJ et al. [109]	1999	Epidemiology is the study of the distribution and change in diseases. The purpose of epidemiology is to identify things in people and their surroundings that affect the occurrence of disease
Moon G et al. [74]	2000	Quoting Lilienfeld, 1980 [103]
Last JM [43]	2001	The study of the distribution and determinants of health-related states or events in specified populations, and the application of this study to control of health problems
Timmreck TC [110]	2002	Epidemiology is an investigative method used to detect the cause or source of diseases, disorders, syndromes, conditions, or perils that cause pain, injury, illness, disability, or death in human populations or groups. . . . Epidemiology has been defined in several ways. One definition is the study of the nature, cause, control, and determinants of the frequency and distribution of disease, disability, and death in human populations.
Friedman G et al.[111]	2003	Epidemiology is the study of disease occurrence in human populations. The primary units of concern are groups of persons, not separate individuals.
Gerstman BB [30]	2003	Modern definitions of epidemiology refer to distributions in populations (statistical), determinants of health and disease (pathophysiological, environmental, behavioral), control of health problems (biological, social, economic, political, administrative, legal)
Coggon D et al. [112]	2003	Epidemiology is the study of how often diseases occur in different groups of people and why. Epidemiological information is used to plan and evaluate strategies to prevent illness and as a guide to the management of patients in whom disease has already developed
Farmer RDT et al. [113]	2004	...In contrast to clinical medicine, epidemiology involves the study of groups of people (populations) rather than direct study of individuals
Friis RH et al. [29]	2004	Epidemiology is concerned with the distribution and determinants of health and diseases, morbidity, injuries, disability, and mortality in populations
Morabia A [40]	2004	The blending of population thinking and group comparisons in an integrated theory to appraise health-related causal relationships characterizes epidemiology
Rossignol A [20]	2005	Epidemiology is the foundational science of public health. Much as a yardstick measures length, epidemiologic investigations measure and compare the frequencies of disease, injury, and other health-related events in human populations
Stedman TL [39]	2005	Epidemiology is defined as the study of the distribution and determinants of health related states or events in human populations and the application of this study to the prevention and the control of health problems

Oakes JM et al. [36]	2006	Epidemiology is the study of the distribution and determinants of states of health in populations
Yarnell J [52]	2007	At the beginning of the twenty-first century, epidemiology is a broad-based population science, drawing on many disciplines from biology and sociology to biostatistics and philosophy of science, which investigates the causes of human disease and methods for their control
Boslaugh S [53]	2008	Epidemiology is the study of frequency and determinants of morbidity and mortality in populations
Bhopal RS [17]	2008	Epidemiology is the science and craft that studies the pattern of disease (and health, though usually indirectly) in populations to help understand both their causes and the burden they impose. This information is applied to prevent, control or manage the problems under study
Porta M [42]	2008	The study of the occurrence and distribution of health-related states or events in specified populations, including the study of the determinants influencing such states, and the application of this knowledge to control health problems
Gordis L [114]	2009	Epidemiology is the study of how disease is distributed in populations and the factors that influence or determine this distribution
Olsen J et al. [115]	2010	Epidemiology is defined by the object of research, "to identify determinants that change the occurrence of health phenomena in human populations."
Saracci R [21]	2010	"Epidemiology is the study of health and disease in populations". The population aspect is the distinctive trait of epidemiology, while health and disease are investigated at other levels as well.
Webb P et al. [35]	2010	Epidemiology . . . is about measuring health, identifying the causes of ill-health, and intervening to improve health. . . Perhaps epidemiology's most fundamental role is to provide a logic and structure for the analyses of health problems both great and small
Carneiro I et al. [51]	2011	Epidemiology is the study of the distribution and determinants of health states or events in specified populations, and the application of this study to control of health problems (Adapted from Porta, 2008). Quoted reference: [116]
Fos PJ [44]	2011	Epidemiology is the study of the distribution and determinants of health-related states and events in defined populations and the application of this study to the control of health problems (Adapted from Last, 2001) Quoted reference: [43] The Greek root of epidemiology and the two definitions have a common theme, namely, the people. The people are considered as a group, which is referred to as a population. This population-centered nature of epidemiology leads to one of the differences between public health services and medical services. . . . Epidemiology is the study of factors affecting the health and illness of populations.
Krickeberg K et al. [117]	2011	. . .epidemiology is about illness in populations, not in individual subjects. More precisely, it is concerned with the frequency of diseases in various parts of the population in which we are interested, the so-called "target population".
Rothman KJ [18]	2012	Often considered the core science of public health, epidemiology involves "the study of the distribution and determinants of disease frequency", or put even more simply "the study of the occurrence of illness
Ward H et al. [22]	2012	Epidemiology (is) the study of the distribution and determinants of health and illness in populations.
Aschengrau A et al. [26]	2013	The study of the distribution and determinants of disease frequency in human populations and the application of this study to control health problems"
Ahrens W et al. [45]	2013	Epidemiology is the study of the distribution and determinants of disease frequency (quoting MacMahon,1970) Epidemiology is the study of the distribution and determinants of health-related states or events in specified populations, and the application of the study to control of health problems (Quoting Last, 2001) Quoted references: [43, 69]
Broeck JV et al. [118]	2013	According to the broadest of views, epidemiology is synonym of community medicine (Miettinen 2011a, b) Quoted references: [119, 120]
Gerstman BB [121]	2013	Quoting several authors and quoting Last, 2001 [43]
Macera CA et al. [46]	2013	Epidemiology is the study of a scientific method of problem solving that helps "disease detectives" understand how people get sick and die, who gets sick and die, and how to avoid getting sick. The current definition of epidemiology is the study of the distribution (who has the problem) and determinants (things that influence the problem) of health-related conditions in human populations and the application of this method to the control of health problem.
Yarnell J et al. [122]	2013	Same definition as Yarnell, 2007
Fletcher RH et al. [123]	2014	Epidemiology is the study of disease occurrence in human populations by counting health related events in people in relation to the naturally accruing groups (populations) of which they are members
Khan AR et al. [47]	2014	The International Epidemiological Association defines epidemiology as "the study of the distribution and determinants of health related states and events in the populations and the application of this study to the control of health problems"
Katz DI et al. [124]	2014	Epidemiology is usually defined as the study of factors that determine the occurrence and distribution of disease in population
Keyes KM et al. [31]	2014	Epidemiology is the science of understanding the causes and distribution of population health so that we may intervene to prevent disease and promote health
Szklo M et al. [37] [33]	2014	Epidemiology is traditionally defined as the study of the distribution and determinants of health-related states or events in specified populations and the application of this study to the control of health problems
Weiss NS et al. [125]	2014	Epidemiology concerns describing and understanding patterns of disease occurrence in human populations, with the ultimate goal of preventing disease
Merrill RM [48]	2015	Epidemiology is the study of the distribution and determinants of health-related states or events in human population and the application of this study to the prevention and control of health problems (Stedman's dictionary. 5th edition, 2005) Quoted reference: [39]
Patten S [126]	2015	Epidemiology is the study of the distribution and determinants of disease in populations
Schneider D et al. [49]	2015	Epidemiology is "The study of the occurrence and distribution of health-related events, states, and processes in specified populations, including the study of the determinants influencing such processes, and the application of this knowledge to control relevant health problems." (Adapted from Lilienfeld, 1980).

Wassertheil-Smoller S et al. [23]	2015	Epidemiology may be defined as the study of the distribution of health and disease in groups of people and the study of the factors that influence this distribution. Modern epidemiology also encompasses the evaluation of diagnostic and therapeutic modalities and the delivery of health-care services.
Bhopal RS [50]	2016	. . .epidemiology will be defined here as the science and practice that describes and explain disease patterns in populations. It uses this knowledge to prevent and control disease, and improve health. The central idea of epidemiology is that patterns of ill health and disease in population may be analyzed systematically to understand their causes and to improve health.
Stewart A [127]	2016	Epidemiology is the study of how often diseases occur in different groups of people, and why (quoting Coggon et al., 1997). Note: same definition of epidemiology as Coggon, 2003 (5th edition) and Coggon, 1997 (4th edition)
Webb P et al. [128]	2016	Same definition as Webb, 2010
Friis RH [129]	2017	Same definition as Friis, 2004
World Health Organization [32] **(WHO)** [22]	No information	Epidemiology is the study of the distribution and determinants of health-related states or events (including disease), and the application of this study to the control of diseases and other health problems. Various methods can be used to carry out epidemiological investigations: surveillance and descriptive studies can be used to study distribution; analytical studies are used to study determinants

https://doi.org/10.1371/journal.pone.0208442.t001

Analysis of the content of definitions from the period 1978 to 2017

Human Medicine, General Epidemiology

A total of 20 terms and concepts were identified in definitions from 1978 to 2017. Of the 20 terms and concepts, only three were not present in the definitions used by Lilienfeld: "*knowledge*" appeared in 1979 [3], "*problems*" appeared in 1991 [15] and "*public health*" appeared in 1999 [16] Table 2 shows the frequencies of appearance of terms in definitions from 1978 to 2017.

Table 2. Frequency of occurrence of the terms identified from 69 definitions from the period 1978–2017.

Terms used*	Definition using terms	
	Number	%
Population	58	84
Disease/Illness	51	74
To study/Studying	47	68
Health	40	58
Distribution	34	49
To determine/ Determinant	30	43
States/Stages/Events/Conditions	34	49
To occur/Occurrence	19	28
Control	22	32
Problems	20	29
Science	12	17
Human populations/Human groups	17	25
To prevent/ Prevention	11	16
Factors	8	12
Knowledge	6	9
To understand/ understanding	7	10
Frequency	7	10
Dynamics	2	3
Process	3	4
Public health	4	6

* "The terms examined by Lilienfeld have been grouped, in the spirit of his article, into categories".

https://doi.org/10.1371/journal.pone.0208442.t002

The term "*population*" or "*group*" was found in 84% of the definitions (58/69). For 25% of definitions (17/69), it was stated that epidemiology applies to human beings.

For 68% of the definitions (47/69), epidemiology was "*the study of something*" (study of ...) and 17% of the definitions (12/69) defined epidemiology as a "*science*". Four definitions (11%) used both terms [16–19].

Epidemiology was associated with "*health*" in 58% of definitions (40/69) and "*disease*" in 74% of definitions (51/69). Both terms were present in 23% of the definitions (16/69) [15–17, 20–32]. In 25% (17/69) of definitions, the term "disease" was not present but the terms "health problem" or "health states" was present [33–42] [43–49].

The concept of "*health control*" or "*disease control*" was present in 32% of the definitions (22/69). The concept of "*disease prevention*" or "prevention of *health problems*" was present in 16% of the definitions (11/69). These notions were associated in 9% (6/69) of the definitions [3, 17, 19, 39] [48, 50].

The term "*problem*" was present in 29% of the definitions (20/69). In 25% of definitions (17/69), the terms "*problem*" and "*health*" were associated ("health *problem*") [15, 26] [28, 30, 32, 35, 37, 39, 42–49, 51].

Epidemiology was associated with the field of statistics in 6% (4/69) of the definitions [3, 30, 32, 52].

Analysis of the evolution of definition content between 1978 and 2017

Table 3 shows a comparison of the content of definitions reviewed by Lilienfeld and those from 1978 to 2017 according to the concepts identified by Evans.

Table 3. Comparison between the contents of definitions reviewed by Lilienfeld and definitions from the period 1978–2017 according to the terms and concepts identified by Evans [3].

Terms	Definition from the period 1978-2017 (N = 69)		Definition reviewed by Lilienfeld (N = 23)		*BF*[*]
	Number	%	Number	%	
Statut of person					
Diseased (see response 2.6)	51	74	21	91	1.1
Infectious disease	0	0	4	17	65.9
Physiological condition	0	0	1	4	2
Injuries	3	4	1	4	0.5
Health	40	58	1	4	18709.8
Group affected					
Population	58	84	11	48	63.1
Community	1	1	3	13	4.8
Mass phenomena/ Outbreak	0	0	3	13	18.3
Total with group affected mentioned	58	84	17	74	0.5
Distribution of disease	14	20	9	39	1.2
Spread (propagation. spread. dynamics)	2	3	3	13	2.1
Incidence. Occurrence	20	29	5	22	0.3
Etiology					
Causes. Determinant. Factors	28	41	5	22	0.8
Circumstances of occurrence	1	1	2	9	1.9
Ecology	0	0	1	4	2
Total with etiology mentioned	29	42	8	35	0.3
Understanding disease					
Natural history or nature	1	1	2	9	1.9
Understanding the process	7	10	1	4	0.5
Total with understanding disease mentioned	3	4	3	13	1.2
Prevention and Control	6	9	3	13	0.5

[*] BF: Bayes Factor

https://doi.org/10.1371/journal.pone.0208442.t003

Several concepts ("*infectious disease*", "*mass phenomena*"…) identified by Evans were not present in definitions from 1978 to 2017 (Table 3).

The term "*health*" was more frequent in definitions from 1978 to 2017 than in the definitions used by Lilienfeld (58% versus 4%, Bayes Factor(BF) = 18709.9).

The term "infectious disease" was less frequent in definitions from 1978 to 2017 than in the definitions used by Lilienfeld, with a BF estimated to 65.9 (0% versus 17%). "Mass phenomena/outbreak" and "infectious disease" can be related. The BF associated with "mass phenomena/outbreak, was estimated at 18, and this also reflects a decrease in the use of the term "infectious disease".

The term "*population*" was more frequent in definitions from 1978 to 2017 than in the definitions used by Lilienfeld (84% versus 48%, Bayes (BF) = 63.1).

Table 4 shows a comparison of the content of definitions used by Lilienfeld and those from 1978 to 2017 according to the terms and concepts we have identified.

Table 4. Comparison between the content of definitions used by Lilienfeld and definitions from the period 1978 to 2017 according to the terms and concepts identified in definitions from the period 1978–2017.

Terms	Definition from the period 1978–2017 (N = 69)		Definition used by Lilienfeld (N = 23)		*BF*[*]
	Number	%	Number	%	
Population	58	84	11	48	63.1
Disease/Illness	51	74	21	91	1.2
To study/Studying	47	68	14	61	0.3
Health	40	58	1	4	18709.8
Distribution	34	49	9	39	0.3
To determine/ Determinant	30	43	5	22	1.2
States/Stages/Events/Conditions	34	49	3	13	31.3
To occur/Occurrence	19	28	3	13	0.6
Control	22	32	1	4	11.5
Problems	20	29	0	0	43.8
Science	12	17	5	22	0.3
Human populations/Human groups	17	25	7	30	0.3
To prevent/ Prevention	11	16	2	9	0.4
Factors	8	12	5	22	0.7
Knowledge	6	9	0	0	0.9
Frequency	7	10	1	4	0.5
To understand/ Understanding	7	10	1	4	0.5
Dynamics	2	3	1	4	0.6
Process	3	4	1	4	0.5
Public health	4	6	0	0	0.7

[*] BF: Bayes Factor

https://doi.org/10.1371/journal.pone.0208442.t004

For the term "*population*" and "*health*", the results shown in Table 3 and Table 4 are identical.

The term "*problems*" was probably more frequent in definitions from 1978 to 2017 than in the definitions used by Lilienfeld (29% versus 0%, BF = 43.8).

Human Medicine, Subspecialities of Epidemiology

The following subspecialties (outcome-oriented or exposition-oriented) were taken into account, according to Boslaugh [53], Rothman [24]and Carter-Pokras, [8], among many others: cancer epidemiology; cardiovascular epidemiology; obesity epidemiology; nutritional epidemiology, psychiatric epidemiology; genetic and molecular epidemiology; molecular epidemiology; genetic epidemiology; infectious diseases epidemiology; reproductive and perinatal epidemiology; environmental epidemiology; occupational epidemiology; social epidemiology. A total of 24 definitions were selected (Table 5).

Table 5. Selected definitions of epidemiology, by subspecialty.

AUTHOR (Reference)	DATE	DEFINITION
Dos Santos Silva I [64] Cancer	1999	"Epidemiology is the study of the distribution and determinants of health-related states or events in specified populations, and the application of this study to control of health problems." Quoting Last, 1995 [62] Cancer epidemiology is the branch of epidemiology concerned with the disease cancer. Therefore, this definition is as valid to cancer epidemiology as it is to epidemiology in general.
Schottenfeld D et al. [60] Cancer	2006	No definition of epidemiology provided
Adami HO et al. [130] Cancer	2008	No definition of epidemiology provided The roles of epidemiology have been to describe the distribution of cancer in populations and to seek deviations from randomness that offer clues as to their explanations, to assess the validity and strength of associations that are suspected causes, and to evaluate the effectiveness of preventive measures by continued monitoring of cancer incidence and mortality.
Loda M et al. [131] Cancer	2016	According to the World health Organization (WHO), epidemiology is "the study of the distribution and determinants of health-related states of events. and the application of this study to the control of diseases and other health problems" Quoting WHO, 2014 http://www.who.int/topics/epidemiology/en/
Luepker RV [84] Cardiovascular	2009	Epidemiology is the study of disease patterns and outcomes in populations. Because of this public health perspective, epidemiologists also play a learning role in the study of disease prevention. Observational epidemiologists define the role of risk factors in predicting, preventing, and treating cardiovascular disease through population studies. Working with laboratory medical and behavioral scientists, they elucidate the underlying mechanisms and initiate large prevention and treatment trials. Our understanding of lipids, hypertension, smoking, diabetes, inflammation, and other factors leading to cardiovascular diseases originated in epidemiological observations.
Labarthe DR [59] Cardiovascular	2010	No definition of epidemiology provided

Hu F [54] Obesity Epidemiology	2008	No definition of epidemiology provided, but referring to classic textbooks in epidemiology: Morris, 1964 [132]; MacMahon, 1970 [69], Lilienfeld, 1976 [133], Miettinen, 1985 [104]; Rothman, 1986 [134]. Obesity epidemiology uses epidemiologic approaches to examine the causes and consequences of obesity in human populations. It includes the study of multiple, broad, interrelated domains such as (a) the distributions, patterns, and dynamics of obesity in populations; (b) health and other consequences of obesity: (c) the determinants or causes of obesity; and (c) the development and validation of body composition measurement methods used in epidemiologic studies. Knowledge gained is eventually applied to public health initiatives to prevent and control obesity and related health conditions.
[55] Nutritional Epidemiology	2012	No definition of epidemiology provided, but recommended books. Although I have not attempted to define or explain basic epidemiologic terms, most of the chapters can be read by someone with elementary statistical knowledge and some epidemiologic background. Readers without exposure to epidemiology would profit by referring to an introductory text such as MacMahon and Trichopoulos' Principles of Epidemiology, Aschengrau and Seage's Essentials of Epidemiology in Public Health, Rothman's Epidemiology: An Introduction, or Rothman, Greenland, and Lash's Modern Epidemiology, the last being the most advanced. Similarly, epidemiologists without formal exposure to nutrition can benefit by reading… Quoted books: [18, 24, 26, 135] The field of nutritional epidemiology has developed from interest in the concept that aspects of diet may influence the occurrence of human disease.
Susser E et al. [61] Psychiatry	2006	No definition of epidemiology provided

Tsuang MT et al. [65] Psychiatry	2011	Quoting Last, 2001 [43]; Wikipedia ; Rothman, 2008 [24] Epidemiology, according to Last's Dictionary of Epidemiology, is 'The study of the distribution and determinants of health-related states or events in specified populations and the application of this study to control of health problems' Wikipedia states 'Epidemiology is the study of factors affecting the health and illness of populations, and serves as the foundation of interventions made in the interest of public health and preventive medicine' Rothman and Greenland: 'the ultimate goal of most epidemiologic research is the elaboration of causes that can explain patterns of disease occurrence' Psychiatric epidemiology is simply the epidemiology of psychiatric disorders–no more, no less. The principles and practice are the same when studying psychiatric disorder as they are when studying other medical conditions
Ioannidis JPA [56] Genetic and molecular epidemiology	2007	No definition of epidemiology provided Genetic and molecular epidemiology, the investigation of genetic and molecular determinants of health and disease…
Schulte PA et al. [57] Molecular Epidemiology	2012	No definition of epidemiology provided Progress in the molecular approach to biology and medicine has simulated and excited both the public and researchers, who now believe these advances can be applied to the study, prevention and control of health risks faced by human populations. The term "molecular epidemiology" may be used to describe such an approach: the incorporation of molecular, cellular, and other biologic measurements into epidemiological research.
Burton PR et al. [66] Genetic epidemiology	2005	Epidemiology is usually defined as "the study of the distribution, determinants [and control] of health-related states and events in populations". Quoting Last, 2001 [43] By contrast, genetic epidemiology means different things to different people (referring six authors) We regard it as a discipline closely allied to traditional epidemiology that focuses on the familial, and in particular genetic, determinants of disease and the joint effects of genes and non-genetic determinants.
Palmer L et al. [67] Genetic epidemiology	2011	Quoting Last, 2001 [43] and Burton, 2005 [66]

Austin M et al. [58] Genetic epidemiology	2013	No definition of epidemiology provided " … an emerging field … that represents an important interaction between the two parent disciplines: genetics and epidemiology. Genetic epidemiology differs from epidemiology by its explicit consideration of genetic factors and family resemblance; it differs from population genetics by its focus on disease; it also differs from medical genetics by its emphasis on population aspects." Quoting Rao, 1984 [90]
Teare MD [136] Genetic epidemiology	2016	Genetic epidemiology. Is the study of the role of genes and environments on markers of health and disease risk in population.
Giesecke J [137] Infectious diseases	2017	Most standard definitions of epidemiology mention something like "the study of distribution and determinants of diseases in the population"… My own practical view is that epidemiology is about putting people into groups. …Epidemiology identifies such groups, ignoring the uniqueness of its members, and tries to discover whether this division of people into groups tells us something more than we could have learnt by just observing each person separately.
Louis GMB et al. [68] Reproductive and Perinatal Epidemiology	2011	Quoting Last, 2001 [43] … the Faculty defines reproductive and perinatal epidemiology for the purposes of this book as the study of the distribution, determinants, and sequelae of reproductive and/or perinatal processes and events. This definition is consistent with the more general definition of epidemiology as defined by John Last (2001) in his textbook A Dictionary of Epidemiology. … the field is essentially defined by reproductive and perinatal outcomes (and not exposures, per se)…
Wilkinson P [138] Environmental Epidemiology	2006	No definition of epidemiology provided

Merrill RM [139] Environmental Epidemiology	2008	In a book compiled by Last the definition of epidemiology provided is "the study of the distribution and determinants of health-related states or events in specified populations, and the application of this study to control for health problems." Quoting Last, 1995 [62] … epidemiology involves more than just the study of disease states (e.g. respiratory illness), but also includes the study of events (e.g., injury) and behaviors and conditions associated with health (e.g., hand washing). In addition, note that epidemiology is concerned with health-related states or events that occur in populations, not a specific individual Environmental epidemiologists study the frequency and pattern of disease and health-related events and attempt to explain the environmental factors that influence these conditions. The study of why and how environmental factors affect peoples' health is the essence of environmental epidemiology. Environmental epidemiology emphasizes the idea that health is largely influenced by environmental factors, and that by identifying these factors and their modes of transmission, the public's health can be better protected.
Baker D [140] Environmental Epidemiology	2008	Environmental epidemiology is a subspecialty of epidemiology, the basic science of public health. Epidemiology is the study of the distribution of health and disease in the population, and of the determinants of this distribution. Environmental epidemiology studies the effects of environmental exposures on health and disease in the population. The subject matter of environmental epidemiology is environmental health: this, in principle, covers all factors external to the human body which may affect health. However, … environmental epidemiology does not try to cover all external factors which may conceivably influence population health, but focuses on physical, chemical and (noninfectious) biological factors in our everyday environment

Venables K [141] Occupational epidemiology	2013	Occupational epidemiology is the study of the distribution and determinants of illness and injury related to the work environment. … Furthermore, harmful work exposures can be conceptualized as a "natural experiment" in the causation of illness and injury, because the workplace usually has much better-defined levels and timing of exposure than in the general community.
O'Campo P et al. [142] Social epidemiology	2011	Epidemiology is the study of the patterns of health and illness in populations, while social epidemiology focuses on the social determinants that shape the risk and occurrence of poor health in these populations. Quoting Berkman and Kawachi, 2000 [143] and James, 2009 [144].
Berkman LF et al. [145] Social epidemiology	2014	Epidemiology is the study of the distribution and determinants of states of health in population. Quoting Susser, 1973 [63]. We define social epidemiology as the branch of epidemiology that studies the social distribution and social determinants of states of health. …We focus on specific social phenomena such as socioeconomic stratification, social networks, discrimination, workplace organization, and public policies rather than on specific disease outcomes.
		Emerging subspecialties
Brown SL [96] Medical device epidemiology	2007	Epidemiology, known as the basic science of public health, aims to study the distribution and determinants of diseases in populations …when applied to medical devices, epidemiology may describe patterns of use or factors associated with use or characterize the risk for certain outcomes in defined subgroups.

Shatin D et al. [97] Implanted medical device epidemiology	2006	Epidemiology, in general, is "the science of occurrence of diseases in human populations. Disease occurrence is measured and related to different characteristics of individuals or their environments". Although clinical trials are sometimes considered a subset of epidemiology, we restrict our discussion here to observational studies of patients who receive their implants as a result of routine care. The epidemiology of implanted devices includes the following: (1) Descriptions of patients with implants in terms of demographics, underlying disease, and concurrent disease. (2) Assessment of efficacy and safety of the implant. (3) Comparison of treatments.
Salathé M [94] Digital epidemiology	2018	The goal of epidemiology, very broadly speaking, is to understand the patterns of disease and health dynamics in populations as well as the causes of these patterns, and to use this understanding to mitigate and prevent disease, and to promote health. The goal of digital epidemiology is exactly the same...Digital epidemiology is epidemiology that uses data that was generated outside the public health system, i.e. with data that was not generated with the primary purpose of doing epidemiology.

Epidemiology, in general, is "the science of occurrence of diseases in human populations. Disease occurrence is measured and related to different characteristics of individuals or their environments". Although clinical trials are sometimes considered a subset of epidemiology, we restrict our discussion here to observational studies of patients who receive their implants as a result of routine care.

The epidemiology of implanted devices includes the following: (1) Descriptions of patients with implants in terms of demographics, underlying disease, and concurrent disease. (2) Assessment of efficacy and safety of the implant. (3) Comparison of treatments.

https://doi.org/10.1371/journal.pone.0208442.t005

The concepts mentioned in the definitions of subspecialty of epidemiology are similar to those found for general epidemiology. Some authors did not provided general definition of epidemiology [54–58], nor specific definition to subspecialty [59–61].

Others took up the main concepts of epidemiology and adapted them to their subspecialty. In this respect, the definition of the epidemiology of obesity speaks for itself: it contains virtually all concepts: "causes and consequences"; "human populations"; "the distributions, patterns, and dynamics"; "the determinants"; "to prevent and control"; "health conditions". This definition also mentions a specificity of the subspecialty: "the development and validation of body composition measurement methods used in epidemiologic studies".

Others referred to authoritative epidemiological literature or to WHO. For this case, the definitions of Last, 1995 and 2001 [43, 62] were most often cited (6 times out of 24 definitions), WHO (1994), Rothman and Greenland, 2008 [24], Susser, 1973 [63] and Wikipedia being quoted once. Different subspecialties referred to the definition of Last: oncology [64], psychiatry [65], genetic epidemiology [66, 67], reproduction [68].

Finally, others Other authors proposed a pure and simple application of the general definitions of epidemiology to their subspecialty [64, 65]. Thus, Dos Santos after recalling Last's definition of epidemiology proposed a definition of cancer epidemiology and stated: "Therefore, this definition is as valid to cancer epidemiology as it is to epidemiology in general" [64]. Tsuang defined psychiatric epidemiology as follow "Psychiatric epidemiology is simply the epidemiology of psychiatric disorders–no more, no less [65].

Some authors provided definitions by highlighting differences. This is the case, for example, for genetic epidemiology. Austin stated "Genetic epidemiology differs from epidemiology by its explicit consideration of genetic factors and family resemblance" [58].

Veterinary Medicine

Table 6 shows the frequencies of appearance of terms in definitions of epidemiology from 1977 to 2015, for veterinary medicine.

Table 6. Definitions of epidemiology from the period 1977–2015.

AUTHOR (Reference)	DATE	DEFINITION
Schwabe CW et al. [73]	1977	The study of the health status of populations" (Schwabe et al., 1977). Quoted reference: [73]
Martin SW et al. [146]	1987	Epidemiology may be defined as the study of the patterns of disease that exist under field conditions. More specifically, epidemiology is the study of the frequency, distribution, and determinants of health and disease in populations.
Toma B [147, 148]	1999	The study of the health status of populations.
Slater MR [149]	2003	What is epidemiology? Most simply, it is the study of diseases and health in populations. Diseases include not only the classic infectious diseases and disease outbreaks but also chronic diseases like arthritis, cancer, and renal failure. Also included are injuries and exposures to environmental contaminants. Factors that maintain good health and quality of life are also within the purview of epidemiology. In this context, questions about the causes of animals being relinquished to animal shelters or about animal well-being in long-term kennel settings may also be addressed by epidemiologists.
Smith RD [147]	2005	Over the years there have been many definitions of epidemiology. Some definitions follow: 1. "The study of the distribution and determinants of disease frequency in man" (MacMahon and Pugh, 1970). 2. "The study of the patterns of disease" (Halpin, 1975). 3. "The study of the health status of populations" (Schwabe et al., 1977). 4. "Epidemiology is nothing more than ecology with a medical and mathematical flavor" (Norman D. Levine, 1990, personal communication). 5. "The study of the distribution and determinants of health-related states or events in specified populations and the application of this study to control of health problems" (Stedman's Medical Dictionary, 2000). Quoted references:[69, 73, 150]

Pfeiffer D [71]	2009	…the science of veterinary epidemiology deals with the investigation of these determinants of disease distribution in animal populations. Productivity and welfare of animals may also be outcomes of interest since disease will usually impact on both, and they may indeed often be what stakeholders will have as their primary focus.
Dohoo I et al. [70]	2009	Epidemiology is largely concerned with disease prevention and therefore, with the "succession of events which result in the exposure of specific types of individual to specific types of environment" (ie exposures) (MacMahon & Pugh, 1970). Thus, epidemiologists strive to identify these exposures and evaluate their associations with various outcomes of interest (eg health, welfare, productivity) so as to improve the lives of animals and their keepers. … associations which are likely to be causal in nature and which, once identified, we can take advantage of to improve the health, welfare and productivity of animals and the quality and safety of foods derived from them. Quoted reference: [69]
Thrusfield M [72]	2013	Epidemiology is the study of disease in populations and of factors that determine its occurrence; the key word being populations. Veterinary epidemiology additionally includes investigation and assessment of others health-related events, notably productivity.
Sergeant et al. [151]	2015	Epidemiology is the study of patterns and causes of disease in populations. Understanding these issues will in turn contribute to identification of options for control and prevention of diseases. As its simplest, epidemiology is about supporting better decision making to ensure appropriate response or preventative measure for population health.

Veterinary medicine.

https://doi.org/10.1371/journal.pone.0208442.t006

A total of 9 definitions were selected

Again, the concepts mentioned in the definitions of veterinary epidemiology are similar to those found for human epidemiology, with several authors using definitions proposed in human medicine and quoting for example MacMahon and Plug [69].

One notable concept, specific to veterinary medicine, is productivity which is mentioned by two authors [70–72].

General Epidemiology (websites)

The definitions of epidemiology are given in Table 7 and are classified according to the nature of the website: websites of international organizations, WHO, CDC, FAO (for veterinary medicine); national health institutes, academies of sciences …; public health schools; medical journals (BMJ); nonprofit organizations; Wikipedia; For-profit organizations; non-medical dictionaries online. The distribution of concepts, by order of frequency, is the following: health (12/30); population (18/30); disease / illness (27/30); study (19/30); distribution (16/30). This distribution is almost that found with the definitions of epidemiology provided by books.

Table 7. Definitions of epidemiology available on websites.

AUTHOR and url	DEFINITION	Comments
WHO http://www.who.int/topics/epidemiology/en/	Epidemiology is the study of the distribution and determinants of health-related states or events (including disease), and the application of this study to the control of diseases and other health problems. Various methods can be used to carry out epidemiological investigations: surveillance and descriptive studies can be used to study distribution; analytical studies are used to study determinants.	
CDC https://www.cdc.gov/ophss/csels/dsepd/ss1978/lesson1/section1.html	Epidemiology is the study of the distribution and determinants of health-related states or events in specified populations, and the application of this study to the control of health problems.	Online version of a printed textbook. Dicker, 2006 [152] Definition based on Last, 2001 [43].

CDC Excellence in Curriculum Innovation through Teaching Epidemiology (EXCITE) CDC a self-study course https://www.cdc.gov/careerpaths/k12teacherroadmap/index.html	Epidemiology is the study of the distribution and determinants of health-related states or events in specified populations, and the application of this study to the control of health problems.(1) (1) Last, 2001	Quoted by Dicker, 2006 [152] And based on Last, 2001 [43].
Australasian Epidemiological Association https://aea.asn.au/about/epidemiology/what-is-epidemiology	...The study of diseases or other health related events in populations. Epidemiology has three main aims: • To describe disease and other health related event patterns in human populations. • To identify the causes of diseases and other health related events (also known as aetiology). • To provide data essential for the management, evaluation and planning of services for the prevention, control and treatment of disease and other health related events.	
FAO (Food and Agriculture Organization of the United Nations) http://www.fao.org/wairdocs/ilri/x5436e/x5436e04.htm	... A useful general definition is that given by Schwabe et al (1977), which defines epidemiology as the study of disease in populations.	Based on Schwabe, 1977 [73].
National Institute on Deafness and Other Communication Disorders https://www.nidcd.nih.gov/health/statistics/what-epidemiology	Epidemiology is the branch of medical science that investigates all the factors that determine the presence or absence of diseases and disorders. Epidemiological research helps us to understand how many people have a disease or disorder, if those numbers are changing, and how the disorder affects our society and our economy.	"Last Updated Date: September 13, 2011"

The National Academies of Sciences, Engineering, and Medicine https://www.nap.edu/read/5804/chapter/2#3	Modern epidemiology, the study of disease patterns in populations, encompasses a broad array of subject matter, including subspecialties that concentrate on such domains as clinical trials of pharmaceutical agents; such outcomes as reproductive and developmental effects, infectious diseases, and chronic diseases; such risk factors as occupation, nutrition, and alcoholism; and special populations	
National Research Council (US). Committee on Environmental Epidemiology, National Research Council (US). Commission on Life Sciences. Environmental Epidemiology http://www.ncbi.nlm.nih.gov/books/NBK233638/	Modern epidemiology, the study of disease patterns in populations, encompasses a broad array of subject matter, including subspecialties that concentrate on such domains as clinical trials of pharmaceutical agents; such outcomes as reproductive and developmental effects, infectious diseases, and chronic diseases; such risk factors as occupation, nutrition, and alcoholism; and special populations	Date of publication: 1997
Encyclopedia of occupational health and safety. 4th edition International Labour Office. http://www.ilocis.org/documents/chpt28e.htm Chapter 28—Epidemiology and Statistics Epidemiological method applied to occupational health and safety. Franco Merletti, Colin L. Soskolne and Paolo Vineis	Several operational definitions of epidemiology have been suggested. The simplest is that epidemiology is the study of the occurrence of disease or other health-related characteristics in human and in animal populations. Epidemiologists study not only the frequency of disease, but whether the frequency differs across groups of people; i.e., they study the cause-effect relationship between exposure and illness. Diseases do not occur at random; they have causes—quite often man-made causes—which are avoidable…	

School of Public Health. University of Alabama at Birmingham. http://www.soph.uab.edu/epi/academics/studenthandbook/what	Many definitions have been proposed; here are two that capture the underlying principles and the public health spirit of epidemiology: "Epidemiology is the study of the distribution and determinants of health-related states or events in specified populations, and the application of this study to the control of health problems." (2). "Epidemiology is the study of the distribution and determinants of disease frequency in man." (3). (2) Last JM, 1995. (3) MacMahon & Pugh, 1970	Definition Based on Last, 1995 [62] and MacMahon, 1970[69].
School of Public Health. Boston University http://sphweb.bumc.bu.edu/otlt/mph-modules/ep/ep713_history/EP713_History9.html#headingtaglink_3	Epidemiology is the study of the distribution and determinants of disease frequency in human populations. Epidemiology focuses on making comparisons in order to establish cause-effect relationships, evaluate information, and make good decisions that will improve outcomes.	
Public Health Social Work Maternal and Child Health Leadership Training Program University of North Carolina at Chapel Hill https://ssw.unc.edu/mch/node/148	Epidemiology is the "study of the distribution and determinants of health-related states or events (including disease), and the application of this study to the control of diseases and other health problems" (World Health Organization, 2011). … Epidemiology examines how health conditions are distributed among a population and seeks to understand the risks or causes associated with those conditions.	Based on WHO (2011) definition

BMJ http://www.bmj.com/about-bmj/resources-readers/publications/epidemiology-uninitiated/1-what-epidemiology	Epidemiology is the study of how often diseases occur in different groups of people and why. Epidemiological information is used to plan and evaluate strategies to prevent illness and as a guide to the management of patients in whom disease has already developed.	
School of Public Health Colorado http://www.ucdenver.edu/academics/colleges/PublicHealth/Academics/departments/Epidemiology/About/Pages/Learn.aspx	Epidemiology is the study of patterns of disease and injury in human populations and the application of this study to the control of health problems. Epidemiologists study the variation of disease in relation to age, sex, race, occupational and social characteristics, place of residence, susceptibility, exposure to specific agents or other pertinent characteristics	
Fulton county, Georgia http://www.fultoncountyga.gov/dhw-epidemiology/8261-what-is-epidemiology	Epidemiology is a branch of science that deals with the incidence, distribution and control of disease in a population. It is a core public health science. Epidemiologists monitor health trends and statistics to identify groups of people who are affected by various diseases. In addition, epidemiologists investigate cases of disease to determine the source, modes of transmission, and risk factors for disease.	
Epidemiology Education Movement http://www.epiedmovement.org/BM3.html	Epidemiology is a fundamental population science and tries to answer questions about health-related behaviors and outcomes in populations. Epidemiologists study how health and disease are distributed in populations and the factors that influence or determine those distributions.	

People's Epidemiology Library http://www.epidemiology.ch/history/PeopleEpidemiologyLibrary.html	Epidemiology is the science of counting health-related events and comparing these counts over time, place and people. Health-related events are the diseases themselves and their causes.	
Merck manual. Veterinary http://www.merckvetmanual.com/public-health/public-health-primer/basic-principles-of-epidemiology	The definition of epidemiology is "the study of disease in populations and of factors that determine its occurrence over time." The purpose is to describe and identify opportunities for intervention. Epidemiology is concerned with the distribution and determinants of health and disease, morbidity, injury, disability, and mortality in populations. For veterinary epidemiology, this intervention is to enhance not only health but also productivity.	Productivity is taken into account
Capella University Minneapolis (Profit Institution) https://www.capella.edu/blogs/cublog/what-is-epidemiology/	We often think about epidemiology in terms of disease; however, it is really the study of how to keep the population healthy. A large of part of epidemiology is determining the cause of disease, providing education about it, and creating plans to prevent and mitigate widespread illnesses	
Wikipedia https://en.wikipedia.org/wiki/Epidemiology	Epidemiology is the study and analysis of the patterns, causes, and effects of health and disease conditions in defined populations. It is the cornerstone of public health, and shapes policy decisions and evidence-based practice by identifying risk factors for disease and targets for preventive healthcare.	

https://www.kullabs.com/classes/subjects/units/lessons/notes/note-detail/3453	Epidemiology is the study and analysis of the patterns, causes and effects of health and disease conditions have defined the population. It is derived from the word epidemic (epi -among; demos-people; logos -study). "The study of distribution and determinants of health-related or events in specified populations and the application of this study to the control of health problems."	Based on WHO, CDC and Wikipedia Following links are provided by this internet site are: •www.who.int/topics/epidemiology/en/ •www.who.int/mediacentre/factsheets/fs355/en/ •www.cdc.gov/ophss/csels/.../section1.htm •www.cdc.gov/.../data_stats/glossary.html •https://en.wikipedia.org/wiki/Non-communicable_disease
healthcare-management http://www.healthcare-management-degree.net/faq/what-is-epidemiology-and-what-does-an-epidemiologist-do / Degree Guide	Epidemiology is a field where trained epidemiologists study patterns of frequency and the causes and effects of diseases in human populations	
Business Dictionary http://www.businessdictionary.com/definition/epidemiology.html	Study of distribution and pattern of diseases in a population to determine or trace the circumstances or events causing them. This information is used by public health authorities in designing appropriate policies and interventions to protect the population.	

Cambrigde Dictionary http://dictionary. cambridge.org/dictionary/ english/epidemiology	The scientific study of diseases and how they are found, spread, andcontrolled in groups of people	
Collins Dictionary https://www. collinsdictionary.com/ dictionary/english/ epidemiology	Epidemiology is a branch of medicine that is concerned with the occurrence, distribution, and control of disease.	
Dictionary.com http://www.dictionary. com/browse/ epidemiology	The branch of medicine dealing with the incidence and prevalence of disease in large populations and with detection of the source and cause of epidemics of infectious disease.	
The Free Dictionary http://www. thefreedictionary.com/ Epidemiolgy	Based on several dictionaries: American Heritage, Collins, Webster.	
The Law Dictionary http://thelawdictionary. org/epidemiology/	Study of distribution, pattern, and cause of diseases in a specific population. The objective is to determine and trace events and reasons causing epidemic-like situations. Resulting information leads to more effective plans, policies, and containments protecting the population by public health authorities.	
Oxford Dictionary https:// en.oxforddictionaries. com/definition/ epidemiology	The branch of medicine which deals with the incidence, distribution, and possible control of diseases and other factors relating to health.	
Webster Dictionary http://www.webster-dictionary.org/definition/ epidemiology	That branch of medicine which studies the incidence and distribution of disease in a population, and uses such information to find the causes, modes of transmission, and methods for control of disease.	

https://doi.org/10.1371/journal.pone.0208442.t007

The majority of websites offer definitions of epidemiology already known, but does not cite the authors.

CDC and FAO based their definition on Last, 2001 [43] and Schwabe, 1977 [73], respectively.

Public health schools do not provided the source of their definition of epidemiology except one (University of Alabama) with Last, 2001 [43] and MacMahon, 1970 [69] being quoted.

Few websites mentioned the date of the last update (e.g. National Institute on Deafness and Other Communication Disorders).

DISCUSSION

This study allowed us to identify, synthesize and analyse the definitions of epidemiology from the period 1978–2017. This evolution is almost absent from books on epidemiology and deserved to be presented. Even though many other fields of epidemiology have grown, only definitions of epidemiology were studied. A significant number of definitions of epidemiology were found despite the short period of study and the inclusion criteria. However, this number must be related to the increase in the number of publications in epidemiology.

Many of the books consulted and the definitions of epidemiology they contained summarized definitions that were already known [74] or definitions of other authors [75] or proposed mnemonics tools "the 3-D definition of epidemiology" [76, 77].

In summary, 69 definitions of epidemiology were retained from 1978 to 2017. All of these came from epidemiology textbooks or published articles. Only one definition (WHO) was not associated with a date of appearance [32]. The fact that this definition was not cited by Lilienfeld was enough to keep it in this study.

The evolution of the content of definitions of epidemiology was studied on the basis of terms and concepts identified by Evans [3] and those identified by us.

The definitions from 1978 to 2017 helped to highlight 20 terms and concepts related to epidemiology. A great majority of concepts were either identified by Evans [3] or present in definitions used by Lilienfeld [1]. Among the terms identified, only three ("*problems*", "*knowledge*" and "*public health*") were not present in Evan's work. Five terms were present in more than 50% of definitions from 1978 to 2017: "*population*", "*study*", "*disease*", "*health*" and "*distribution*". These five terms can summarize epidemiology as being the study of the distribution of disease and health in the population. The term "*study*" may not seem specific enough for epidemiology. But this term is linked to the concept of "*science*". Epidemiology was considered a "*science*" during both periods. In fact, the term "*science*" was used in more than 15% of definitions for the two periods.

Several developments have occurred. (i) the emergence of the term "*problem*" in the definitions of epidemiology; (ii) strengthening of the terms "*control*" and "*health*" already used; (iii) other terms and concepts ("*infectious* diseases", "*mass phenomenon*") identified by Evans [3] were no longer used in definitions from 1978 to 2017. But these terms were rare in the definitions used by Lilienfeld: 4/23 (17%), 1/23 (4%), 3/23 (13%) and 3/23 (13%), respectively.

The prevalence and distribution of diseases, infectious or not, also influence epidemiology. Thus, in low-income countries, new infectious diseases have emerged such as HIV infection. Nevertheless, the prevalence of these infectious diseases has declined over time. In low-income countries, infectious diseases are still present [78], but there is an increase in the incidence of chronic diseases [79]. Epidemiological definitions have evolved in this direction, with a decrease in the presence of infectious disease terms.

Indeed, for definitions from 1978 to 2017, only one definition associated the terms "epidemiology" and "epidemic" [80]. Currently, the term "epidemic" is associated with both infectious disease and the growing prevalence and incidence of disease states in the population, for example, obesity (obesity epidemic) [81]. Thus, the concept of "disease" used in epidemiology is no longer limited to infectious diseases.

Epidemiology is currently associated with the study of disease, but also more broadly, with the study of health phenomena. In 23% (16/69) of the definitions retrieved, the concepts of "*disease*" and the concept of "*health*" were related. The term "disease" has a rather negative connotation, while the term "health" has a positive connotation. They reflect two opposing views of the same phenomenon. Some definitions suggested this phenomenon was dynamic and not static and evolved in the population, in space and in time [3, 33, 38, 82, 83].

In the definitions from 1978 to 2017, the link between epidemiology and public health was rare [16, 18, 20, 44, 84]. However, these definitions describe epidemiology as an essential part of public health. Other developments, nonetheless, were related to public health, like the expansion of the field of epidemiology to areas such as health promotion [18] and the use of statistical tools [3, 30, 32, 52].

Finally, 69 definitions (general epidemiology) were analyzed. This is greater than the number of definitions used by Lilienfeld. The fact that the definitions retained came from different sources, both grey and published literature, suggests that the definitions retained are representative of the available definitions of epidemiology, thus limiting the risk of selection bias. The evolution of the content of definitions of epidemiology was studied by using the concepts identified by Evans [3] and those identified by us. The comparison with the concepts cited by Evans allowed an objective analysis of the evolution of the content of definitions of epidemiology from 1978 to 2017. The statistical analysis revealed the differences observed between the two sets of definitions of epidemiology, even if the definitions of epidemiology from 1978 to 2017 were not exhaustively from this period.

The analysis of the evolution of the definitions of human epidemiology by subspecialty and veterinary epidemiology was not possible, due to the low number of definitions.

The number of subspecialties of epidemiology increases from year to year and almost all fields of medicine are concerned ([85]). As early as 2007, Lillienfeld already raised the question of the general epidemiologist [86]. As for general epidemiology, they are several definitions for each subspecialty. Austin [58], quoting Khoury's book published in 1993 [87], mentions that there were at least eight different definitions of genetic epidemiology at that time.

This increase in the number of definitions of epidemiological subspecialties can be linked to several factors including technological advances. The significant development of data analysis in epidemiology (epidemiological methods) could be related to the emergence of informatics. Molecular and genetic epidemiology is also undoubtedly linked to major advances in molecular and genomic biology [88]. Advances in genome sequencing techniques have led to a better understanding of the genetic determinants of disease occurrence. They have therefore promoted the emergence of genetic epidemiology as a subspecialty of epidemiology [89], with a specific definition [90].

The prevalence of some cancers has also changed over time. Epidemiology has made it possible to identify some risk factors related to the environment and lifestyles. Nevertheless, the etiology of many types of cancer is still poorly understood. Indeed, it is generally accepted that most cancers result from the combined effects of environmental and genetic factors and that only a few cancers are only "of genetic origin" [91]. The integration of molecular techniques into epidemiological studies has also led to the emergence of molecular epidemiology [92], with a definition proposed by Schulte et al. in 2012 [57].

Ecology and environment are a concern of human populations. The same applies to their impact on health. The emergence of environmental epidemiology is a particular reflection of this. The consideration of environment in the definition of epidemiology appeared in the 2000s, with the definition of Gerstman [30], which includes the term "modern" (with reference to old definitions). In previous years, only Cole, 1979 [82] refers to the environment in his definition of epidemiology.

The digital revolution, linked to the rapid and unprecedented increase in the availability of data from various digital sources, has created new opportunities for collecting and analyzing data produced outside the health system. This led Salathé et al. [93] to introduce the term "digital epidemiology" and to propose a definition in 2018 [94]. Some tools such as Google Trends are available and their use has led to mixed results [95].

The use of medical devices, whether implanted or not, has become crucial in many medical specialties for the diagnosis and treatment of certain diseases. Although their use is frequently beneficial to the patient, side effects may occur sometimes (e. g. infection on a vascular catheter). Monitoring these side effects and studying the risk factors for their occurrence is a necessity that has led to the individualization of the epidemiology of medical devices, with a specific definition [96, 97]. The methodological aspects specific to this field of epidemiology (definition of "real" exposure; choice of short- or long-term assessment criteria; various biases) were reported by Jalbert et al. [98].

Faced with the numerous specialties and the numerous books about on the topics, we had to make a selection. We recall that the main objective was general epidemiology, the description of the definitions of subspecialties of epidemiology being a complement.

The concepts found in the definitions of subspecialties of human epidemiology and those of veterinary epidemiology are generally the same. Consideration of economic aspects (productivity) is a specific feature of veterinary epidemiology.

The concepts found in the definitions of human epidemiology available on the websites are generally the same as those contained in the books or articles. Nevertheless, bibliographic references are often unavailable and access of information may change over time, with the disappearance of pages [99] or the existence of dead links ("error 404"…). The posting date is also rarely available.

Publications about the definitions of epidemiology exist. The fields and the aim are for example: introductory epidemiology textbooks [10]; description of definitions epidemiology over 50 years [9]; epidemiology and the methods needed for public health assessments [100]; evolution of epidemiological methods and concepts [11]. But our review examines the evolution of definitions of epidemiology, taking into account more than 100 definitions of epidemiology retrieved from books or articles, introductory texts or not, general epidemiology or subspecialties, and a non-temporal description and analysis of the definitions of veterinary medicine. We also describe 29 definitions of epidemiology retrieved from websites.

This study has several limitations. Only English definitions were retained. National definitions of epidemiology not translated into English were eliminated [101]. These definitions may have contained terms and concept related to epidemiology that were different from those in the definitions retained. But the probability that we have missed an important concept seems quite low. The terms and concepts identified depend on the content of the definitions selected. The definitions of epidemiology available in online dictionaries and other media were easy to find but the quality of the definitions was heterogeneous. To determine the quality of definitions, quality criteria need to be used. For the definitions from 1978 to 2017, the quality criteria may seems subjective. Moreover, no weights were assigned to definitions. The choice of weighting criteria may be subjective.

The purpose of the study was to identify concepts related to epidemiology and not to prioritize them. Given the variety of definitions, the fact that the method used to identify terms and concepts was easy to use and the fact that the majority of terms and concepts were present in more than 10% of definitions suggests that for the definitions retained all of the terms and concepts related to epidemiology were identified.

For future studies, we propose some weighting criteria such as inverse-variance weighting (meta-analysis…) which are widely accepted. We also propose that some of the following criteria be taken into account in the weighting grid: the number of concepts in the definition; the book type (collective dictionary as those of Last/ Porta; advanced texts; introductory texts).

CONCLUSIONS

In summary, this work led to a synthesis of different concepts related to epidemiology proposed during the period 1978–2017 and highlighted the evolution of the content of definitions of epidemiology over time. Most of the terms and concepts identified by us had already been used in the definitions of Lilienfeld while several terms and concepts identified by Evans were no longer used in definitions from 1978 to 2017. Increased usage of the terms "control" and "health" was found in definitions of epidemiology from 1978 to 2017. A thematic analysis of definitions of epidemiology could be conducted to complete this study, in order to improve our understanding of the changes observed.

ACKNOWLEDGMENTS

The authors wish to thank Philippe Bastable for his assistance with editing the manuscript.

REFERENCES

1. Lilienfeld DE. Definitions of epidemiology. Am J Epidemiol. 1978;107(2):87–90. pmid:341693
2. Abramson JH. Re: "definitions of epidemiology". Am J Epidemiol. 1979;109(1):99–102 pmid:433921
3. Evans AS. Re: "Definitions of epidemiology". Am J Epidemiol. 1979;109(3):379–82. pmid:453174
4. Coughlin SS, Beauchamp TL, Weed DL. Ethics and Epidemiology: Oxford University Press; 2009 2009. 328 p. https://doi. org/10. 1097/ EDE. 0b013e3181a90ff1
5. Broadbent A. Philosophy of Epidemiology: Springer; 2013 2013/06/10/. 238 p.
6. Amsterdamska O. Demarcating Epidemiology. Science, Technology, & Human Values. 2005;30(1):17–51.
7. Morabia A. Enigmas of Health and Disease: How Epidemiology Helps Unravel Scientific Mysteries: Columbia University Press; 2014 2014/06/24/. 297 p.
8. Carter-Pokras OD, Spirtas R, Bethune L, Mays V, Freeman VL, Cozier YC. The training of epidemiologists and diversity in epidemiology: findings from the 2006 Congress of Epidemiology survey. Annals of Epidemiology. 2009;19(4):268–75. pmid:19344867
9. Krieger N. Epidemiology and the People's Health: Theory and Context. 1 ed. Oxford: OUP USA; 2013 2013/09/12/. 400 p.
10. Bhopal R. Which book? A comparative review of 25 introductory epidemiology textbooks. Journal of Epidemiology and Community Health. 1997;51(6):612–22. pmid:9519123
11. Zhang FF, Michaels DC, Mathema B, Kauchali S, Chatterjee A, Ferris DC, et al. Evolution of epidemiologic methods and concepts in selected textbooks of the 20th century. Sozial- Und Praventivmedizin. 2004;49(2):97–104. pmid:15150860
12. Kass RE, Raftery AE. Bayes Factors. Journal of the American Statistical Association. 1995;90(430):773–95.
13. R. Core Team. R: A Language and Environment for Statistical Computing. Vienna, Austria: R Foundation for Statistical Computing; 2017.
14. Wasserstein RL, Lazar NA. The ASA's Statement on p-Values: Context, Process, and Purpose. The American Statistician. 2016;70(2):129–33.

15. Buck C. The Challenge of Epidemiology: Issues and Selected Readings: Pan American Health Org; 1988 1988. 1046 p.
16. Schoenbach VJ, Rosamond WD. Understanding the fundamentals of epidemiology: an evolving text. Chapel Hill: North Carolina. 1999.
17. Bhopal R. Concepts of Epidemiology: Integrating the Ideas, Theories, Principles and Methods of Epidemiology. 2 edition ed. Oxford; New York: Oxford University Press, USA; 2008 2008/10/15/. 456 p.
18. Rothman KJ. Epidemiology: An Introduction: OUP USA; 2012 2012/06/21/. 281 p.
19. Savitz DA, Poole C, Miller WC. Reassessing the role of epidemiology in public health. American Journal of Public Health. 1999;89(8):1158–61. pmid:10432898
20. Rossignol A. Principles and Practice of Epidemiology: An Engaged Approach: McGraw-Hill Companies,Incorporated; 2005 2005/12/14/. 336 p.
21. Saracci R. Epidemiology: A Very Short Introduction: OUP Oxford; 2010 2010/02/25/. 143 p.
22. Ward H, Toledano MB, Shaddick G, Davies B, Elliott P. Oxford Handbook of Epidemiology for Clinicians: OUP Oxford; 2012 2012/05/24/. 417 p.
23. Wassertheil-Smoller S, Smoller JW. Biostatistics and epidemiology: a primer for health and biomedical professionals. Fourth edition ed. New York: Springer; 2015 2015. 260 p.
24. Rothman KJ, Greenland S, Lash TL. Modern Epidemiology: Lippincott Williams & Wilkins; 2008 2008. 776 p.
25. Kelsey JL, Thompson WD, Evans AS. Methods in Observational Epidemiology: Oxford University Press; 1986 1986. 366 p.
26. Aschengrau A, Seage GR. Essentials of Epidemiology in Public Health: Jones & Bartlett Publishers; 2013 2013/06/03/. 547 p.
27. Kleinbaum DG, Kupper LL, Morgenstern H. Epidemiologic Research: Principles and Quantitative Methods: John Wiley & Sons; 1982 1982/05/15/. 552 p.
28. Vaughan P, Morrow RH. Manual of epidemiology for district health management. Vaughan P, Morrow RH, editors. Geneva: World Health Organization; 1989 1989. 198 p.
29. Friis RH, Sellers TA. Epidemiology for Public Health Practice: Jones and Bartlett; 2004 2004. 650 p.

30. Gerstman BB. Epidemiology Kept Simple: An Introduction to Classic and Modern Epidemiology, Second Edition. 2 edition ed. Hoboken, N. J: Wiley-Liss; 2003 2003/09/04/. 436 p.
31. Keyes KM, Galea S. Epidemiology Matters: A New Introduction to Methodological Foundations: Oxford University Press; 2014 2014. 257 p.
32. WHO. WHO | Epidemiology [updated 2017/08/27/15:40:58]. Available from: http://www. who. int/topics/epidemiology/en/.
33. Frerichs RR, Neutra R. RE: "definitions of epidemiology". Am J Epidemiol. 1978;108(1):74–5. pmid:685977
34. Berkman LF, Macintyre S. The measurement of social class in health studies: old measures and new formulations. IARC scientific publications. 1997;(138):51–64. pmid:9353663
35. Webb P, Bain C. Essential Epidemiology: An Introduction for Students and Health Professionals. 2 edition ed. Cambridge, UK; New York: Cambridge University Press; 2010 2010/12/16/. 466 p.
36. Oakes JM, Kaufman JS. Methods in social epidemiology. San Francisco, Calif. : Jossey-Bass; 2006 2006. 400 p.
37. Szklo M, Nieto J. Epidemiology: Jones & Bartlett Publishers; 2014 2014. 530 p.
38. Rich H. More on definitions of epidemiology. Am J Epidemiol. 1979;109(1):102–.
39. Stedman . Stedman's Medical Dictionary. Twenty-Eighth edition ed. Philadelphia: Lippincott Williams and Wilkins; 2005 2005/12/01/. 2100 p.
40. Morabia A. A history of epidemiologic methods and concepts. Morabia A, editor. Basel; Boston: Springer Science & Business Media; 2004 2004. 405 p.
41. Schwartz S, Susser E, Susser M. A future for epidemiology? Annual Review of Public Health. 1999;20:15–33. pmid:10352847
42. Porta M. Dictionary of Epidemiology. 5 ed. Oxford; New York: OUP USA; 2008 2008//03/juillet. 320 p.
43. Last JM, International Epidemiological Association. A dictionary of epidemiology. New York: Oxford University Press; 2001 2001.
44. Fos PJ. Epidemiology Foundations: The Science of Public Health. Pap/Psc ed. San Francisco, CA: John Wiley & Sons; 2011 2011/01/25/. 304 p.

45. Ahrens W, Pigeot I. Handbook of Epidemiology. 2nd ed. 2014 ed. New York: Springer-Verlag New York Inc. ; 2013 2013/11/14/. 2498 p.
46. Macera CA, Shaffer R, Shaffer PM. Introduction to Epidemiology: Distribution and Determinants of Disease: Cengage Learning; 2013 2013/06/25/. 498 p.
47. Khan AR, Narayan KA. Lecture Notes on Epidemiology: Penang Medical College; 2014 2014/08/29/. 195 p.
48. Merrill RM. Introduction to Epidemiology. 7th Revised edition ed. Burlington, MA: Jones and Bartlett Publishers, Inc; 2015 2015//18/ décembre. 340 p.
49. Schneider D, Lilienfeld DE. Lilienfeld's Foundations of Epidemiology. 4 ed: Oxford University Press; 2015 2015//06/août. 333 p.
50. Bhopal RS. Concepts of Epidemiology: Integrating the ideas, theories, principles, and methods of epidemiology: Oxford University Press; 2016 2016/09/08/. 481 p.
51. Carneiro I, Howard N. Introduction to epidemiology. 2nd ed ed. Maidenhead, Berkshire; New York: Open University Press; 2011 2011. 183 p. https://doi. org/10. 1097/EDE. 0b013e3182087634
52. Yarnell J. Epidemiology and Prevention: A Systems Based Approach: OUP Oxford; 2007 2007. 310 p.
53. Boslaugh S. Encyclopedia of epidemiology. Boslaugh S, editor. Los Angeles: Sage Publications; 2008 2008. 1240 p.
54. Hu F. Obesity Epidemiology. 1 ed: Oxford University Press; 2008 2008/03/21/. 498 p.
55. Willett W. Nutritional Epidemiology. 3 ed: Oxford University Press; 2012 2012/11/07/. 552 p. https://doi. org/10. 1097/EDE. 0b013e31825afb0b
56. Ioannidis JPA. Genetic and molecular epidemiology. Journal of Epidemiology and Community Health. 2007;61(9):757–8. pmid:17699527
57. Schulte PA, Perera FP. Molecular Epidemiology: Principles and Practices: Academic Press; 2012 2012/12/02/. 609 p.
58. Austin M, Thornton T, Stanford J, Stamatoyannopolous J, Schwartz S, Psaty B, et al. Genetic Epidemiology: Methods and Applications. 1 ed. Wallingford, Oxfordshire, UK: CABI Publishing, 2013 2013//27/ juin. 216 p.

59. Labarthe DR. Epidemiology and Prevention of Cardiovascular Diseases: A Global Challenge: Jones and Bartlett Publishers, Inc; 2010 2010//01/mai.
60. Schottenfeld D, Fraumeni JF. Cancer epidemiology and prevention. Oxford; New York: Oxford University Press; 2006 2006.
61. Susser E, Schwartz S, Morabia A, Bromet E. Psychiatric Epidemiology: Searching for the Causes of Mental Disorders. 1 ed. Oxford; New York: OUP USA; 2006 2006//06/juillet. 544 p.
62. Last JM, Abramson JH, International Epidemiological A. A dictionary of epidemiology. 3rd ed ed. Last JM, Abramson JH, International Epidemiological A, editors. New York: Oxford University Press; 1995 1995. 180 p.
63. Susser M. Causal thinking in the health sciences: concepts and strategies of epidemiology: Oxford University Press; 1973 1973/08//. 208 p.
64. Dos Santos Silva I. Cancer Epidemiology: Principles and Methods: IARC; 1999 1999. 466 p.
65. Tsuang MT, Tohen M, Jones P. Textbook of Psychiatric Epidemiology: John Wiley & Sons; 2011 2011/03/25/. 666 p.
66. Burton PR, Tobin MD, Hopper JL. Key concepts in genetic epidemiology. The Lancet. 2005;366(9489):941–51.
67. Palmer L, Burton PR, Smith GD. An introduction to genetic epidemiology. Palmer L, Burton PR, Smith GD, editors. Bristol: Policy Press; 2011 2011. 230 p.
68. Louis GMB, Platt RW. Reproductive and Perinatal Epidemiology. 1 ed. Oxford; New York: OUP USA; 2011 2011//26/mai. 360 p. https://doi. org/10. 1097/EDE. 0b013e3181fe759f
69. MacMahon B, Pugh TF. Epidemiology: Principles and Methods: Little, Brown; 1970 1970. 408 p.
70. Dohoo I, Martin W, Stryhn H. Veterinary Epidemiologic Research. 2nd edition ed. Charlottetown, CA: VER Inc. ; 2009 2009. 865 p.
71. Pfeiffer D. Veterinary Epidemiology: An Introduction. 1 ed. Chichester, West Sussex: Wiley-Blackwell; 2009 2009//31/décembre. 150 p.
72. Thrusfield M. Veterinary Epidemiology: John Wiley & Sons; 2013 2013/04/24/. 1289 p.
73. Schwabe CW, Riemann H, Franti CE. Epidemiology in Veterinary Practice: Lea & Febiger; 1977 1977. 303 p.

74. Moon G, Gould M. Epidemiology: An Introduction: McGraw-Hill Education (UK); 2000 2000/08/01/. 210 p.
75. Bonita R, Beaglehole R, Kjellström T, Organization WH. Basic Epidemiology: World Health Organization; 2006 2006. 227 p. https:// doi. org/10. 1097/01. ede. 0000229155. 05644. 43
76. Merrill RM. Introduction to epidemiology. 6th ed ed. Burlington, Mass: Jones & Bartlett Learning; 2013 2013. 434 p.
77. Oleckno WA. Epidemiology: Concepts and Methods: Waveland Press; 2008 2008/01/18/. 664 p.
78. Semenza JC, Lindgren E, Balkanyi L, Espinosa L, Almqvist MS, Penttinen P, et al. Determinants and Drivers of Infectious Disease Threat Events in Europe. Emerg Infect Dis. 2016;22(4):581–9. Epub 2016/03/18. pmid:26982104; PubMed Central PMCID: PMC4806948.
79. Nugent R. Chronic Diseases in Developing Countries: Health and Economic Burdens. Annals of the New York Academy of Sciences. 2008;1136(1):70–9.
80. Kuller LH. Epidemiology is the study of "epidemics" and their prevention. Am J Epidemiol. 1991;134(10):1051–6. pmid:1746513
81. Gurwitz AA. The spreading worldwide obesity epidemic. Journal of Community Health. 2014;39(5):827. pmid:24938851
82. Cole P. The evolving case-control study. Journal of Chronic Diseases. 1979;32(1–2):15–27. pmid:312804
83. Shy CM. Epidemiologic principles and methods for occupational health studies: The National Institute for Occupational Safety and Health (NIOSH); [updated 2017/08/27/15:55:52]. Available from: https://www. cdc. gov/niosh/nioshtic-2/00171282. html.
84. Luepker RV. Careers in Cardiovascular Disease Epidemiology and Prevention. Circulation. 2009;120(6):533–8. pmid:19667244
85. Freeman MD, Zeegers MP. Forensic epidemiology: principles and practice. Freeman MD, Zeegers MP, editors. Amsterdam: Elsevier/ Academic Press; 2016 2016. 416 p.
86. Lilienfeld DE. The general epidemiologist: is there a place in today's epidemiology? Am J Epidemiol. 2007;166(1):1–4. pmid:17496312
87. Khoury MJ, Beaty TH, Cohen BH. Fundamentals of genetic epidemiology. New York: Oxford University Press; 1993 1993. 383 p.

88. Hofman A. New studies, technology, and the progress of epidemiology. Eur J Epidemiol. 2010;25(12):851–4. pmid:21153562
89. Kaprio J. Genetic epidemiology. BMJ. 2000;320(7244):1257–9. pmid:10797040
90. Rao DC. Genetic epidemiology of coronary heart disease: past, present, and future: proceedings of a workshop held in St. Louis, Missouri, August 10–12, 1983. Rao DC, National Heart L, Blood I, editors. New York: Liss; 1984 1984. 575 p.
91. Le Marchand L. The predominance of the environment over genes in cancer causation: implications for genetic epidemiology. Cancer Epidemiol Biomarkers Prev. 2005;14(5):1037–9. pmid:15894649
92. Chen Y-C, Hunter DJ. Molecular epidemiology of cancer. CA Cancer J Clin. 2005;55(1):45–54; quiz 7. pmid:15661686
93. Salathé M, Bengtsson L, Bodnar TJ, Brewer DD, Brownstein JS, Buckee C, et al. Digital Epidemiology. PLOS Computational Biology. 2012;8(7):e1002616. pmid:22844241
94. Salathé M. Digital epidemiology: what is it, and where is it going? Life Sci Soc Policy. 2018;14. pmid:29302758
95. Cervellin G, Comelli I, Lippi G. Is Google Trends a reliable tool for digital epidemiology? Insights from different clinical settings. J Epidemiol Glob Health. 2017;7(3):185–9. pmid:28756828
96. Brown SL, Bright RA, Tavris DR. Medical Device Epidemiology and Surveillance. New York, NY: John Wiley & Sons; 2007 2007.
97. Shatin D, Bright RA, Astor B. Databases for Studying the Epidemiology of Implanted Medical Devices. In: Johnson FE, Virgo KS, Lairmore TC, Audisio RA, editors. The Bionic Human: Health Promotion for People With Implanted Prosthetic Devices. Totowa, NJ: Humana Press; 2006. p. 115–32.
98. Jalbert JJ, Ritchey ME, Mi X, Chen C-Y, Hammill BG, Curtis LH, et al. Methodological Considerations in Observational Comparative Effectiveness Research for Implantable Medical Devices: An Epidemiologic Perspective. Am J Epidemiol. 2014;180(9):949–58. pmid:25255810
99. Klein M, Nelson ML. Moved but not gone: an evaluation of real-time methods for discovering replacement web pages. International Journal on Digital Libraries. 2014;14(1–2):17–38.

100. Gouda HN, Powles JW. The science of epidemiology and the methods needed for public health assessments: a review of epidemiology textbooks. BMC Public Health. 2014;14:139. pmid:24507570
101. Valleron A-J, Schwartz D, Goldberg M, Salamon R, Collectif. L'épidémiologie humaine: Conditions de son développement en France, et rôle des mathématiques. édition 2006 ed. Paris: EDP Sciences; 2006 2006//15/juin. 424 p.
102. Barker DJP, Rose GA. Epidemiology in medical practice. 3rd ed ed. Edinburgh; New York: Churchill Livingstone; 1984 1984. 158 p.
103. Lilienfeld AM, Lilienthal DE. Foundations of Epidemiology: Oxford University Press; 1980 1980. 416 p.
104. Miettinen OS. Theoretical epidemiology: principles of occurrence research in medicine. New York: Wiley; 1985 1985. 359 p.
105. Weiss NS. Clinical epidemiology: the study of the outcome of illness. New York: Oxford University Press; 1986 1986. 144 p.
106. Hennekens CH, Buring JE. Epidemiology in Medicine. Boston: Lippincott Williams and Wilkins; 1987 1987//01/août. 400 p.
107. Ahlbom A, Norell S. Introduction to Modern Epidemiology: Epidemiology Resources; 1990 1990. 126 p.
108. Kelsey JL. Methods in Observational Epidemiology: Oxford University Press; 1996 1996. 458 p.
109. Vetter NJ, Matthews I. Epidemiology and Public Health Medicine. 5th Revised edition ed. Edinburgh: Churchill Livingstone; 1999 1999/09/07/. 272 p.
110. Timmreck TC. An Introduction to Epidemiology: Jones & Bartlett Learning; 2002 2002. 532 p.
111. Friedman G, Friedman GD. Primer of Epidemiology. 5 edition ed. New York: McGraw-Hill Education / Medical; 2003 2003/10/16/. 401 p. https://doi. org/10. 1097/01. ede. 0000091680. 23304. 21
112. Coggon D, Barker D, Rose G. Epidemiology for the Uninitiated: Bmj Publishing Group; 2003 2003. 81 p.
113. Farmer RDT, Lawrenson R. Lecture Notes: Epidemiology and Public Health Medicine. 5th Revised edition ed. Malden, Mass: Wiley; 2004 2004/11/12/. 204 p.
114. Gordis L. Epidemiology. Philadelphia: Elsevier/Saunders; 2009 2009. 400 p.

115. Olsen J, Christensen K, Murray J, Ekbom A. An Introduction to Epidemiology for Health Professionals: Springer Science & Business Media; 2010 2010/06/14/. 152 p.

116. Porta M. A Dictionary of Epidemiology. 6 ed. Oxford: OUP USA; 2014 2014//20/juin. 376 p.

117. Krickeberg K, Pham VT, Pham TMH. Epidemiology: Key to Prevention: Springer Science & Business Media; 2011 2011/11/09/. 256 p.

118. Broeck JVd, Brestoff JR. Epidemiology: Principles and Practical Guidelines: Springer Science & Business Media; 2013 2013/04/22/. 622 p.

119. Miettinen OS. Up from Clinical Epidemiology & EBM. 1. , Edition ed. Berlin: Springer Netherland; 2011 2011. 175 p.

120. Miettinen OS. Epidemiological research: terms and concepts. Dordrecht: Springer; 2011 2011. 175 p.

121. Gerstman BB. Epidemiology kept simple: an introduction to traditional and modern epidemiology. 3rd ed ed. Chichester, West Sussex: John Wiley & Sons; 2013 2013. 461 p.

122. Yarnell J, O'Reilly D. Epidemiology and Disease Prevention: A Global Approach: OUP Oxford; 2013 2013/05/23/. 450 p.

123. Fletcher RH, Fletcher SW, Fletcher GS. Clinical epidemiology: the essentials, 5th ed. Philadelphia: Wolters Kluwer; 2014. 253 p.

124. Katz DL, Jekel JF. Jekel's epidemiology, biostatistics, preventive medicine, and public health. 4th ed ed. Katz DL, Jekel JF, editors. Philadelphia, Pa. ; London: Saunders; 2014 2014. 405 p.

125. Weiss NS, Koepsell TD. Epidemiologic Methods: Studying the Occurrence of Illness: Oxford University Press; 2014 2014/07/03/. 636 p.

126. Patten S. Epidemiology for Canadian Students: Principles, Methods and Critical Appraisal: Brush Education; 2015 2015/03/23/. 297 p.

127. Stewart A. Basic statistics and epidemiology: a practical guide. Fourth edition ed. Boca Raton: CRC Press, Taylor & Francis Group; 2016 2016. 218 p.

128. Webb P, Bain C, Page A. Essential Epidemiology: An Introduction for Students and Health Professionals: Cambridge University Press; 2016 2016/11/30/. 864 p.

129. Friis RH. Epidemiology 101: Jones & Bartlett Learning; 2017 2017/03/02/. 317 p.

130. Adami H-O, Hunter DJ, Trichopoulos D. Textbook of Cancer Epidemiology: Oxford University Press; 2008 2008. 783 p.
131. Loda M, Mucci L, Mittelstadt ML, Hemelrijck MV, Cotter MB. Pathology and Epidemiology of Cancer: Springer; 2016 2016/09/01/. 664 p.
132. Morris JN. Uses of Epidemiology. 3rd Revised edition ed. Edinburgh; New York: Churchill Livingstone; 1976 1976 août. 332 p.
133. Lilienfeld AM. Foundations of epidemiology. New York: Oxford Univ. Press; 1976 1976. 283 p.
134. Rothman KJ. Modern epidemiology. 1st ed ed. Boston: Little, Brown; 1986 1986. 358 p.
135. MacMahon B, Trichopoulos D. Epidemiology: Principles and Methods: Little, Brown; 1996 1996. 347 p.
136. Teare MD. Genetic Epidemiology: Humana Press; 2016 2016//15/mai. 256 p.
137. Giesecke J. Modern Infectious Disease Epidemiology, Third Edition. 3 ed. Boca Raton, FL: CRC Press; 2017 2017/03/08/. 192 p.
138. Wilkinson P. Environmental Epidemiology. 1 edition ed. Wilkinson P, editor. Maidenhead, Berkshire, England: Open University Press; 2006 2006/03/01/. 224 p.
139. Merrill RM. Environmental epidemiology: principles and methods. Sudbury, Mass: Jones and Bartlett Publishers; 2008 2008. 483 p.
140. Baker D, Nieuwenhuijsen MJ. Environmental Epidemiology: Study methods and application. 1 ed. Oxford; New York: OUP Oxford; 2008 2008//18/août. 416 p.
141. Venables K. Current Topics in Occupational Epidemiology. Oxford, United Kingdom: OUP Oxford; 2013 2013/09/15/. 288 p.
142. O'Campo P, Dunn JR. Rethinking Social Epidemiology: Towards a Science of Change. 2012 edition ed. O'Campo P, Dunn JR, editors. Dordrecht; New York: Springer; 2011 2011/10/04/. 350 p.
143. Berkman LF, Kawachi I. Social Epidemiology: Oxford University Press, USA; 2000 2000/03/09/. 428 p.
144. James SA. Epidemiologic research on health disparities: some thoughts on history and current developments. Epidemiologic Reviews. 2009;31:1–6. pmid:19822533

145. Berkman LF, Kawachi I, Glymour M. Social Epidemiology. 2 edition ed. Berkman LF, Kawachi I, Glymour M, editors. Oxford: Oxford University Press; 2014 2014/08/06/. 640 p.
146. Martin SW, Meek AH, Willeberg P. Veterinary Epidemiology: Principles and Methods: Iowa State University Press; 1987 1987. 343 p.
147. Smith RD. Veterinary Clinical Epidemiology, Third Edition: CRC Press; 2005 2005/09/07/. 288 p.
148. Toma B. Dictionary of veterinary epidemiology. 1st English language ed ed. Iowa State University P, editor. Ames: Iowa State University Press; 1999 1999. 284 p.
149. Slater MR. Veterinary Epidemiology: Butterworth-Heinemann; 2003 2003. 244 p.
150. Halpin B. Patterns of Animal Disease: Baillière Tindall; 1975 1975. 184 p.
151. Sergeant E, Perkins N. Epidemiology for Field Veterinarians: An Introduction. First ed: CABI; 2015 2015//28/juillet. 320 p.
152. Dicker R, Coronado F, Koo D, Parrish RG. Principles of Epidemiology in Public Health Practice, 3rd Edition. 3rd edition ed: CDC; 2006 2006.

CHAPTER 2

The Application of Epidemiology in Aquatic Animal Health -Opportunities and Challenges

Edmund J Peeler and Nicholas GH Taylor

Centre for Environment, Fisheries and Aquaculture Science (Cefas), Barrack Road, Weymouth, Dorset, DT4 8UB, UK

ABSTRACT

Over recent years the growth in aquaculture, accompanied by the emergence of new and trans boundary diseases, has stimulated epidemiological studies of aquatic animal diseases. Great potential exists for both observational and theoretical approaches to investigate the processes driving emergence but, to date, compared to terrestrial systems, relatively few studies exist in aquatic animals. Research using risk methods has assessed routes of introduction of

Citation: Peeler, E.J., Taylor, N.G. The application of epidemiology in aquatic animal health -opportunities and challenges. Vet Res 42, 94 (2011). https://doi.org/10.1186/1297-9716-42-94

aquatic animal pathogens to facilitate safe trade (e.g. import risk analyses) and support biosecurity. Epidemiological studies of risk factors for disease in aquaculture (most notably Atlantic salmon farming) have effectively supported control measures. Methods developed for terrestrial livestock diseases (e.g. risk-based surveillance) could improve the capacity of aquatic animal surveillance systems to detect disease incursions and emergence. The study of disease in wild populations presents many challenges and the judicious use of theoretical models offers some solutions. Models, parameterised from observational studies of host pathogen interactions, have been used to extrapolate estimates of impacts on the individual to the population level. These have proved effective in estimating the likely impact of parasite infections on wild salmonid populations in Switzerland and Canada (where the importance of farmed salmon as a reservoir of infection was investigated). A lack of data is often the key constraint in the application of new approaches to surveillance and modelling. The need for epidemiological approaches to protect aquatic animal health will inevitably increase in the face of the combined challenges of climate change, increasing anthropogenic pressures, limited water sources and the growth in aquaculture.

INTRODUCTION

Epidemiology was originally considered as the study of pathogen lifecycles, but is now more comprehensively defined in the Oxford English dictionary as "*the branch of medicine which deals with the incidence, distribution, and possible control of diseases and other factors relating to health*". The terms "epizootic" and "epizootiology" have been used in the veterinary literature but in keeping with current practice we refer to "epidemics" and "epidemiology" in this paper [1]. The birth of modern veterinary epidemiology can be traced back to work by Schwabe [2], who identified that traditional approaches to disease control were failing to resolve animal health problems occurring in intensive livestock production systems. From its inception, modern epidemiology was primarily an applied discipline, practised in the field, with the purpose of preventing and resolving animal health problems. The requirement for inter-disciplinary approaches was central to modern epidemiology, and it encompasses aspects of biostatistics, animal health economics, risk analysis and theoretical modelling studies [3]. A founding principle of modern epidemiology is that disease is caused by multiple interacting factors, which was formulated in the early 1970s as the causal triad of pathogen, host and environment [4]. However, it has since been argued that pathogens should be considered as a component of the environment [5].

Until relatively recently, epidemiological research focused on human and terrestrial animal systems with comparatively little effort directed towards aquatic animal health. However, with 71% of the Earth's surface being covered in water, we live on a wet planet. Life began in water, and the diversity of aquatic organisms is greater than on land. For example, viruses are the most abundant life form in the oceans. There are estimated to be millions of virus-like particles in every millilitre of seawater [6] and that every second 10^{23} viral infections occur in the marine environment [7].

Humans use the aquatic environment for a range of services: potable water, wild capture fisheries, aquaculture, irrigation, travel, recreation and waste disposal. The interaction between humans and aquatic environment means that the study of pathogens in aquatic animals is important to protect a valuable food and recreational resource, and becaise diseases of humans may originate in the aquatic environment [8]. Globally aquaculture and fisheries supplied 110 million tonnes of food in 2006, of which 47% was accounted for by aquaculture [9], compared with only 3.9% in 1970 [10]. The majority of wild capture fisheries are over-exploited [9], growth in consumer demand and new technologies has led to a 10% per annum growth in aquaculture over between 1970 to 2008 [9]. Over 230 aquatic animal species are now farmed [10]. The development of aquaculture has, to a significant extent, depended on the use of non-native species [11] (e.g. marine culture of Altantic salmon in Chile and tilapia and sea-bass in land-based systems in the UK). The diversity and number of species being produced has also resulted in new culture systems and practices, which themselves could influence the emergence and spread of pathogens and occurrence of disease [12]. For example, polyculture (keeping several species on the same site) provides the opportunity for pathogen transmission between species [13]. Farmed fish may be exposed to wild aquatic animals and their pathogens, to which the they would not normally exposed. Similarly, aquaculture may expose wild fish to novel pathogens. Both cases are illustrated by tuna ranching in southern Australia. Juvenile tuna are caught in the open ocean and moved inshores, where they come in to contact with parasites to which they would normally have little or no exposure, resulting in disease emergence [14–16]. Additionally, the tuna are fed imported frozen fish which was very likely to have been the route of introduction of pilchard herpes virus, which has caused epidemics in wild populations [17].

Fish are one of the most commonly kept companion animals (in the UK there are estimate to be 135 million [18]) and as a consequence the global trade in ornamental aquatic animals is massive. Over 1 billion ornamental fish comprising more than 4000 freshwater and 1400 marine species are traded internationally each year [19, 20]. Trade in live fish (for aquaculture, food and as ornamental animals) and changing climates has expanded the geographic range of many aquatic animal species (and their pathogens), facilitating disease emergence through host-switching [21–23].

The uses of aquatic animals produce interactions between industry sectors that are not obvious and may drive disease emergence and pathogen spread. For example, within the UK cyprinid fish (e.g. carp and goldfish) industry there are two main interacting sectors: the ornamental and coarse (sport) fish (Figure 1). In summary, cyprinid fish may be imported from both European and non-European countries, the majority of which will be intended for the ornamental (fish keeping) sector. In addition cyprinid farm production with the UK supplies both sectors. Fish are also moved between fisheries either directly or via dealers. Data on fish movement in this network is patchy. It is believed that many movements between fisheries are non-consented and therefore not officially recorded [24], and additionally hobbyists are known to stock fish into lakes and rivers [19, 25–27]. The cross-over that occurs between the ornamental and coarse fishing sectors has the potential to contribute substantially to pathogen spread in fisheries, and is believed to have been one of the main drivers in the early stages of the koi herpesvirus epidemic in the UK [24].

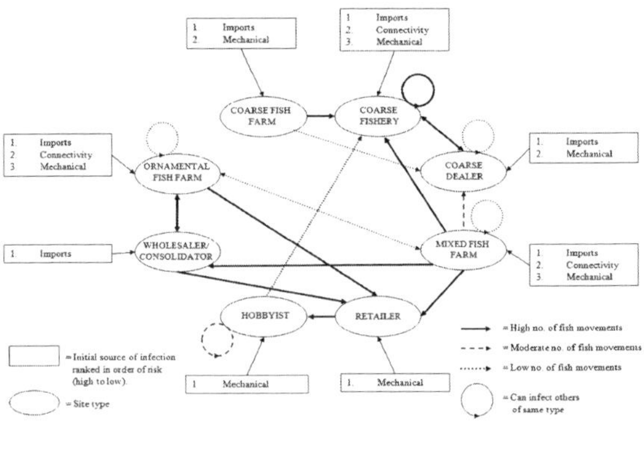

Figure 1. Contact network structure occurring between businesses in the UK cyprinid fish sector, demonstrating the potential routes of pathogen introduction and spread (connectivity refers to sites being connected by the river network, and mechanical refers to transmission via vectors and fomites).

The diversity and interactions of aquatic systems presents a series of unique challenges and complexities that requires the holistic approach which is fundamental to modern epidemiology. This review aims to summarise the application of aquatic animal epidemiology in three key areas of aquatic animal health: i) transboundary and emerging aquatic animal diseases, ii) disease monitoring, surveillance and surveys, and iii) spread and impact of aquatic animal pathogens.

In addressing these questions, epidemiological approaches needed for diseases of livestock are compared with those required by studies of aquatic animal disease. Examples of where these approaches have been applied, their success and limitations, are discussed. Unique challenges and constraints that arise in epidemiological studies of aquatic systems are identified. The underlying theme running through this paper is the role of epidemiology in improving aquatic animal health.

THE DEVELOPMENT OF AQUATIC EPIDEMIOLOGY

It has been argued that fish health research has focused on isolating and characterising pathogens, at the expense of host and environmental factors [28]. Only relatively recently have epidemiological studies investigated causes of aquatic animal diseases under field conditions. Inevitably the development of epidemiological methodology has focused firstly human, and secondly terrestrial livestock health issues. As a consequence the development of aquatic epidemiology has lagged behind its terrestrial counterparts, however in recent years the field has developed rapidly. Epidemiological studies of aquatic animals systems are becoming more common place and the techniques used have increased in complexity.

The beginning of aquatic animal epidemiology mirrored the evolution of modern epidemiology in terrestrial livestock. The first risk factor study was published by Thorburn [29], who applied approaches previously used in terrestrial systems to investigate mortality due to vibriosis in farmed rainbow trout. Intensive Atlantic salmon (the first large scale commercially produced fish) production was pioneered in Norway in the 1970s. Inevitably, disease problems emerged, most notably infectious salmon anaemia (ISA) [30]. ISA and other diseases stimulated epidemiological research to support cost-effective control strategies in farmed Atlantic salmon [31–33]. Subsequently, other important diseases have emerged in aquaculture, some with serious economic and environmental consequences. *Gyrodacytlus salaris* emerged in Norway in the early 1970s causing serious declines in many wild Atlantic salmon populations [34]. Farmed Atlantic salmon may act as a reservoir for infection, thus contributing to populations declines in wild salmon [35]. A number of viral diseases, notably white spot syndrome virus (WSSV), have emerged in the nascent paenid shrimp farming sector, with devastating economic consequences in some countries [36]. Cold water strawberry disease emerged in 2006 in the UK, causing financial losses to trout producers [37]. Disease emergence has clearly provided an important spur to the application of epidemiology.

The growth of aquatic animal epidemiology is illustrated by an analysis of the number and type of studies presented at the International Society of Veterinary Epidemiology and Economics (ISVEE) conferences, held every three years since 1976 (Table 1). The number of aquatic abstracts and the species being studied have increased, the questions addressed and the techniques used have become more diverse (the location of the conference may influence the number of aquatic animal submissions). Shellfish (molluscs and crustacea) production accounts for over a quarter of the aquatic animals consumed annually [9], but have been the subject of comparatively few studies (which are predominantly of cultured shrimp). This may be explained by the limited options for disease control.

Table 1 Analysis of oral presentations at meetings of the International Society for Veterinary Epidemiology and Economics (ISVEE), 1976-2009

Year	Total papers presented	Aquatic papers presented	Species (number of presentations)	Subject area (number of presentations)
1976 - 1982	246	0 (0%)		
1985	117	1 (0.85%)	Fish (1)	Economics/Control (1)
1988	164	1 (0.61%)	Fish (1)	Diagnostic tools (1)
1991	207	2 (0.97%)	Fish (2)	Control (1) Pathogen dynamics (1)
1994	228	0 (0%)	NA	NA

1997	423	9 (1.83%)	Fish (7) Shellfish (2)	Baseline/monitoring/investigation (6) Control (1) Risk analysis (1) Risk factors (1)
2000	603	15 (2.49%)	Fish (13) Shellfish (2)	Baseline/monitoring/investigation (6) Risk factors (4) Control (2) Diagnostic testing x2 Risk analysis (1)
2003	691	19 (2.75%)	Fish (16) Shellfish (3)	Baseline/monitoring/investigation (7) Risk factors (4) Risk analysis (2) Zoonosis (2) Epidemiological tools (2) Disease modelling (1) Control (1)
2006	911	67 (7.35%)	Fish (41) Shellfish (12) Amphibians (4) General (10)	Baseline/monitoring/investigation (25) Risk factors (13) Epidemiological tools (9 Control (6) Risk analysis (5) Diagnostic testing (4) Pathogen dynamics (2) Disease modelling (1) Zoonosis (1) Economic (1)

2009	755	26 (3.44%)	Fish (24) Shellfish (1) General (1)	Baseline/monitoring/investigation (11) Risk factors (6) Epidemiological tools (2) Control (2) Risk analysis (2) Diagnostic testing (1) Pathogen dynamics (1) Disease modelling (1)

TRANSBOUNDARY AND EMERGING DISEASES

A key role for epidemiology is to identify and assess exotic and emerging disease threats. Transboundary diseases are a major threat to both farmed and wild terrestrial and aquatic animal populations [21]. In recent years, a number of aquatic animal diseases have emerged and spread rapidly across international political boundaries, causing serious financial and ecological impacts. WSSV emerged in shrimp (*Paenus monodon*) culture in China (Fuijan Province) in 1992 and subsequently spread to Taiwan and Japan and then became panzootic [36]. The main route of transmission was the movement of live animals [36]. *G. salaris* emerged in Norway in the early 1970s, and is hypothesised to have been introduced from Sweden with the introduction of a Baltic strain of Atlantic salmon [34]. A new variant of oyster herpes virus recently emerged in European oyster production [38] and has spread rapidly through live animals movements. Koi herpesvirus achieved widespread distribution across the globe within a few years of emerging through movements of carp, mainly destined for the ornamental trade industry [39]. The movements of bullfrogs (as pets, food and for laboratory use) has been implicated in the spread of the fungal disease, chytridiomycosis in wild amphibia [40]. Movements of animals drive disease emergence not only through extending the geographic range of known diseases but by moving species outside of their natural ranges, thus providing the opportunity for putative pathogens to find new hosts [22]. The scale and diversity of aquatic animal movements, mainly for the ornamental trade but also for food and aquaculture, far outweighs international terrestrial movements, and presents important challenges to epidemiologists.

Import Risk Analysis (IRA)

IRA assesses the risk of disease spread through trade in animals or their products. In aquatic animal health they have been undertaken for a variety of reasons (reviewed by Peeler et al., [41]), but mainly to support animal health policy making with respect to trade and biosecurity. Epidemiological approaches are used within the import risk analysis framework [42]. Consistent, defensible and scientifically sound assessments of disease threats and disease freedom are required to fulfil the requirement of the Agreement on the Application of Sanitary and Phytosanitary (the SPS agreement) of the World Trade Organization (WTO) [43] (which was enacted to allow Members to adopt measures to protect human, animal or plant life or health, whilst preventing restrictions on international trade disguised as sanitary measures). Thus this application of epidemiology, for both terrestrial and aquatic animal diseases, have developed in response to both requirements of international trade agreements and the real threat of transboundary diseases. One of first aquatic animal IRA was Biosecurity Australia's assessment of the risk of disease introduction with the import of wild caught salmon carcasses from Canada [44]. IRA requires information on the epidemiology of the diseases in the exporting country, biophysical characteristics of the pathogen and product information to be integrated. As part of the consequence assessment epidemiological and economic modelling are needed to assess disease spread and impact on local industries and wild populations. More broadly, an assessment of routes of introduction (through trade, movement of shipping, lorries etc.) is the necessary basis on which to review a country's biosecurity measures. However, there are constraints to IRA, as currently practised. Frequently, relevant information is not available. IRA cannot account for unidentified hazards; this is particularly relevant for aquatic animal health in general [45] and the trade in ornamental aquatic animals in particular [19].

Aquaculture and Disease Emergence

The same risk framework which has been used for examining disease spread through trade in live animals and animal products have also been applied to disease emergence. Murray and Peeler [12] and Bridges et al. [46] used risk models to investigate drivers for the emergence of disease in aquaculture. However, disease emergence in terrestrial systems has been studied in greater depth [47–49], in particular the evolution of virulence [50–52], the importance of animal reservoirs [53] and multiple hosts

[54–56]. These studies have in large part been driven by the emergence of zoonotic pathogens [57, 58]. Epidemiological studies need to more fully explore the conditions under which putative pathogens spread between wild and farmed aquatic animals. The introduction of non-native species has been fundamental to the expansion of aquaculture. Rainbow trout, a native of North America, is the most important freshwater finfish species in Europe and is cultured across the globe [59]. *Paenus monodon* and other species have been shipped internationally to establish shrimp aquaculture. A recent review demonstrated that the introduction of non-native aquatic animal species underpinned many instances of disease emergence in Europe [22] and this is likely to be true for other parts of the world. It is likely that the culture of new species, the expansion of aquaculture into new areas and the global trade in both live aquatic animals and commodity will continue to increase as the demand for animal protein increases with the growth in human population and the decline of wild capture fisheries [10]. Diseases which emerge in farmed populations invariably originate in wild populations. Thus interaction between wild and farmed populations facilitated by the open design of many aquaculture systems [60, 61] and the current fish farming practices [62] is fundamental to disease emergence. Additionally, restrictions in the use of chemicals or the development of resistance could lead to the re-emergence of old diseases that had previously be controlled. Applying the epidemiological modelling approaches which have been used to study disease emergence in terrestrial systems may provide insights into the processes driving disease emergence in the aquatic environment, and inform the development of improved mitigation measures.

Climate Change and Disease Emergence

Climate change will influence the development of aquaculture and the epidemiology of aquatic animal diseases. Many important aquatic animal pathogens have temperature thresholds, above or below which disease or infection does not occur. A risk assessment model to screen pathogens, to identify those whose threat will increase under established climate change scenarios has been developed for freshwater fish diseases and applied to the UK [23], which identified *Lactococcus garvieae* and *Anphanomyces invadans* as the exotic diseases whose threat will increase most with rising water temperatures. Epidemiological modelling and the application of geographic information systems (GIS) have been used to good effect to assess the likely impact of climate change scenarios on terrestrial livestock diseases (especially vector borne diseases such as blue tongue virus and trypanosomiasis) [63], and huge scope exists to apply these approaches to aquatic animal diseases.

Outbreak Investigations

Outbreak investigation is a core activity for epidemiologists and uses key methods of modern epidemiology: measurement of disease frequency in time and space and in different populations. More recently molecular analyses have been incorporated into outbreak investigations in order to aid source tracking. Outbreak investigations of transboundary diseases are generally undertaken by the veterinary services and generally remain unpublished. However, research papers investigating the origin of transboundary diseases have been published. Investigations of infectious haematopoietic necrosis virus (IHNV) in Europe have attributed its introduction to imported rainbow trout eggs from the US [64]. This hypothesis has been supported by molecular epidemiology studies showing that all European IHNV isolates belong to one of the four N. American genogroups [65]. Conversely, genotyping demonstrated that viral haemorrhagic septicaemia virus (VHSV) in N. America strain had emerged separately from a marine reservoir with no epidemiological link to Europe [66, 67] (where the disease first emerged). Similarly genotyping of ISAV demonstrated emergence from the marine reservoir on separate occasions in Norway, Scotland and N. America [68, 69]. However, genetic analysis of the isolate from the recent outbreak in Chile is identical to Norwegian isolates suggesting spread via imported eggs in 1996 [70, 71].

Despite its growing use and popularity, aquatic molecular epidemiology is yet to reach the level of sophistication applied to terrestrial systems. Examples of the potential power of molecular epidemiology can be seen in studies such as Cottam et al. [72] who, in studying the transmission of foot and mouth disease virus (FMDV) used small mutations found in the entire viral sequence to identify possible previously unidentified contacts in the 2001 epidemic. The availability of technologies that allow for rapid, large-scale sequencing, such techniques provide exciting potential to develop aquatic epidemiology, and may provide a powerful tool to enhance our ability to accurately trace the source and course of infections.

SURVEILLANCE AND SURVEYS

Investigation of Disease Prevalence

Epidemiological textbooks [73, 74] generally define disease monitoring as ongoing efforts to assess health and disease in a specified population. The term surveillance is reserved for a more active system and implies action will be taken based on the results. Surveys collect information for a specific aim or to test a hypothesis. Monitoring and surveillance for aquatic animal disease is undertaken by governments to demonstrate disease freedom or progress in disease control, and generally remains unpublished (the exception being work by East et al. [75] demonstrating freedom in Australia from WSSV). In the scientific literature there are a number of reports of surveys of disease in farmed and wild fish.

Studies of gyrodactylid infections in farmed rainbow trout in Denmark [76, 77] showed considerable between farm and seasonal variation. Investigations of the prevalence of renal myxosporidiosis in wild [78–81] and farmed salmonids [82] were undertaken to assess the impact of these diseases. Studies of *Renibacterium salmoninarum* (Rs) revealed a low prevalence in wild salmonid fish populations [83] and farms [84], and considerable variation between rivers and farms. More recent a sero-survey of koi herpesvirus found that the virus was geographically widespread in fisheries in E&W, but largely absent from farms [25].

Relatively few papers have used more sophisticated spatial and statistical analytical methods However, analysis of data from a large-scale survey of Atlantic salmon in Scotland [85]) modelled the within and between river prevalences to correct the bias that arose from low sample sizes (in some rivers) and pooling [86].

Peeler et al. [81] used multi-level modelling approaches to analyse prevalence data of renal myxosporidiosis and hepatitis in wild brown trout, and concluded that site (compared with factors at the level of the fish or river) level factors exerted most influence.

Developments in Surveillance Methodology

In terrestrial animal epidemiology there has been considerable progress in the development of methods and approaches to support the design and analysis of surveillance systems and data [73]. In this section, the application of these methods to aquatic animal diseases is discussed.

Risk-based surveillance and scenario tree modelling

Risk-based surveillance (RBS) and scenario tree modelling are powerful tool for improving the design and analysis of surveillance systems, not least because complex data sources can be used and historic data discounted [87, 88]. To date there are only publications on the use of RBS and scenario tree modelling in terrestrial systems, however, these method have started to be used in aquatic animal production systems. The more sophisticated approaches used in terrestrial animal surveillance have not yet been applied to aquatic systems, in part, because of an inevitable lag that will occur for methods developed for one system to be refined and used elsewhere. The EU aquatic animal health directive 2006/88/EC requires member states to use RBS, and so will drive the application of this approach in aquatic systems.

Spatial and temporal analysis

The use of for example the scan statistic [89] to detect spatial and temporal clusters or the nearest neighbour technique (first developed by Clark and Evans [90]) to detect spatial clusters, helps identify confounding and risk factors for disease, and generate research hypotheses [91, 92]. These techniques could serve the same role in aquatic animal health research, though spatial analysis would need to account for hydrodynamic connections between populations (e.g. for farms in the same river catchment). In the terrestrial field, data from long term data collection exercises have been used. These have often been collected for statutory purposes (e.g. monitoring for bovine spongiform encephalopathy). Arguably, it is the lack of suitable datasets which have held back the adoption of these approaches. The main exception to this are datasets of sealice infections of farmed Atlantic salmon [93–95] and a single dataset of 14 years of monitoring of two cyprinid populations and their parasites in three rivers in north east England [96], which could be used to assess parasite burdens over time.

Test Validation

Advances in Bayesian statistical methods to determine test characteristics (sensitivity and specificity) in the absence of a gold standard (TAGS) have also contributed to the development of surveillance [97, 98], and are of great importance in aquatic systems as few true gold standard methods currently exist. However, the test characteristics of relatively few diagnostic tests have been validated under field conditions and in the absence of a gold standard, the exception being tests for infectious salmon anaemia [99, 100]. The techniques and expertise exists; once again the main constraint is the lack of resources to test a sufficient number of populations (with different levels of disease). Pooling diagnostic samples occurs more frequently with aquatic compared with terrestrial animals, with many of the tests approved by the World Organisation for Animal health [101] for aquatic animal diseases allow samples from up to 10 animals to be pooled. The effect of pooling on the characteristics of these test have in general not been considered, however, methods to adjust prevalence estimates for pooled samples have been developed [102]. Epidemiologists have an important role to play in designing trials to generate information on test characteristics and to assess the impact of pooling [97].

SPREAD, ESTABLISHMENT AND IMPACT OF PATHOGENS

Identifying Routes of Spread

Understanding the routes of disease spread is of key importance to containment and control. Transmission mechanisms must first be identified and understood, and then the importance of routes of transmission between areas or sites quantified. Studies of the UK 2001 foot and mouth disease (FMD) epidemic provide examples of techniques which may be of use in the study of aquatic systems: farmed based stochastic modelling [103]; social network analysis [104] and space time interaction [105]. As in terrestrial systems, the main routes of transmission in aquatic systems are the movement of live animals, transmission via vectors and fomites and the contiguous nature of some sites (by connecting water sources). The specific processes involved for each route and their importance can be investigated ex-ante by theoretical probabilistic approaches (e.g. risk assessment), and ex-post observational studies of patterns of disease spread.

Ex-ante studies of disease spread

Peeler et al. [41] reviewed the use of risk assessment to study the spread of aquatic animal pathogens between countries, river catchments and farms. A number of risk assessments focused on *G. salaris*. These ex-ante studies tended to focus on one pathway and addressed specific management questions. Paisley et al. [106] used Monte-Carlo simulation methods to study the likelihood of *G. salaris* being introduced to the river Tana in Norway from infected Atlantic salmon smolts in marine cages. The study identified the components of this route that had the most influence in determining the final probability of introduction (salinity at the cage site followed by the time of year fish escaped from the farm). Høgasen et al. [107] used similar methods to assess the risk associated with infected wild Atlantic salmon smolts moving to uninfected catchments and successfully transmitting *G. salaris*. Peeler, Gardiner and Thrush [108] conducted a qualitative risk assessment to establish the most important routes by which *G. salaris* would spread through England and Wales should it be introduced. This study concluded that in England and Wales the most likely transmission routes were the movement of live rainbow trout followed by the movement of other fish species. Such studies provide valuable information for contingency plans to prevent spread should an introduction occur.

Once the contact pathways have been established, their importance in an epidemic and the overall rate of spread can be modelled between sites (or individuals as is often the case with human pathogens). This however relies on an accurate picture of the total population size of susceptible, potential reservoirs, and the degree of contact between them. If the question is simply to determine the rate of pathogen spread under different conditions, simple compartment based epidemic models, in which the host moves from susceptible to infected states [109], may be applied. Murray [110] and Ruane et al. [111] used such models to study the spread and dynamics of infectious pancreatic necrosis virus (IPNV) between salmon farms in Scotland and Ireland, respectively. Both studies suggested that the stocking of infected smolts from freshwater sites drove the epidemic in the marine sites, but Ruane et al. [111] found that endogenous spread between marine sites took over as the dominant transmission route shortly after introduction from freshwater sites. Taylor et al. [112] also found that the role of infection pressures in spreading a pathogen could change as an epidemic progressed: the stocking of imported ornamental fish drove the initial phase of the koi herpesvirus epidemic in England and Wales (E&W), but that live fish movements between lakes became the main driving force after a few years.

Network analysis may be employed to investigate which sites are most likely to be exposed to pathogens. The technique is used to investigate the number and direction of contacts (edges) made by one site to another, and allows several useful measures that may facilitate an understanding of pathogen spread to be derived. In their simplest form they simply model contacts and not transmission. However, they can be combined with epidemic models, and simulations run to determine how a disease might spread through a population. Such approaches are powerful, but rely on a full knowledge of the network structure, which is also assumed to be stable over time, i.e. sites always contact the same sites. In aquatic systems, network approaches have received relatively little attention, as the required contact data is often not available. However, some systems do lend themselves to this approach, and two good examples of this are demonstrated by Thrush and Peeler [113], who looked at live trout movements in E&W, and Green [114] who studied live fish movements between Scottish trout and salmon sites. In the UK, data on live fish movements from farms exists farmers are obliged to keep these records. Both studies used these data to construct contact networks. Thrush and Peeler developed a stochastic model to examine how rapidly a pathogen may spread on the network. Green et al. [115] used a network model to identify sites where movement controls should be targeted to most effectively slow the spread of disease. Sharkey et al. [116] further developed the work of Thrush and Peeler [113] by adding other pathways such as connectivity via the river network to the analysis. Hydrographical connections between farm sites have also been studied in the marine environment using particle tracking, advection decay models. These have been used to good effect in the study of the dispersal of sea lice [117], ISAV [118], pancreas disease virus [119] and IPNV [120]. If network and hyodrographical models were accurately parameterised, for a particular pathogen, they could be used during an outbreak to estimate the likely spread of the pathogen. In addition they can be used to develop outbreak scenarios which to test contingency plans and specifically competing control measures. Such studies are, however, often limited by a lack of available data on the contact network and key epidemiological parameters such as transmission rate, latent periods, duration of shedding and immunity.

Ex-post observational studies

Retrospective observational studies (e.g. case control studies) of spread are a natural extension of outbreak investigations. They have been used extensively in terrestrial epidemiology to identify routes of spread, e.g. analysis of data from the FMD 2001 outbreak [121]. Again *G. salaris* provides an example of a case-control study [122] which used logistic regression to assess routes of spread (whilst accounting for confounding variables); and concluded that the amount of freshwater inflow into fjords and the distance travelled by infected fish were the most important factors. A number of retrospective studies of ISA have provided important insights into routes of spread. Early work in Norway demonstrated that spread from neighbouring infected farms and slaughter houses were important [32, 123]. Mechanical routes of transmission (e.g. divers moving between sites) were also significant [123]. Investigations of the ISA outbreak in Scotland [124] and Canada [125] identified the use of well boats, for moving and harvesting fish, as an important factor in the spread of the disease.

The limitation of studies such as case-control are that it is often not known if the pathogen has been introduced to a control site but has not be detected as conditions at the site were not conducive to disease expression or establishment. Serological testing may provide insight into whether this may have been the case, and was used by Taylor et al. [24] to identify carp fishery sites that may have been exposed to koi herpesvirus. Subsequent contact tracing studies also using such tests subsequently suggested that live fish movements were the main route of transmission [126]. Unfortunately, at present few validated serological test methods exist for aquatic animal diseases.

Identifying Risk Factors for Disease Establishment

Classical epidemiological studies such as cross-sectional, case-control and cohort are commonly applied to identify risk factors associated with pathogen establishment and disease. In terrestrial systems individual animals were often the unit of study. In aquatic systems, although these study types are now commonly used by aquatic epidemiologists, they tend to be conducted at the level of the site (i.e. farm or fishery) or batch/cohort of fish due to the logistics of tracking individual fish.

There are a growing number of observational epidemiological studies that have identified risk factors for aquatic animal diseases. Most commonly these studies have undertaken to support the development of management advice; examples include diseases of farmed salmon [31–33], including sealice [127], and white spot syndrome virus (WSSV) infection in paenid shrimp [128]. The same approaches have also been applied to investigate risk factors for non-infectious causes of disease, for example fin erosion in rainbow trout [129], cataracts [130] and skin lesions [131] in Atlantic salmon. To date, few studies have focused on ornamental fish or fishery populations. With some exceptions these studies have investigated diseases in managed stillwater fisheries (e.g. *Argulus* spp. [132, 133] and KHV [25]). Most of these studies use standard generalised linear modelling techniques such as logistic regression to determine risk factors, but due to the difficulties in accurately measuring many factors of interest, latent variable modelling techniques may provide an approach that should be considered in future studies. The lack of studies in ornamental fish and fishery systems is likely a reflection of difficulties in collecting data from these systems. However, with a growing focus on conservation of wild stocks, their study and interactions with other sectors is likely to be of increased interest and should be the focus of future research efforts.

Most of the risk factor studies have been cross-sectional, which have the advantage over case-control studies that they provide an estimate of disease prevalence, but are not suitable for studying factors which change over time. Few prospective longitudinal studies of disease frequency in aquatic animals have been published, possibly because these studies face a number of constraints not encountered in terrestrial systems. Animals are not individually identified, and on farms groups of fish are split and mixed during the production cycle. These factors make repeated observations over time problematic (possible solutions were discussed by Thorburn et al. [134]). However, successful prospective studies of WSSV in shrimp [128] and a cohort studies in Atlantic salmon [131, 135, 136] have been completed.

Assessing Impact at the Population Level

It is important to have accurate assessments of the impact of disease at the population level to justify and prioritise expenditure on aquatic animal health. However, a range of impacts need to be considered. Most classically one thinks of mortality and the economic impact attributed to it; however there may also be indirect financial impacts. The key question about the impact of pathogens on wild populations is whether mortality, reduced fecundity, recruitment etc. actually reduce the overall population size, or whether their influence is mitigated by other density-dependent processes, i.e. can impacts at the level of the individual be scaled to the population.

Recording mortality

Although mortality is a commonly used measure of impact, in the aquatic environment even this may not be easy to measure. In the fish farm environment dead and moribund fish are relatively easy to observe compared to wild fish or fishery populations, and there is generally a legal obligation on farmers to record mortalities. However due to the large number of fish that may die (especially in the case of juvenile fish), accurate recording may be a problem. Additionally many farms experience relatively high and variable levels of "natural" background mortality compared to terrestrial farms, and multiple infections may occur at the same time. Consequently, attributing a cause of death to each individual and identifying when a disease first occurs is not always easy (although with a time-series of mortality data simple statistical techniques such as cumulative sum plots are useful). Fish farmers will often record the most obvious potential cause (often an ectoparasite, as they are easy to see and identify), to which all mortalities are attributed. The quality and value of diagnoses by farmers is likely to be highly variable, and this must be taken into consideration when using it for the purpose of epidemiological studies.

Assessing mortality attributed to a disease in wild or even recreational fisheries is more difficult, and as a consequence has been little studied. Even small recreational fisheries are relatively extensive and deep compared to farm systems. The likelihood of detecting mortality or capturing moribund fish for the purpose of diagnosis is therefore low, especially in the early stages of and epidemic. Methods for assessing impact wild populations is discussed in section 5.3.3.

Farm health and production records

Some farmers use production software that records mortalities and their causes. Such information is potentially of great use to the epidemiologist, but commercial sensitivities may constrain its use. Wheatley et al. [137] used production records from Irish salmon production to assess the impact of endemic diseases. Attempts have been made to create a national database to store production and mortality information for the UK trout industry (Development of a scheme for monitoring sentinel farms in the UK trout industry, Scottish Aquaculture Research Forum, project number 028). This database was designed to compile a national picture that would allow farmers to compare their sites performance with their sectors baseline performance (a common practice in cattle production where extensive health and production databases exist). Currently this system is of limited use to the research scientist, but a similar database, MonAqua which holds Norwegian salmon production data has successfully been used by epidemiologists to assess health related impacts [138]. The Faroe Islands have also successfully implemented such a database created following outbreaks of infectious salmon anaemia between 2000 and 2005, which could be used for similar purposes. This is a government initiative which requires by law that farmers report production figures, fish movements, mortalities and their causes on a regular basis for each cage. Where such systems are available, care must be taken to note the purpose of the database when compiling and analysing data, as data collected by the industry for their own monitoring purposes is likely to be subject to a different set of biases to that collected to comply with government regulations.

Of the few databases used to compile disease information in aquatic systems the most extensive are of sea lice infections in farmed Atlantic salmon. This is possibly due to the serious impact of this pathogen across the industry, and the fact that the majority of farms are owned by only a few companies that want to tightly monitor their lossess and the efficacy of treatments across sites. These data have been extensively used to build statistical and mathematical models of the epidemiology of *Lepeophtheirius salmonis* [139], *Caligus elongatus* [140] and other sea lice [141]. The studies include the identification of risk factors [127], time series analysis for trends [142] the impact of treatment timing [143], and economic impacts [144].

Assessing the impact of disease in wild populations

Although the incidence and dynamics of disease in wild aquatic animal populations has been studied, relatively few have adequately assessed impact at the population level. A few observational studies have provided assessments of pathogen induced. Johnsen and Jensen [145] compared Atlantic salmon catch rates and parr densities in Norwegian rivers where *G. salaris* was known to be present and absent. Although no formal statistical associations were provided, the data clearly demonstrated a decline in infected compared with uninfected salmon populations. Several longitudinal studies have been conducted to establish statistical associations between pathogen prevalence/abundance and impact. Taylor et al. [146] for example, studied the impact of *Argulus foliaceus* abundance and other risk factors on fish capture rates in several recreational trout fisheries and demonstrated a statistically significant reduction in capture rates when the parasite burden was high but it was not possible to quantify host mortality. Murray et al. [147] used mark recapture methods combined with non-lethal pathogen sampling methods in a longitudinal study to assess the impact of chytridiomycosis, demonstrating a clear decline in frog survival attributable to the infection.

The effort required to collect the data sets required to conduct observational studies is costly and labour intensive, especially if the host under study is not of interest in terms of commercial fishing, and a dedicated survey is therefore needed. Due to their extensive nature, such surveys are difficult in freshwater systems, but exceptionally demanding in marine systems. Consequently, theoretical modelling approaches that specifically model the host-pathogen interaction may be more appropriate (given knowledge of host-pathogen life cycles and processes), since they provide a method of extrapolating data on individual host level impacts to the population level. The application of such techniques is reflected in the scientific literature. The majority of these modelling studies are based on deterministic differential equation based models, but the model structure can vary greatly in nature. Patterson [148] extended a fisheries capture model to incorporate different mortality rates for infected and uninfected fish. This was used to study the influence of *Ichthophonus hoferi* infections in herring and demonstrated a 21% reduction in stock size. The study highlighted one of the short comings of standard fisheries models, in that they normally do not account for dynamic natural mortality processes such as disease.

The impact of pathogens reproducing on fish farms and effecting wild stocks is a major area of concern, and a promising subject for theoretical studies. Krkošek et al. [35] modelled the impact of sea lice (*L. salmonis*) emerging from Canadian fish farms on the survival of two species of migrating wild Pacific salmon. Deterministic differential equation based models were combined with several different model types. They used an advection-diffusion-decay model to track parasite dispersal and transmission, and coupled this to a survival model for the two host species. The study showed that the sea lice substantially reduced host survival, and that fish farms produced the number of parasite required to induce large scale population effects.

Other approaches to modelling have also been used to good effect in studying the impact of pathogens in wild aquatic animal populations. Borsuk et al. [149] used a Bayesian probabilistic network to study declining brown trout populations in Swiss rivers. This network incorporated many factors thought to be influential in the life-cycle of this fish. The parasite *T. bryosalmonae* (the aetiological agent of proliferative kidney disease) was one of the factors included, and the analysis demonstrated that this was one of the most influential factors in driving population declines, especially at higher temperatures. The approach provides evidence to support recommendations for risk mitigation and the identification of sites that are likely to be at high risk, thus allowing monitoring efforts to be targeted.

One of the major challenges in modelling the impact of epidemics and their impact in wild aquatic animal populations is they can occur over vast areas in many distinct populations and subpopulations. Murray and O'Callagan [150] addressed some of these issues in their review of the different approaches used to study the pilchard herpesvirus epidemic in

Australian stocks in 1998 to 1999. The authors demonstrated the need to apply several different model types to understand the processes occurring at different scales; from the epidemic within a shoal, to between shoal spread, and the large scale transmission around the southern Australian coastline.

In summary, epidemiological studies to study pathogen impact in wild aquatic animal populations are sparse compared to farmed species. This is in part likely to be due to the challenges of data collection, the scales over which these processes occur, and the many interacting factors. Adjusting for the effect of density-dependent processes may pose a significant challenge in accurately assessing impact, however, in exploited populations such as wild capture fisheries it may be argued that such processes are of less concern. Sub-lethal effects of pathogens (such as reduced growth or fecundity) that are often observed at the individual level are also difficult to assess at the population level, and modelling approaches provide and important tool in their assessment.

A lack of research into the impact of pathogens on wild fish populations may be because it is difficult to influence events, and some fisheries scientists may argue that pathogen-induced mortality is of little consequence in the presence of high fishing mortality. However, with increasing pressures on fish stocks, understanding the influence of these combined pressures on a population may be of great importance when trying to identify tipping points and set catch limits, in order to prevent population collapses. Many of the challenges described here for the study of impacts in fisheries and wild aquatic animals are similar to those faced by researchers studying wildlife epidemiology and population ecology. These disciples may provide further tools helpful to the study of aquatic systems (e.g research on the impact of parasites on gamebirds [120, 151] and rabies in wildlife populations [20]).

CONCLUSIONS

Georgiadis et al. [152] reviewed how epidemiology was and could be used in aquatic animal health. Peeler et al. [41] reviewed the application of risk analysis in aquatic animal health management. In this review we have tried to demonstrate the wide breadth of epidemiological studies and the range of tools, approaches currently being applied to study disease aquatic systems, and the constraints and challenges facing epidemiologists. The single unifying actor is the purpose of epidemiology, namely the improvement of animal health. Although epidemiological studies in aquatic systems have, and continue to lag behind those applied to terrestrial systems in terms of scope and complexity, rapid development in the application of epidemiology to aquatic systems over a relatively short period has occurred. Consequently, epidemiological studies have made a significant contribution to the health of both wild and farmed aquatic animals, through improved biosecurity and surveillance of exotic diseases and control of endemic diseases.

The role for epidemiology in aquatic animal health became clear as pathogens emerged and rapidly spread in intensive aquaculture systems and disease interactions between wild and farmed populations became increasingly problematic. These complex aquatic animal health issues required inter-disciplinary approaches and epidemiologists are best placed to integrate and analyse diverse data sources and lead teams of pathogens specialists, economists, modellers and statisticians. Despite its success to date many challenges remain. Difficulties in obtaining data on host populations, especially in extensive systems require epidemiologists to work with fishery scientists and ecologists to better integrate wild fish demographics into epidemiological studies and models. Economic, population and disease model needed to be better integrated to improve assessments of disease spread to support aquatic animal health policy making. Surveillance methods developed for terrestrial systems need to be adapted for aquatic animals, and the lack of diagnostic test methods suited to surveillance (as opposed to disease confirmation) is currently a major limitation. More sophisticated models of the processes underlying disease emergence that take into account, *inter-alia*, the role of multi-host pathogens and reservoir populations and a greater understanding of the way pathogens are transmitted in aquatic systems are needed to support measures to minimise the number and impact of new diseases. The volume of both legal and illegal trade in live aquatic animals far surpasses that of terrestrial animals, and is likely to be the biggest challenge in trying to prevent the spread of emerging aquatic diseases. Additionally, over the next few decades, not only is such trade set to increase, but aquaculture is likely to further grow, and pressures of limited water resources and climate change effects are likely to combine to increase the rate at which diseases emerge and spread. In the face of these combined challenges the demand for epidemiological approaches to support measures to protect aquatic animal health will undoubtedly increase.

ACKNOWLEDGEMENTS

This work was supported by research funding (FC 1185 and FB001) from the Department for Environment, Food and Rural Affairs (Defra, UK).

REFERENCES

1. Dohoo IR, Morris RS, Martin SW, Perry BD, Bernardo T, Erb H, Thrusfleld M, Smith R, Welte VR: Epidemiology. Nature. 1994, 368: 284-
2. Schwabe C: The current epidemiological revolution in veterinary medicine. Part 1. Prev Vet Med. 1982, 1: 5-15. 10.1016/0167-5877(82)90003-4.
3. Thrusfield M: Veterinary Epidemiology. 2005, Blackwell Science Ltd, 3
4. MacHohan B, Pugh TF: Epidemiology: Principles and Methods. 1970, Boston: Little Brown
5. Martin SW, Meek AH, Willeberg P: Veterinary Epidemiology - principles and methods. 1987, Iowa: Iowa Statue University Press/Ames, 1
6. Bergh O, Borsheim KY, Bratbak G, Heldal M: High abundance of viruses found in aquatic environments. Nature. 1989, 340: 467-468. 10.1038/340467a0.
7. Suttle CA: Marine viruses [mdash] major players in the global ecosystem. Nat Rev Micro. 2007, 5: 801-812. 10.1038/nrmicro1750.
8. Taylor NGH, Verner-Jeffreys DW, Baker-Austin C: Aquatic systems: maintaining, mixing and mobilising antimicrobial resistance?. Trends Ecol Evol. 2011, 26: 278-284. 10.1016/j.tree.2011.03.004.
9. F.A.O: State of the World Fisheries and Aquaculture. Book State of the World Fisheries and Aquaculture. 2008, City: Food and Agriculture Organisation Fisheries and Aquaculture Department, 196-
10. Subasinghe RP: Epidemiological approach to aquatic animal health management: opportunities and challenges for developing countries to increase aquatic production through aquaculture. Prev Vet Med. 2005, 67: 117-124. 10.1016/j.prevetmed.2004.11.004.
11. Gozlan RE: Introduction of non-native freshwater fish: is it all bad?. Fish Fish. 2008, 9: 106-115.
12. Murray AG, Peeler EJ: A framework for understanding the potential for emerging diseases in aquaculture. Prev Vet Med. 2005, 67: 223-235. 10.1016/j.prevetmed.2004.10.012.
13. McInerney J: Old economics for new problems - livestock disease: presidential address. J Agric Econ. 1996, 47: 295-314. 10.1111/j.1477-9552.1996.tb00695.x.

14. Aiken HM, Hayward CJ, Nowak BF: An epizootic and its decline of a blood fluke, Cardicola forsteri, in farmed southern bluefin tuna, Thunnus maccoyii. Aquaculture. 2006, 254: 40-45. 10.1016/j.aquaculture.2005.10.013.
15. Deveney MR, Bayly TJ, Johnston CJ, Nowak BF: A parasite survey of farmed Southern bluefin tuna, *Thunnus maccoyii* (Castelnau). J Fish Dis. 2005, 28: 279-284. 10.1111/j.1365-2761.2005.00629.x.
16. Nowak BF: Parasitic diseases in marine cage culture - An example of experimental evolution of parasites?. Int J Parasitol. 2007, 37: 581-588. 10.1016/j.ijpara.2007.01.003.
17. Whittington RJ, Jones JB, Hine PM, Hyatt AD: Epizootic mortality in the pilchard *Sardinops sagax neopilchardus* in Australia and New Zealand in 1995. 1. Pathology and epizootiology. Dis Aquat Organ. 1997, 28: 1-16. Ornamental fish. [http://www.ornamentalfish.org/aquanautmarket/petfishstats.php]
18. Whittington RJ, Chong R: Global trade in ornamental fish from an Australian perspective: the case for revised import risk analysis and management strategies. Prev Vet Med. 2007, 81: 92-116. 10.1016/j.prevetmed.2007.04.007.
19. Lembo T, Hampson K, Haydon DT, Craft M, Dobson A, Dushoff J, Ernest E, Hoare R, Kaare M, Mlengeya T, Mentzel C, Cleaveland S: Exploring reservoir dynamics: a case study of rabies in the Serengeti ecosystem. J Appl Ecol. 2008, 45: 1246-1257. 10.1111/j.1365-2664.2008.01468.x.
20. Hedrick RP: Movements of pathogens with the international trade of live fish: problems and solutions. Rev Sci Tech. 1996, 15: 523-531.
21. Peeler EJ, Oidtmann BC, Midtlyng PM, Miossec L, Gozlan RE: Non-native aquatic animals introductions have driven disease emergence in Europe. Biol Invasions.
22. Marcos-López M, Gale P, Oidtmann BC, Peeler EJ: Assessing the impact of climate change on disease emergence in freshwater fish in the United Kingdom. Transbound Emerg Dis. 2010, 57: 293-304. 10.1111/j.1865-1682.2010.01150.x.
23. Taylor NGH, Dixon PF, Jeffery KR, Peeler EJ, Denham KL, Way K: Koi herpesvirus: distribution and prospects for control in England and Wales. J Fish Dis. 2010, 33: 221-230. 10.1111/j.1365-2761.2009.01111.x.

24. Schikorski D, Faury N, Pepin JF, Saulnier D, Tourbiez D, Renault T: Experimental ostreid herpesvirus 1 infection of the Pacific oyster Crassostrea gigas: Kinetics of virus DNA detection by q-PCR in seawater and in oyster samples. Virus Res. 2011, 155: 28-34. 10.1016/j.virusres.2010.07.031.
25. Copp G, Villizi L, Gozlan R: The demography of introduction pathways, propagule pressure and non-native freshwater fishes occurrence in England. Aquatic Conserv: Mar Freshw Ecosyst. 2010, 20: 595-601. 10.1002/aqc.1129.
26. Copp GH, Templeton M, Gozlan RE: Propagule pressure and the invasion risks of non-native freshwater fishes: a case study in England. J Fish Biol. 2007, 71: 148-159.
27. Smith P: An overview of current fish disease research; Hyper-reality in a post-modern world?. Bull Eur Assoc Fish Pathol. 1999, 19: 310-312.
28. Thorburn MA: Factors influencing seasonal vibriosis mortality rates in Swedish pen reared rainbow trout. Aquaculture. 1987, 67: 79-85. 10.1016/0044-8486(87)90010-X.
29. Thorud K, Djupvik H: Infectious salmon anaemia in Atlantic salmon (*Salmo salar* L.). Bull Eur Assoc Fish Pathol. 1988, 8: 109-111.
30. Jarp J, Gjevre AG, Olsen AB, Bruheim T: Risk factors for furunculosis, infectious pancreatic necrosis and mortality in post-smolt Atlantic salmon, *Salmo salar* L. J Fish Dis. 1994, 18: 67-78.
31. Jarp J, Karlsen E: Infectious salmon anaemia (ISA) risk factors in sea-cultured Atlantic salmon Salmo salar. Dis Aquat Organ. 1997, 28: 79-86.
32. Jarp J, Tangen K, Willumsen FV, Dyupvik HO, Tveit AM: Risk factors for infection with Aeromonas salmonicida subsp. salmonicida in Norwegian freshwater hatcheries. Dis Aquat Organ. 1993, 17: 81-86.
33. Johnsen BO, Jensen AJ: The *Gyrodactylus* story in Norway. Aquaculture. 1991, 98: 289-302. 10.1016/0044-8486(91)90393-L.
34. Krkosek M, Ford JS, Morton A, Lele S, Myers RA, Lewis MA: Declining wild salmon populations in relation to parasites from farm salmon. Science. 2007, 318: 1772-1775. 10.1126/science.1148744.
35. Walker PJ, Mohan CV: Viral disease emergence in shrimp aquaculture: origins, impacts and the effectiveness of health management strategies. Rev Aquac. 2009, 1: 125-154. 10.1111/j.1753-5131.2009.01007.x.

36. Verner-Jeffreys DW, Pond MJ, Peeler EJ, Rimmer GSE, Oidtmann B, Way K, Mewett J, Jeffrey K, Bateman K, Reese RA, Feist SW: Emergence of cold water strawberry disease of rainbow trout Oncorynchus mykiss in England and Wales: Outbreak investigations and transmission studies. Dis Aquat Organ. 2008, 79: 207-218.
37. Segarra A, Pépin JF, Arzul I, Morga B, Faury N, Renault T: Detection and description of a particular Ostreid herpesvirus 1 genotype associated with massive mortality outbreaks of Pacific oysters, Crassostrea gigas, in France in 2008. Virus Res. 2010, 153: 92-99. 10.1016/j.virusres.2010.07.011.
38. Haenen OLM, Way K, Bergmann SM, Ariel E: The emergence of koi herpesvirus and its significance to European aquaculture. Bull Eur Ass Fish Pathol. 2004, 24: 293-307.
39. Daszak P, Strieby A, Cunningham AA, Longcore JE, Brown CC, Porter D: Experimental evidence that the bullfrog (Rana catesbeiana) is a potential carrier of chytridiomycosis, an emerging fungal disease of amphibians. Herpetolog J. 2004, 14: 201-207.
40. Peeler EJ, Murray AG, Thebault A, Brun E, Giovaninni A, Thrush MA: The application of risk analysis in aquatic animal health management. Prev Vet Med. 2007, 81: 3-20. 10.1016/j.prevetmed.2007.04.012.
41. O.I.E: Section 2.2. Risk analysis. Aquatic Animal Health Code. 2009, Paris: World Animal Health Organisation, 1: 10
42. WTO: Agreement on the Application of Sanitary and Phytosanitary Measures. Book Agreement on the Application of Sanitary and Phytosanitary Measures. 1995, City: World Trade Organisation, 21-
43. Kahn SA, Beers PT, Findlay VL, Peebles IR, Durham PJ, Wilson DW, Gerrity SE: Import Risk Analysis on Non-viable Salmonids and Non-salmonids Marine Finfish. Book Import Risk Analysis on Non-viable Salmonids and Non-salmonids Marine Finfish. 1999, City: Australian Quarantine and Inspection Service, 409-
44. Gaughan DJ: Disease-translocation across geographic boundaries must be recognized as a risk even in the absence of disease identification: the case with Australian Sardinops. Rev Fish Biol Fish. 2002, 11: 113-123.
45. Bridges VE, Akkina J, Grannis J, Johnson C, Johnson R, Tuszynski C: A qualitative assessment tool for the potential of infectious disease emergence and spread. Prev Vet Med. 2007, 81: 80-91. 10.1016/j.prevetmed.2007.04.008.

46. Woolhouse MEJ: Population biology of emerging and re-emerging pathogens. Trends Microbiol. 2002, 10: 3-7. 10.1016/S0966-842X(02)02428-9.
47. Woolhouse MEJ: Where do emerging pathogens come from?. Microbe. 2006, 1: 511-515.
48. Woolhouse MEJ: Epidemiology: emerging diseases go global. Nature. 2008, 451: 898-899. 10.1038/451898a.
49. Brown SP, Hochberg ME, Grenfell BT: Does multiple infection select for virulence?. Trends Microbiol. 2002, 10: 401-405. 10.1016/S0966-842X(02)02413-7.
50. Davies CM, Webster JP, Woolhouse MEJ: Trade-offs in the evolution of virulence in an indirectly transmitted macroparasite. Proc Biol Sci. 2001, 268: 251-257. 10.1098/rspb.2000.1367.
51. Frank SA: Models of parasite virulence. Q Rev Biol. 1996, 71: 37-78. 10.1086/419267.
52. Haydon DT, Cleaveland S, Taylor LH, Laurenson MK: Identifying reservoirs of infection: A conceptual and practical challenge. Emerg Infect Dis. 2002, 8: 1468-1473.
53. Woolhouse MEJ, Gowtage-Sequeria S: Host range and emerging and reemerging pathogens. Emerg Infect Dis. 2005, 11: 1842-1847. 10.3201/eid1112.050997.
54. Woolhouse MEJ, Haydon DT, Antia R: Emerging pathogens: the epidemiology and evolution of species jumps. Trends Ecol Evol. 2005, 20: 238-244. 10.1016/j.tree.2005.02.009.
55. Woolhouse MEJ, Taylor LH, Haydon DT: Population biology of multihost pathogens. Science. 2001, 292: 1109-1112. 10.1126/science.1059026.
56. Taylor LH, Latham SM, Woolhouse MEJ: Risk factors for human disease emergence. Philos Trans R Soc Lond B Biol Sci. 2001, 356: 983-989. 10.1098/rstb.2001.0888.
57. Woolhouse M, Gaunt E: Ecological origins of novel human pathogens. Crit Rev Microbiol. 2007, 33: 231-242. 10.1080/10408410701647560.
58. Maccrimmon HR: World distribution of rainbow trout (*Salmon gairdneri*). J Fish Res Board Can. 1971, 28: 663-704. 10.1139/f71-098.
59. Amos K, Thomas J: Disease interactions between wild and cultured fish: observation and lessons learned in the Pacific Northwest. Bull Eur Ass Fish Pathol. 2002, 22: 95-102.

60. Coutant CC: What is normative for fish pathogens? A perspective on the controversy over interactions between wild and cultured fish. J Aquat Anim Health. 1998, 10: 101-106. 10.1577/1548-8667(1998)010<0101:WINFFP>2.0.CO;2.
61. Mennerat A, Nilsen F, Ebert D, Skorping A: Intensive Farming: Evolutionary Implications for Parasites and Pathogens. Evol Biol. 2010, 37: 59-67. 10.1007/s11692-010-9089-0.
62. Rogers DJ, Randolph SE: Studying the global distribution of infectious diseases using GIS and RS. Nat Rev Microbiol. 2003, 1: 231-237. 10.1038/nrmicro776.
63. Bovo G, Giorgetti G, Jørgensen PEV, Olesen NJ: Infectious hematopoietic necrosis: first detection in Italy. Bull Eur Ass Fish Pathol. 1987, 7: 124-
64. Enzmann PJ, Kurath G, Fichtner D, Bergmann SM: Infectious hematopoietic necrosis virus: monophyletic origin of European IHNV isolates from North-American genogroup M. Dis Aquat Organ. 2005, 66: 187:195-
65. Basurco B, Benmansour A: Distant strains of the fish rhabdovirus VHSV maintain a sixth functional cistron which codes for a nonstructural protein of unknown function. Virology. 1995, 212: 741-745. 10.1006/viro.1995.1534.
66. Basurco B, Vende P, Monnier AF, Winton JR, de Kinkelin P, Benmansour A: Genetic diversity and phylogenetic classification of viral hemorrhagic septicemia virus (VHSV). Vet Res. 1995, 26: 460-463.
67. Cunningham CO, Snow M: Genetic analysis of infectious salmon anaemia virus (ISAV) from Scotland. Dis Aquat Organ. 2000, 41: 1-8.
68. Kibenge FSB, Lyaku JR, Rainnie D, Hammell KL: Growth of infectious salmon anaemia virus in CHSE-214 cells and evidence for phenotypic differences between virus strains. J Gen Virol. 2000, 81: 143-150.
69. Kibenge FS, Godoy MG, Wang Y, Kibenge MJ, Gherardelli V, Mansilla S, Lisperger A, Jarpa M, Larroquete G, Avendão F, Lara M, Gallardo A: Infectious salmon anaemia virus (ISAV) isolated from the ISA disease outbreaks in Chile diverged from ISAV isolates from Norway around 1996 and was disseminated around 2005, based on surface glycoprotein gene sequences. Virol J. 2009, 6: 88-10.1186/1743-422X-6-88.

70. Godoy MG, Aedo A, Kibenge MJT, Groman DB, Yason CV, Grothusen H, Lisperguer A, Calbucura M, Avendaño F, Imilán M, Jarpa M, Kibenge FS: First detection, isolation and molecular characterization of infectious salmon anaemia virus associated with clinical disease in farmed Atlantic salmon (Salmo salar) in Chile. BMC Vet Res. 2008, 4: 28-10.1186/1746-6148-4-28.
71. Cottam EM, Thébaud G, Wadsworth J, Gloster J, Mansley L, Paton DJ, King DP, Haydon DT: Integrating genetic and epidemiological data to determine transmission pathways of foot-and-mouth disease virus. Proc Biol Sci. 2008, 275: 887-895. 10.1098/rspb.2007.1442.
72. Salman MD: Animal disease surveillance and survey systems - methods and applications. 2003, Iowa: Iowa State Press, 1
73. Cameron A: Survey toolbox for aquatic animal diseases. A practical Manual and Software Package. Book Survey toolbox for aquatic animal diseases. A practical Manual and Software Package. 2002, City: ACIAR, 375-
74. East IJ, Black PF, McColl KA, Hodgson RAJ, Bernoth EM: Survey for the presence of White Spot Syndrome virus in Australian crustaceans. Aust Vet J. 2004, 82: 236-240. 10.1111/j.1751-0813.2004.tb12688.x.
75. Lindenstroem T, Nielsen B, Buchmann K: Gyrodactylids on salmonids from Danish streams. Systematics and Phylogeny of Platyhelminthes; St Petesburg, Russia. 1999, 62-64.
76. Nielsen CV, Buchmann K: Occurrence of *Gyrodactylus* parasites in Danish fish farms. Bull Eur Assoc Fish Pathol. 2001, 21: 19-25.
77. Wahli T, Bernet D, Steiner PA, Schmidt-Posthaus H: Geographic distribution of *Tetracapsuloides bryosalmonae* infected fish in Swiss rivers: an update. Aquat Sci. 2007, 69: 3-10. 10.1007/s00027-006-0843-4.
78. Wahli T, Knuesel R, Bernet D, Segner H, Pugovkin D, Burkhardt-Holm P, Escher M, Schmidt-Posthaus H: Proliferative kidney disease in Switzerland: current state of knowledge. J Fish Dis. 2002, 25: 491-500. 10.1046/j.1365-2761.2002.00401.x.
79. Feist SW, Peeler EJ, Gardiner R, Smith E, Longshaw M: Proliferative kidney disease and renal myxosporidiosis in juvenile salmonids from rivers in England and Wales. J Fish Dis. 2002, 25: 451-458. 10.1046/j.1365-2761.2002.00361.x.

80. Peeler EJ, Feist SW, Longshaw M, Thrush MA, St-Hilaire S: An assessment of the variation in the prevalence of renal myxosporidiosis and hepatitis in wild brown trout, *Salmo trutta* L., within and between rivers in South-West England. J Fish Dis. 2008, 31: 719-728. 10.1111/j.1365-2761.2008.00942.x.
81. Seagrave CP, Bucke D, Hudson EB, McGregor D: A survey of the prevalence and distribution of proliferative kidney disease (PKD) in England and Wales. J Fish Dis. 1981, 4: 437-439. 10.1111/j.1365-2761.1981.tb01155.x.
82. Chambers E, Gardiner R, Peeler EJ: An investigation into the prevalence of *Renibacterium salmoninarum* in farmed rainbow trout (*Onchorhynchus mykiss*) and wild fish populatins in selected river catchments in England and Wales between 1998 and 2000. J Fish Dis. 2008, 31: 89-96. 10.1111/j.1365-2761.2007.00868.x.
83. Austin B, Rayment JN: Epizootiology of Renibacterium salmoninarum, the causal agent of bacterial kidney disease in salmonid fish. J Fish Dis. 1985, 8: 505-509. 10.1111/j.1365-2761.1985.tb00965.x.
84. Rodger HD, Turnbull T, Muir F, Millar S, Richards RH: Infectious salmon anemia (ISA) in the United Kingdom. Bull Eur Assoc Fish Pathol. 1998, 18: 115-116.
85. Raynard RS, Murray AG, Gregory A: Infectious salmon anaemia virus in wild fish from Scotland. Dis Aquat Organ. 2001, 46: 93-100.
86. Martin PAJ: Current value of historical and ongoing surveillance for disease freedom: Surveillance for bovine Johne's disease in Western Australia. Prev Vet Med. 2008, 84: 291-309. 10.1016/j.prevetmed.2007.12.002.
87. Martin PAJ, Cameron AR, Barfod K, Sergeant ESG, Greiner M: Demonstrating freedom from disease using multiple complex data sources. 2: Case study-Classical swine fever in Denmark. Prev Vet Med. 2007, 79: 98-115. 10.1016/j.prevetmed.2006.09.007.
88. Weinstock MA: A generalised scan statistic test for the detection of clusters. Int J Epidemiol. 1981, 10: 289-293. 10.1093/ije/10.3.289.
89. Clark PJ, Evans FC: Distance to nearest neighbor as a measure of special relationship in populations. Ecology. 1954, 159-163.
90. Carpenter TE: Methods to investigate spatial and temporal clustering in veterinary epidemiology. Prev Vet Med. 2001, 48: 303-320. 10.1016/S0167-5877(00)00199-9.

91. Ward MP, Carpenter TE: Techniques for analysis of disease clustering in space and in time in veterinary epidemiology. Prev Vet Med. 2000, 45: 257-284. 10.1016/S0167-5877(00)00133-1.
92. Heuch PA, Revie CW, Gettinby G: A comparison of epidemiological patterns of salmon lice, Lepeophtheirus salmonis, infections on farmed Atlantic salmon, Salmo salar L., in Norway and Scotland. J Fish Dis. 2003, 26: 539-551. 10.1046/j.1365-2761.2003.00490.x.
93. Revie CW, Hollinger E, Gettinby G, Lees F, Heuch PA: Clustering of parasites within cages on Scottish and Norwegian salmon farms: Alternative sampling strategies illustrated using simulation. Prev Vet Med. 2007, 81: 135-147. 10.1016/j.prevetmed.2007.04.004.
94. Lees F, Gettinby G, Revie CW: Changes in epidemiological patterns of sea lice infestation on farmed Atlantic salmon, Salmo salar L., in Scotland between 1996 and 2006. J Fish Dis. 2008, 31: 259-268. 10.1111/j.1365-2761.2007.00897.x.
95. Longshaw M, Frear PA, Nunn AD, Cowx IG, Feist SW: The influence of parasitism on fish population success. Fish Manag Ecol. 2010, 17: 426-434. 10.1111/j.1365-2400.2010.00741.x.
96. Georgiadis MP, Johnson WO, Gardner IA: Sample size determination for estimation of the accuracy of two conditionally independent tests in the absence of a gold standard. Prev Vet Med. 2005, 71: 1-10. 10.1016/j.prevetmed.2005.04.004.
97. Georgiadis MP, Johnson WO, Gardner IA, Singh R: Correlation-adjusted estimation of sensitivity and specificity of two diagnostic tests. Appl Stat. 2003, 52: 63-76. 10.1111/1467-9876.00389.
98. Nérette P, Stryhn H, Dohoo I, Hammell L: Using pseudogold standards and latent-class analysis in combination to evaluate the accuracy of three diagnostic tests. Prev Vet Med. 2008, 85: 207-225. 10.1016/j.prevetmed.2008.01.011.
99. Nérette P, Dohoo I, Hammell L: Estimation of specificity and sensitivity of three diagnostic tests for infectious salmon anaemia virus in the absence of a gold standard. J Fish Dis. 2005, 28: 89-99. 10.1111/j.1365-2761.2005.00612.x.
100. O.I.E: Manual of Diagnostic Tests for Aquatic Animals. 2009, Paris: Office International des Epizooties, 5
101. Williams CJ, Moffitt CM: Estimation of fish and wildlife disease prevalence from imperfect diagnostic tests on pooled samples with varying pool sizes. Ecol Inform. 2010, 5: 273-280. 10.1016/j.

ecoinf.2010.04.003.

102. Keeling MJ, Woolhouse MEJ, May RM, Davies G, Grenfell BT: Modelling vaccination strategies against foot-and-mouth disease. Nature. 2003, 421: 136-142. 10.1038/nature01343.

103. Ortiz-Pelaez A, Pfeiffer DU, Soares-Magalhaes RJ, Guitian FJ: Use of social network analysis to characterize the pattern of animal movements in the initial phases of the 2001 foot and mouth disease (FMD) epidemic in the UK. Prev Vet Med. 2006, 76: 40-55. 10.1016/j.prevetmed.2006.04.007.

104. Picado A, Guitian FJ, Pfeiffer DU: Space-time interaction as an indicator of local spread during the 2001 FMD outbreak in the UK. Prev Vet Med. 2007, 79: 3-19. 10.1016/j.prevetmed.2006.11.009.

105. Paisley LG, Karlsen E, Jarp J, Mo TA: A Monte Carlo simulation model for assessing the risk of introduction of *Gyrodactylus salaris* to the Tana river, Norway. Dis Aquat Organ. 1999, 37: 145-152.

106. Høgåsen HR, Brun E: Risk of inter-river transmission of *Gyrodactylus salaris* by migrating Atlantic salmon smolts, estimated by Monte Carlo simulation. Dis Aquat Organ. 2003, 57: 247-254.

107. Peeler EJ, Gardiner R, Thrush MA: Qualitative risk assessment of routes of transmission of the exotic fish parasite *Gyrodactylus salaris* between river catchments in England and Wales. Prev Vet Med. 2004, 64: 175-189. 10.1016/j.prevetmed.2004.05.005.

108. Anderson RM, May RM: Population biology of infectious diseases: Part 1. Nature. 1979, 280: 361-367. 10.1038/280361a0.

109. Murray AG: A model of the emergence of infectious pancreatic necrosis virus in Scottish salmon farms 1996-2003. Ecol Modell. 2006, 199: 64-72. 10.1016/j.ecolmodel.2006.06.010.

110. Ruane NM, Murray AG, Geoghegan F, Raynard RS: Modelling the spread of Infectious Pancreatic Necrosis Virus (IPNV) in the Irish salmon farming industry: the role of inputs. Ecol Modell. 2009, 220: 1369-1374. 10.1016/j.ecolmodel.2009.02.016.

111. Taylor NGH, Norman RA, Way K, Peeler EJ: Modelling the koi herpesvirus (KHV) epidemic highlights the importance of active surveillance within a national control policy. J Appl Ecol. 2011, 48: 348-355. 10.1111/j.1365-2664.2010.01926.x.

112. Thrush M, Peeler E: Stochastic simulation of live salmonid movement in England and Wales to predict potential spread of exotic pathogens. Dis Aquat Organ. 2006, 72: 115-123.

113. Green DM: A strategic model for epidemic control in aquaculture. Prev Vet Med. 2010, 94: 119-127. 10.1016/j.prevetmed.2009.12.004.

114. Green DM, Gregory A, Munro LA: Small- and large-scale network structure of live fish movements in Scotland. Prev Vet Med. 2009, 91: 261-269. 10.1016/j.prevetmed.2009.05.031.

115. Sharkey KJ, Fernandez C, Morgan KL, Peeler E, Thrush M, Turnbull JF, Bowers RG: Pair-level approximations to the spatio-temporal dynamics of epidemics on asymmetric contact networks. J Math Biol. 2006, 53: 61-85. 10.1007/s00285-006-0377-3.

116. Murray AG, Gillibrand PA: Modelling dispersal of salmon lice in Loch Torridon, Scotland. Mar Pollut Bull. 2006, 53: 128-135. 10.1016/j.marpolbul.2005.09.013.

117. Murray AG, Amundrud TL, Gillibrand PA: Models of hydrodynamic pathogen dispersal affecting Scottish salmon production: modelling shows how Scotland eradicated ISA, but not IPN. Bull Aquacult Assoc Can. 2005, 105: 79-86.

118. Viljugrein H, Staalstrøm A, Molvær J, Urke H, Jansen P: Integration of hydrodynamics into a statistical model on the spread of pancreas disease (PD) in salmon farming. Dis Aquat Organ. 2009, 88: 35-44.

119. Murray AG, Amundrud TL, Gillibrand PA: Models of hydrodynamic pathogen dispersal affecting Scottish salmon production: modelling shows how Scotland eradicated ISA, but not IPN. Bull Aquacult Assoc Can. 2005, 105: 80-87.

120. Keeling MJ, Woolhouse MEJ, Shaw DJ, Mattthews L, Chase-Topping M, Haydon DT, Cornell SJ, Kappey J, Wilesmith JW, Grenfell BT: Dynamics of the 2001 foot and mouth epidemic: stochastic dispersal and a hetergenous landscape. Science. 2001, 294: 812-817. 10.1126/science.1064627.

121. Jansen PA, Matthews L, Toft N: Geographic risk factors for inter-river dispersal of Gyrodactylus salaris in fjord-systems in Norway. Dis Aquat Organ. 2007, 74: 139-149.

122. Vågsholm I, Djupvik HO, Willumsen FV, Tveit AM, Tangen K: Infectious salmon anaemia (ISA) epidemiology in Norway. Prev Vet Med. 1994, 19: 277-290. 10.1016/0167-5877(94)90095-7.

123. Murray AG, Smith RJ, Stagg RM: Shipping and the spread of infectious salmon anemia in Scottish aquaculture. Emerg Infect Dis. 2002, 8: 1-5. 10.3201/eid0801.010144.
124. McClure CA, Hammell KL, Dohoo IR: Risk factors for outbreaks of infectious salmon anemia in farmed Atlantic salmon, Salmo salar. Prev Vet Med. 2005, 72: 263-280. 10.1016/j.prevetmed.2005.07.010.
125. Taylor NGH, Way K, Jeffery KR, Peeler EJ: The role of live fish movements in spreading koi herpesvirus throughout England and Wales. J Fish Dis. 2010, 33: 1005-1007. 10.1111/j.1365-2761.2010.01198.x.
126. Revie CW, Gettinby G, Treasurer JW, Wallace C: Identifying epidemiological factors affecting sea lice *Lepeophtheirus salmonis* abundance on Scottish salmon farms using general linear models. Dis Aquat Organ. 2003, 57: 85-89.
127. Corsin F, Turnbull JF, Hao NV, Mohan CV, Phi TT, Phuoc LH, Tinh NTN, Morgan KL: Risk factors associated with white spot syndrome virus infection in a Vietnamese rice-shrimp farming system. Dis Aquat Organ. 2001, 47: 1-12.
128. St-Hilaire S, Ellis T, Cooke A, North BP, Turnbull JF, Knowles T, Kestin S: Fin erosion on rainbow trout on commercial trout farms in the United Kingdom. Vet Rec. 2006, 159: 446-451. 10.1136/vr.159.14.446.
129. Ersdal C, Midtlyng PJ, Jarp J: An epidemiological study of cataracts in seawater farmed Atlantic salmon Salmo salar. Dis Aquat Organ. 2001, 45: 229-236.
130. Vaagsholm I, Djupvik HO: Risk factors for skin lesions in Atlantic salmon, Salmo salar L. J Fish Dis. 1998, 21: 449-453. 10.1046/j.1365-2761.1998.00123.x.
131. Thrush MA, Murray AG, Brun E, Wallace S, Peeler EJ: The application of risk and disease modelling to emerging freshwater diseases in wild aquatic animals. Freshwat Biol. 2011, 56: 658-675. 10.1111/j.1365-2427.2010.02549.x.
132. Taylor NGH, Sommerville C, Wootten R: The epidemiology of Argulus spp. (Crustacea: Branchiura) infections in stillwater trout fisheries. J Fish Dis. 2006, 29: 193-200. 10.1111/j.1365-2761.2006.00704.x.
133. Thorburn MA: Variables associated with the use of chemotherapeutic agents on trout farms in Ontario, Canada. Prev Vet Med. 1995, 24: 1-14. 10.1016/0167-5877(95)00472-9.

134. Vågsholm I, Djupvik HO: Risk factors for abdominal adhesions in Atlantic salmon, *Salmo salar* L. J Fish Dis. 1999, 22: 53-58. 10.1046/j.1365-2761.1999.00137.x.

135. Vågsholm I, Djupvik HO: Risk factors for spinal deformities in Atlantic salmon, *Salmo salar* L. J Fish Dis. 1998, 21: 47-53. 10.1046/j.1365-2761.1998.00069.x.

136. Wheatley SB, McLoughlin MF, Menzies FD, Goodall EA: Site management factors influencing mortality rates in Atlantic salmon (*Salmo salar* L.) during marine production. Aquaculture. 1995, 136: 195-207. 10.1016/0044-8486(95)01058-0.

137. Aunsmo A, Valle PS, Sandberg M, Midtlyng PJ, Bruheim T: Stochastic modelling of direct costs of pancreas disease (PD) in Norwegian farmed Atlantic salmon (Salmo salar L.). Prev Vet Med. 2010, 93: 233-241. 10.1016/j.prevetmed.2009.10.001.

138. Revie CW, Gettinby G, Treasurer JW, Rae GH, Clark N: Temporal, environmental and management factors influencing the epidemiological patterns of sea lice (*Lepeophtheirus salmonis*) infestations on farmed Atlantic salmon (*Salmo salar*) in Scotland. Pest Manag Sci. 2002, 58: 576-584. 10.1002/ps.476.

139. Revie CW, Gettinby G, Treasurer JW, Rae GH: The epidemiology of the sea lice, *Caligus elongatus* Nordmann, in marine aquaculture of Atlantic salmon, *Salmo salar* L. in Scotland. J Fish Dis. 2002, 25: 391-400. 10.1046/j.1365-2761.2002.00388.x.

140. Zagmutt-Vergara FJ, Carpenter TE, Farver TB, Hedrick RP: Spatial and temporal variations in sea lice (Copeda: Caligidae) infestations of three salmonid species farmed in net pens in southern Chile. Dis Aquat Organ. 2005, 64: 163-173.

141. McKenzie E, Gettinby G, McCart K, Revie CW: Time-series models of sea lice Caligus elongatus (Nordmann) abundance on Atlantic salmon Salmo salar L. in Loch Sunart, Scotland. Aquacult Res. 2004, 35: 764-772. 10.1111/j.1365-2109.2004.01099.x.

142. Tucker CS, Norman R, Shinn AP, Bron JE, Sommerville C, Wooten R: A single cohort time delay model of the life-cycle of the salmon louse *Lepeophtheirus salmonis* on Atlantic salmon *Salmo salar*. Fish Pathol. 2002, 37: 107-118. 10.3147/jsfp.37.107.

143. Pike AW, Wadsworth SL: Sea lice on salmonids: their biology and control. Adv Parasitol. 2000, 44: 231-337.

144. Johnsen BO, Jensen AJ: Infestations of Atlantic salmon, *Salmo salar*, by *Gyrodactylus salaris* in Norwegian rivers. J Fish Biol. 1986, 29: 233-241. 10.1111/j.1095-8649.1986.tb04941.x.
145. Taylor NGH, Wootten R, Sommerville C: The influence of risk factors on the abundance, egg laying habits and impact of Argulus foliaceus in stillwater trout fisheries. J Fish Dis. 2009, 32: 509-519. 10.1111/j.1365-2761.2009.01007.x.
146. Murphy PJ, St-Hilaire S, Bruer S, Corn PS, Peterson CR: Distribution and pathogenicity of Batrachochytrium dendrobatidis in boreal toads from the Grand Teton area of western Wyoming. Ecohealth. 2009, 6: 109-120. 10.1007/s10393-009-0230-4.
147. Patterson KR: Modelling the impact of disease-induced mortality in an exploited population of the fungal parasite Icthyophonus hoferi in the North Sea herring (*Clupea harengus*). Can J Fish Aquat Sci. 1996, 53: 2870-2877.
148. Borsuk ME, Reichert P, Peter A, Schager E, Burkhardt-Holm P: Assessing the decline of brown trout (*Salmo trutta*) in Swiss rivers using a Bayesian probability network. Ecol Model. 2006, 192: 224-244. 10.1016/j.ecolmodel.2005.07.006.
149. Murray AG, O'Callaghan M: The numerical and programming methods used to implement models of the spread and impact of a major epidemic disease: pilchard herpesvirus, Australia 1995 and 1998/1999. Environ Model Software. 2005, 20: 575-585. 10.1016/j.envsoft.2004.03.011.
150. Barber I, Scharsack JP: The three-spined stickleback-*Schistocephalus solidus* system: an experimental model for investigating host-parasite interactions in fish. Parasitology. 2010, 137: 411-424. 10.1017/S0031182009991466.
151. Georgiadis MP, Gardner IA, Hedrick RP: The role of epidemiology in the prevention, diagnosis and control of infectious disease of fish. Prev Vet Med. 2001, 48: 287-302. 10.1016/S0167-5877(00)00202-6.

CHAPTER 3

Application of Loop Analysis for the Qualitative Assessment of Surveillance and Control in Veterinary Epidemiology

Lucie Collineau[1], Raphaël Duboz[1,4], Mathilde Paul[2], Marisa Peyre[1], Flavie Goutard[1,5], Sinel Holl[3] and François Roger[1,5]

[1] CIRAD, AGIRs (Animal and Integrated Risks Management Unit, UPR22), Montpellier, France.

[2] UMR 1225 IHAP INRA-ENVT, National Veterinary School, Toulouse, France.

[3] National Veterinary Research Institute, Phnom Penh, Cambodia.

[4] Asian Institute of Technology, Computer Sciences and Information Management, Pathumthani, Thailand.

[5] Faculty of Veterinary Medicine, Kasetsart University, Bangkok, Thailand.

Citation: Collineau, L., Duboz, R., Paul, M. et al. Application of loop analysis for the qualitative assessment of surveillance and control in veterinary epidemiology. Emerg Themes Epidemiol 10, 7 (2013). https://doi.org/10.1186/1742-7622-10-7

ABSTRACT

Background

Systems for animal disease mitigation involve both surveillance activities and interventions to control the disease. They are complex organizations that are described by partial or imprecise data, making it difficult to evaluate them or make decisions to improve them. A mathematical method, called loop analysis, can be used to model qualitatively the structure and the behavior of dynamic systems; it relies on the study of the sign of the interactions between the components of the system. This method, currently widely used by ecologists, has to our knowledge never been applied in the context of animal disease mitigation systems. The objective of the study was to assess whether loop analysis could be applied to this new context. We first developed a generic model that restricted the applicability of the method to event-based surveillance systems of endemic diseases, excluding the emergence and eradication phases. Then we chose the mitigation system of highly pathogenic avian influenza (HPAI) H5N1 in Cambodia as an example of such system to study the application of loop analysis to a real disease mitigation system.

Results

Breaking down the generic model, we constructed a 6-variables model to represent the HPAI H5N1 mitigation system in Cambodia. This construction work improved our understanding of this system, highlighting the link between surveillance and control which is unclear in traditional representations of this system. Then we analyzed the effect of the perturbations to this HPAI H5N1 mitigation system that we interpreted in terms of investment in a given compartment. This study suggested that increasing intervention at a local level can optimize the system's efficiency. Indeed, this perturbation both decreases surveillance and intervention costs and reduces the disease's occurrence.

Conclusion

Loop analysis can be applied to disease mitigation systems. Its main strength is that it is easy to design, focusing on the signs of the interactions. It is a simple and flexible tool that could be used as a precursor to large-scale quantitative studies, to support reflection about disease mitigation systems structure and functioning.

Keywords: Efficiency, Mitigation system, Loop analysis, Interaction, HPAI H5N1

BACKGROUND

The Context of Animal Disease Mitigation Systems

An animal disease surveillance system consists of all individuals or agencies organized to ensure surveillance in a given region of one or more hazards [1]. Surveillance consists of the continuous collection, recording and analysis of data that are then disseminated in order to control the disease [2], i.e. to reduce the disease prevalence, to early detect cases or to demonstrate the freedom from infection or disease. An intervention aims to control the disease. As suggested by Häsler et al., surveillance and intervention activities are part of a mitigation strategy to make the effect of the disease less severe [3]. The term disease mitigation system is further used in the text as a synonym of disease surveillance and control systems. The activities of surveillance and intervention vary according to the disease being studied and the objectives of the system. Vaccination, culling (test-and-slaughter), vector control or biosecurity improvement are measures that can be applied to control an animal disease. These activities vary in terms of their cost (economic and social), their effectiveness and their feasibility [4, 5]. Disease mitigation systems are complex organizations, and the data they collect are partial, incomplete or imprecise, particularly in developing countries [6]. The evaluation of surveillance programs is crucial to i) assess the limitations and usefulness of the data generated [7], ii) ensure that limited resources are effectively used for protecting health [8] and iii) comply with requirements of international organizations in the era of trade globalization [9]. Different methods and tools exist for the qualitative and quantitative evaluation of animal disease surveillance systems: e.g. OASIS (Outil d'analyse de système d'information en santé) tool [7]; SERVAL (SuRveillance EVALuation framework) [10]; tree scenarios [11], or capture-recapture techniques [12]. Current representations of animal surveillance systems are mostly organigrammes or flowcharts. They provide only an organizational snapshot of these systems, and a limited understanding of their process dynamics (in particular the link between surveillance and intervention activities). This hinders both the evaluation of these systems and any decision on how to improve their efficiency, defined as how effective they are in reducing the prevalence of the disease by making the best use of the financial, human and material resources they mobilize [13].

The Method of Loop Analysis

Feedback loop analysis, referred to simply as loop analysis is a method that applies qualitative models to dynamic systems [14]. It allows i) to represent the system's components and interactions between components, and ii) from the study of the sign of the interactions, to predict the response of the system to the modification of the level of one of its variable; this modification is called perturbation. While the loop analysis method is frequently applied in ecology [15, 16] and has been used for exploring associations between various variables and epidemiological risks [17, 18], it has to our knowledge never been used to study animal disease mitigation systems. In this section, we first present the approach one has to follow to conduct loop analysis.

Let us consider a dynamic system composed of n continuous variables. For i=1 to n, X_i represents the ith system's component. Loop analysis is conducted following seven successive steps:

1. First, the system's components (or variables) are described.
2. Loop analysis requires making the following hypothesis: i) the system is close to a state of equilibrium at every moment and ii) interactions are considered continuous, monotone and of the same order of magnitude.
3. The interactions between the system's variables are defined and represented by a signed directed graph, i.e. a graph in which the arrows representing interactions have a given direction and sign (positive, negative or no interaction from one variable to another or to itself).
4. The directed graph is converted into a unique interaction matrix labelled A, in which the elements a_{ij} represent the effect of a modification of X_j on X_i. The value of elements a_{ij} can be any of {+1; -1; 0}, depending on whether the effect of modifying X_j on X_i is positive, negative, or null, respectively. To get the interaction matrix from the directed graph, only the direction and the sign of the interactions are taken into account, and not its amplitude.
5. We verify that the system meets both the Hurwitz criteria for stability, and is therefore stable. These criteria, based on matrix determinants calculations, mean that the system always returns to its equilibrium after a perturbation. For further details about the calculation method, please see the work from Dambacher et al. [19].

6. We study the response of the system to perturbations to its different variables, by calculating the conjugate transpose (also called adjoint matrix) labelled Adj(-A) for interaction matrix A (see [20] for calculation details). The signs of the adjoint matrix coefficients give the direction of change (increase or decrease) of all the system's variables after the increase (respectively decrease) of the level of one variable of the system (called positive, respectively negative perturbation of this variable). Perturbations are represented in the columns and variables' responses in the rows of the matrix Adj(-A). The response of the system to perturbations can also be displayed using a directed graph in which decreasing variables are represented by small size circles, increasing variables by big size circles and variables having undergone no effect or ambiguous effect by an intermediate size circle.
7. We calculate the matrix W of weighted predictions which allows for a level of confidence to be attributed to the predictions [20], ranging from unambiguous at level $w_{ij}=1$ (certain prediction), to totally ambiguous at $w_{ij}=0$ (very poor confidence in the prediction). A value of $w_{ij}=0.5$ is generally accepted as a validation threshold.

Loop analysis is a dynamic system modeling method. To our knowledge, dynamic modeling has been little used to study disease mitigation systems. An example is provided by Howe et al. who developed a generic model to study the overall economic optimum for disease mitigation [21]. In the wider context of epidemiology, dynamic modeling has been mainly applied to represent the spread of diseases [22]. For example, SIR model is a compartmental model representing the dynamic of the disease between three compartments (or states): susceptible, infectious and recovered. Compartmental models can also be used to study the impact of certain control measures (e.g. vaccination) on the dynamic of infection. The main difference with loop analysis is that time is considered (e.g. we observe the moment at which the outbreak goes extinct). Another difference is that compartmental models require quantifying the interactions between compartments, which is sometimes difficult or impossible; that is not necessary with loop analysis which only requires signs to be defined. This is the strength of loop analysis (easy to design), but also its main limit (the model only predicts whether variables increase or decrease, but not how much and how fast). In addition, loop analysis allows the integration of variables of very different natures, which in turn enables the development of interdisciplinary approaches [23]. The whole approach can also be computerized using for example the Loop analysis website [24].

The objective of this study was to assess whether loop analysis could be applied to the context of animal disease mitigation systems. More especially, we evaluated whether loop analysis allows i) to model the components of a mitigation system and the interactions between them, and ii) to study the mitigation system's behavior after a modification of the level of one of its components (perturbation), that we interpreted in terms of system's efficiency. We first applied loop analysis to a generic disease mitigation system. This study, presented in the next section, allowed both to determine the limits of the method applicability and the conditions necessary for the application of loop analysis to a disease mitigation system. It also provided a starting point for the construction of more complex models. Breaking down this generic model, we applied loop analysis to a real mitigation system: the HPAI H5N1 mitigation system in Cambodia. This work is described in the results section of this paper.

Application of Loop Analysis to A Generic Disease Mitigation System

We considered disease mitigation systems as dynamic systems. Following the 7-steps approach described in the previous section, we applied loop analysis to a simplified and generic model of an animal disease mitigation system, unconnected to any particular disease.

1. In the case of a country affected by an infectious animal disease, the simplest model that can describe its surveillance and intervention activities is a generic one with three intensity variables, called model 3I. The intensity of the disease (variable I_D) represents the level of disease occurrence detected by the surveillance system. I_S and I_C represent the intensity of surveillance and intervention, respectively, in other words, the extent of the financial, human and material resources deployed for the surveillance and control of the disease. We excluded systems designed for the early detection of outbreaks, where I_D and I_C are null.
2. We assumed that the 3I system is close to a state of equilibrium at every moment. This implies that the disease intensity (variable I_D) is close to or oscillates around the equilibrium: this corresponds to an endemic disease situation. Thus, the 3I model is further

restricted to the case when the disease is present and endemic. We also considered the 3I model interactions being continuous, monotone and of the same order of magnitude.

3. We defined the interactions of our generic model as follows: first, I_D has a positive influence on I_S. This is only true if I_S is restricted to event-based surveillance (i.e. surveillance based on the reporting of the disease cases spontaneously or by key stakeholders of the system). On the contrary, active surveillance can stay at a constant intensity level if the disease intensity increases (e.g. systematic surveillance). Consequently, I_S was further restricted to the intensity of event-based surveillance. We also excluded the study of the period of eradication because in this case the interaction I_D-I_S would be negative. Indeed, if the prevalence of the disease approaches zero, surveillance needs to be increased to detect the last, sparser, remaining cases. I_S has a positive action on I_C (we assumed in our model that if more cases are detected, more cases will be later controlled). Finally, I_C has the effect of reducing I_D (we supposed that the deployed control measures effectively reduce the level of the disease). I_S and I_C are self-limiting (each having a negative action on themselves), as they are subject to limited financial, material and human resources. I_D is also self-limiting as the disease is endemic (if I_D increases, I_D later decreases and returns to its level of equilibrium). Finally, we excluded the emergence phase where the interaction I_D–I_D is positive. The directed graph representing the generic system is presented in Figure 1A.

A

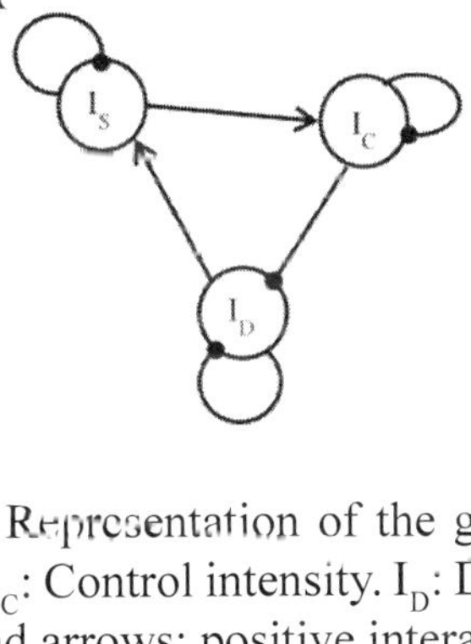

B

$A_1=$

	I_S	I_C	I_D
I_S	-1	0	1
I_C	1	-1	0
I_D	0	-1	-1

Figure 1. Representation of the generic system. I_S. Event-based surveillance intensity. I_C: Control intensity. I_D: Detected disease intensity. **A**: Directed graph. Pointed end arrows: positive interaction. Circular end arrows: negative interaction. **B**: Matrix A_1 of interactions. **A** and **B** are two equivalent representations

of the 3I system. To get the interaction matrix A_1 from the directed graph, only the direction and the sign of the interactions are taken into account, and not their amplitude.

1. We converted this directed graph into an interaction matrix A_1 represented in Figure 1B.
2. The 3I system meets both the Hurwitz stability criteria, and can therefore be considered stable.

A

$Adj(-A_1) =$

	I_S	I_C	I_D
I_S	1	-1	1
I_C	1	1	1
I_D	-1	-1	1

B

C

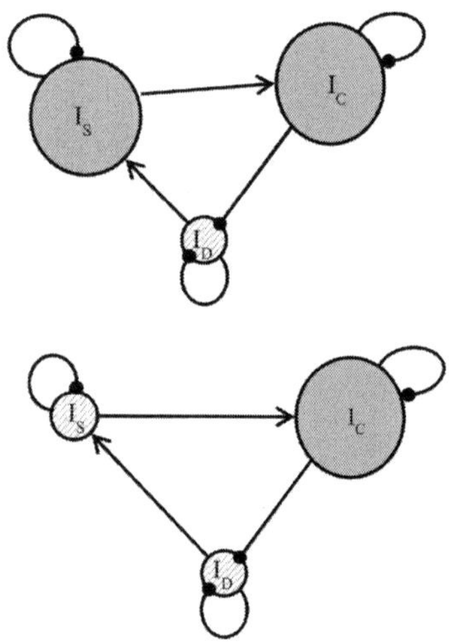

Figure 2. Study of the dynamic of the generic system. I_S: Intensity of event-based surveillance. I_C: Intensity of control. I_D: Detected intensity of the disease. **A**: Matrix adjoint to the generic system. Perturbations are read in the columns and the variables' responses to perturbations are read in the corresponding rows of the matrix $Adj(-A_1)$. **B** and **C**: Directed graphs representing the response of the generic system to a positive perturbation on I_S**(B)** and I_C**(C)**. Greyed-out and enlarged circle: increasing variable. Hashed-out and reduced circle: decreasing variable.

3. The adjoint matrix $Adj(-A_1)$ describes the behavior of the 3I system after a positive perturbation (i.e. increase) of the variables in the columns headers (Figure 2A). The effect of a positive perturbation

to I_S and I_C (read in the first and second columns of Adj($-A_1$)) are also graphically displayed in the Figures 2B and 2C. By creating a positive perturbation on I_S or I_C, the model shows a negative influence on I_D, therefore a reduction in the detected disease level. However, by increasing I_C (Figure 2B), we can note that I_S also diminishes; the intensity of surveillance is reduced.

1. The predictions from the adjoint matrix are certain because all the coefficients of the matrix W_1 associated with A_1 have a value of 1.

Finally, the application of loop analysis to the generic model restricted the applicability of the method to event-based surveillance systems of endemic diseases, excluding the emergence and eradication phases. The HPAI H5N1 mitigation system in Cambodia fulfils these requirements: HPAI H5N1can be considered as endemic in Cambodia, as it has been present since 2004 according to the OIE, circulating mostly at a low level among humans and domestic poultry [25]. Moreover, this mitigation system is mainly based on event-based surveillance [26]. Thus, we used the HPAI H5N1 mitigation system in Cambodia as an example to study the application of loop analysis to a real mitigation system. To apply loop analysis to this particular case, we considered each variable of the generic model and redefined it according to this specific context. For example, the intensity of the surveillance was modeled by combining several interacting variables representing the intensity of surveillance performed by each of the stakeholders of the surveillance system, including the farmers themselves. Thus to elaborate the specific model, we broke down the generic one into one with more variables and more interactions. Building this more complex model therefore required the prior understanding of the HPAI H5N1 mitigation system in Cambodia. By improving our knowledge of the system, we also improved our ability to build a more precise model, that is to say, one with more specific interactions.

RESULTS

We present in this section the application of the 7-steps loop analysis approach to the HPAI H5N1 mitigation system in Cambodia. The first part of the section focuses on the steps 1 to 4, and the second part on the steps 5 to 7.

Description of the HPAI H5N1 Mitigation System in Cambodia

Our knowledge of the HPAI H5N1 mitigation system in Cambodia was based on bibliographic data [26, 27] and a field study conducted in April 2012 in the Cambodian provinces of Kompong Cham and Takeo. In the course of this study, we interviewed key stakeholders of the system to collate different points of view on the functioning of the system. Key stakeholders consisted of 21 farmers, 21 designated Village Animal Health Workers (VAHWs), 6 district veterinarians, 3 province-level veterinarians and 3 NaVRI (National Veterinary Research Institute) personnel. The VAHWs are village volunteers, usually one per village, who have been trained to recognize, treat and report certain animal diseases to the health authorities, in particular cases of HPAI H5N1. In our selection of local stakeholders we targeted active VAHWs working in zones affected by previous HPAI H5N1 outbreaks, whose names were put forward by the district veterinarians. We then interviewed farmers in the village of each selected VAHW. Backyard farming makes up the majority of poultry production in Cambodia [28], and it is reasonable to assume that the behavior of poultry farmers is generally homogeneous throughout the country. Thus, the selected sample gave us a satisfactory representation of the role of local stakeholders within the network, in particular of their surveillance and control activities, and interactions with other stakeholders.

1. We called I_D the intensity of the HPAI H5N1 in Cambodia. Further to the introduction of this disease on the territory in 2004, the Cambodian government, with the support of various international organizations, has put in place a system for the surveillance and control of the disease [26]. This system is principally based on event-based surveillance, organized into three levels. The local level consists of farmers and VAHWs, who monitor and control HPAI H5N1. We called I_{Sl} the intensity of farmers and VAHWs surveillance, measured as the extent of the human, financial and material resources these stakeholders use to monitor livestock (principally by observation), and to report cases to the intermediate level if appropriate. Similarly, I_{Cl} represents all the financial, material and human resources implicated in the control activities by the local stakeholders. It includes disinfecting

premises, or separating and treating diseased animals. For farmers and VAHWs, I_D represents the number of dead or diseased birds.

The intermediate level, consisting of district and province-level veterinarians, also plays a role in event-based surveillance, by passing on information from the local level to the national level. The I_{Si} variable is the extent of the material, human and financial resources required by the district and province veterinarians for the surveillance of the sanitary situation in farms, and its reporting if required. The intensity of the surveillance by the CDC (Centre for Disease Control and Prevention), working under the aegis of the Ministry of Health, can also be integrated into the variable I_{Si}. Indeed, this intermediate actor transfers information acquired via its own hotline, from local stakeholders to the NaVRI. In addition, the NaVRI conducts event-based surveillance of suspected cases of HPAI H5N1 at national level. It receives calls to the hotline, and if HPAI H5N1 is suspected, it intensifies its surveillance by collecting and analyzing samples. The intensity of this surveillance, labeled I_{Sn}, represents the extent of human, financial and material resources deployed by the NaVRI to evaluate the disease's level of occurrence. If HPAI H5N1 is confirmed, the NaVRI implements control measures. It organizes in particular the selective and uncompensated slaughter of animals that test positive for HPAI H5N1, as well as the disinfection of the area and the prevention of animal movements. I_{Cn} corresponds to the human, material and financial resources put in place for the slaughtering, disinfection and blocking the transport of animals. Intermediate stakeholders take few decisions relating to control. Rather, they are more involved in the application of national control measures, as decided by the NaVRI. We therefore chose to include the intensity of the control by intermediate stakeholders in the variable I_{Cn}. The NaVRI also carries out active surveillance of HPAI H5N1 in live poultry markets and sentinel free-range duck farms. This surveillance is conducted by scheduled spot checks, and its intensity is independent from the epidemiological dynamics of the disease. We were therefore unable to integrate it into our model. Indeed, if a variable describing the intensity of active surveillance was included, it would be isolated and have no interaction with the other variables. However, this is of no consequence for the study of the system's dynamics as such. A summary table of the stakeholders involved at different levels, and the corresponding variables is presented in the Table 1.

Table 1. Summary table of the stakeholders involved in the HPAI H5N1 mitigation system in Cambodia and the corresponding variables included in the model

Level	Stakeholders	Variables: surveillance intensity	Variables: control intensity
National	NaVRI	I_{Sn}	I_{Cn}
Intermediate	District and provincial-level veterinarians, CDC	I_{Si}	I_{Cn}
Local	Farmers, VAHWs	I_{Sl}	I_{Cl}

The surveillance (respectively control) intensity variables represent the extent of the human, financial and material resources deployed by the stakeholders involved in the HPAI H5N1 surveillance (respectively control) in Cambodia.

To adapt the generic model to the context of the mitigation system of HPAI H5N1 in Cambodia, we broke down the variable I_S into three variables (I_{Sl}, I_{Si} and I_{Sn}), corresponding to the different surveillance levels. At the same time, we broke down the variable I_C into two new variables (I_{Cl} and I_{Cn}), to reflect the two main levels of control. This breakdown was discussed and validated with two members of the NaVRI and one of the authors of this article, an expert on the surveillance network for HPAI H5N1 in Cambodia.

2. We considered here that HPAI H5N1 is endemic in Cambodia. This hypothesis seems acceptable, as HPAI H5N1 has been present in Cambodia since 2004 according to the OIE, circulating mostly at a low level among humans and domestic poultry [25]. We also decided to represent here the national scale, meaning that I_{Sl} and I_{Cl} include (aggregate) the intensities of surveillance and control of all farmers and VAHWs in Cambodia. Thus we considered that the intensity of national surveillance and control was of the same order of magnitude as the aggregated intensities of surveillance and control at local level.
3. We then described the interactions between the system's variables. The link between I_D and I_{Sl} is positive: the farmers and VAHWs intensify surveillance when they observe affected birds; they

monitor their flocks more frequently and cautiously, and are more diligent about reporting cases. We supposed that if the intensity of the local surveillance increases, the intensity of the local control increases (positive I_{Sl}-I_{Cl} link). Assuming that local control is effective in reducing the incidence of the disease, I_{Cl} in turn then has a negative influence on I_D (negative I_{Cl}- I_D link). I_{Sl} influences I_{Si} positively (positive I_{Sl}-I_{Si} link). Indeed, the resource investment increases when an intermediate-level veterinarian is contacted by a farmer or VAHW, for example, if the veterinarian then makes a field visit to collect samples. I_{Sn} increases when the intensity of the surveillance by local (positive I_{Sl}-I_{Sn} link) and intermediary stakeholders (positive I_{Si}-I_{Sn} link) increases, due to hotline reports. We chose to examine a scenario where a case of HPAI H5N1 is confirmed. The I_{Sn}-I_{Cn} link is therefore positive – without confirmation, there is no such link. I_{Cn} has a negative action on I_D: we assumed that the control at national level effectively reduces the occurrence of the disease. I_{Cn} also has a negative influence on I_{Sl} and I_{Si} (negative I_{Cn}-I_{Sl} and I_{Cn}-I_{Si} links). Indeed, increasing the number of animals slaughtered without compensation reduces the reporting of cases by local and intermediate stakeholders, who are affected by these measures [27]. Figure 3A shows a proposed representation of the directed graph representing this system.

4. This directed graph was then converted into the interaction matrix A_2 showed in Figure 3B.

Thus, loop analysis offered a new representation of the HPAI H5N1 mitigation system in Cambodia. Representing the components of the system and the interactions between them improved our understanding of the system's structure and functioning. It was especially interesting to understand how surveillance and control are connected, as this is usually unclear with classical representations of this system (hierarchical organigramme). Breaking down the generic model to build a more complex model allows for keeping surveillance components and control components together, making the link between surveillance and control clear.

A

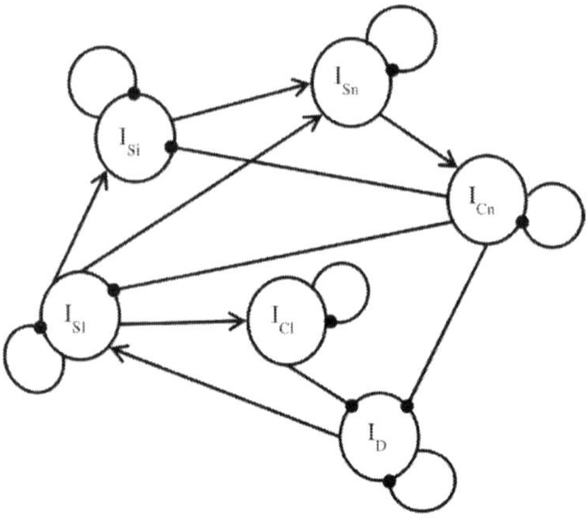

B

$A_2=$

	I_{Sl}	I_{Si}	I_{Sn}	I_{Cl}	I_{Cn}	I_D
I_{Sl}	-1	0	0	0	-1	1
I_{Si}	1	-1	0	0	-1	0
I_{Sn}	1	1	-1	0	0	0
I_{Cl}	1	0	0	-1	0	0
I_{Cn}	0	0	1	0	-1	0
I_D	0	0	0	-1	-1	-1

Figure 3. Representation of the HPAI H5N1 mitigation system in Cambodia. I_{Sl}: Intensity of surveillance by local stakeholders (farmers, VAHWs), I_{Si}: Intensity of surveillance by intermediate stakeholders (district and province-level veterinarians, CDC), I_{Sn}: Intensity of surveillance by national-level stakeholders (NaVRI), I_{Cl}: Intensity of control by local stakeholders, I_{Cn}: Intensity of control by intermediate and national stakeholders, I_D: Detected occurrence level of the disease (number of dead or diseased birds). **A**: Directed graph. Pointed end arrows: positive interaction. Circular end arrows: negative interaction. **B**: Matrix A_2 of interactions. **A** and **B** are two equivalent representations of the HPAI H5N1 mitigation system in Cambodia. To get the interaction matrix A_2 from the directed graph, only the direction and the sign of the interactions are taken into account, and not their amplitude.

Study of the dynamic of the HPAI H5N1 mitigation system in Cambodia

This section focuses on the steps 5 to 7 of the loop analysis approach applied to the HPAI H5N1 mitigation system in Cambodia.

5. The system for the surveillance and control of HPAI H5N1 in Cambodia is considered stable in that it meets both Hurwitz stability criteria.
6. Thus, we studied the predictability of this system by calculating its conjugate transpose presented in Table 2.

Table 2. Matrix adjoint to the HPAI H5N1 mitigation system in Cambodia

		I_{SI}	I_{Si}	I_{Sn}	I_{Cl}	I_{Cn}	I_D
	I_{SI}	2	−2	−2	−2	−2	2
	I_{Si}	0	4	−4	0	−4	0
Adj $(-A_2)$ −	I_{Sn}	2	2	2	−2	−6	2
	I_{Cl}	2	−2	−2	6	−2	2
	I_{Cn}	2	2	2	−2	2	2
	I_D	−4	0	0	−4	0	4

1. I_{SI}: Intensity of surveillance by local stakeholders (farmers, VAHWs), I_{Si}: Intensity of surveillance by intermediate stakeholders (district and province-level veterinarians, CDC), I_{Sn}: Intensity of surveillance by national-level stakeholders (NaVRI), I_{Cl}: Intensity of control by local stakeholders, I_{Cn}: Intensity of control by national-level and intermediate stakeholders, I_D: Detected level of disease occurrence (number of dead or diseased birds).
2. Perturbations are read in the columns and the variables' responses to perturbations are read in the corresponding rows of the matrix Adj$(-A_2)$.

We elected to focus on the positive perturbations on I_{SI}, I_{Cl} and I_{Cn} (read in the first, fourth and fifth columns of the matrix Adj$(-A_2)$). To facilitate the reader's understanding, the behavior of the system in these three cases was represented in three distinct directed graphs (Figure 4).

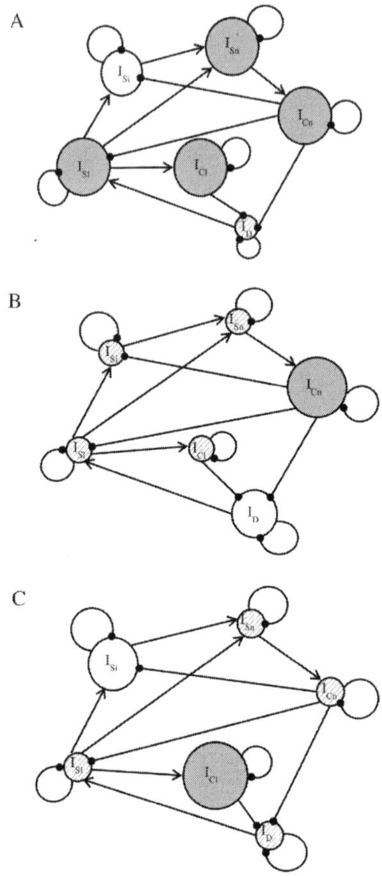

Figure 4. Study of the dynamic of the HPAI H5N1 mitigation system in Cambodia. Impact of a positive perturbation on I_{Sn}(A), I_{Cn}(B) and I_{Cl}(C). I_{Sl}: Intensity of the surveillance by local stakeholders (farmers, VAHWs), I_{Si}: Intensity of the surveillance by intermediate stakeholders (district and province-level veterinarians, CDC), I_{Sn}: Intensity of the surveillance by national-level stakeholders (NaVRI), I_{Cl}: Intensity of the control by local stakeholders, I_{Cn}: Intensity of the control by national-level and intermediary stakeholders, I_D: Detected level of incidence of the disease (number of dead or diseased birds). Greyed-out and enlarged circle: increasing variable. Hashed-out and reduced circle: decreasing variable. White circle: ambiguous direction of change of the variable (impossible to conclude if it increases or decreases).

The results show that a positive perturbation on I_{Sl} (Figure 4A) increases the intensity of surveillance (except for I_{Si}, which is ambiguous) and of the control across the different stakeholders, and reduces the intensity of the disease. Furthermore, a positive perturbation of the intensity of national control (Figure 4B) reduces the intensity of surveillance at all levels, and that of control at local level. On other hand, the impact on I_D is ambiguous: the model does not indicate whether this national-level control is effective in reducing the level of the disease. Finally, results indicate that increasing the intensity of local control (see Figure 4C for the positive impact on I_{Cl}) reduces I_D unambiguously while lowering the intensity of local and national-level surveillance (with ambiguity for I_{Si}).

7. We calculated the matrix W_2 associated to the system; it is presented in Table 3. We note that six predictions out of the 36 are completely ambiguous, which indicates that only a feeble level of confidence is attributed to them. This is linked to the presence of feedback loops of opposite signs within the system [19], which reduces the net number of complementary feedbacks.

Table 3. Matrix W_2 **of the predictions relating to HPAI H5N1 mitigation system** in Cambodia

		I_{Sl}	I_{Si}	I_{Sn}	I_{Cl}	I_{Cn}	I_D
	I_{Sl}	1	1	1	1	1	1
	I_{Si}	0	1	1	0	1	0
$W_2 =$	I_{Sn}	1	1	1	1	1	1
	I_{Cl}	1	1	1	1	1	1
	I_{Cn}	1	1	1	1	1	1
	I_D	1	0	0	1	0	1

1. I_{Sl}: Intensity of surveillance by local stakeholders (farmers, VAHWs), I_{Si}: Intensity of surveillance by intermediate stakeholders (district and province-level veterinarians, CDC), I_{Sn}: Intensity of surveillance by national-level stakeholders (NaVRI), I_{Cl}: Intensity of control by local stakeholders, I_{Cn}: Intensity of control by national-level and intermediate stakeholders, I_D: Detected level of disease occurrence (number of dead or diseased birds).
2. The matrix coefficients represent the level of confidence to be attributed to the predictions, ranging from unambiguous at level $w_{ij}=1$ (certain prediction), to totally ambiguous at $w_{ij}=0$ (very poor confidence in the prediction). A value of $w_{ij}=0.5$ is generally accepted as a validation threshold.

DISCUSSION

Loop Analysis Conditions for Application and Limits

We have described the principle of the loop analysis method, and illustrated its application, first to a minimal, generic model (3I), then to a specific model of the HPAI H5N1 mitigation system in Cambodia. The loop analysis method imposes certain restraints which limit its applicability to the context of disease mitigation systems. The study of the 3I model showed that loop analysis can be applied to any disease subject to event-based surveillance and control. However, the application is limited to endemic diseases; this corresponds to the epidemiological phase when the system is close to a state of equilibrium, which means that the intensities of the disease, its surveillance and its control vary little over time. We can accept this hypothesis for a disease that has been endemic for several years, as in the case of HPAI H5N1 in Cambodia. On contrary, the method cannot be applied to transition phases such as emergence and eradication phases.

The generic and the HPAI H5N1 models are, furthermore, based on the assumption that control measures are effective in reducing the occurrence of the disease, which is debatable. At national level, only cases detected by the system are controlled, effectively making the effectiveness of such controls dependent on the sensitivity of the surveillance system. The field implementation of local control measures can also be questioned, for example with some farmers selling their diseased animals, thereby contributing to the spread of HPAI H5N1 [29].

The method is also based on the assumption that the rate of variation is of the same order of magnitude for all variables. To consider this hypothesis in our model for Cambodia, we assumed that the intensity of national surveillance and control is of the same order of magnitude as the aggregated intensities of surveillance and control at local level. However, further works would be needed to confirm this hypothesis. Another limitation of the model is that some interactions between the compartments of the system are subject to threshold effects. For example, a farmer may tend to wait until he has observed a certain number of dead animals before reporting the case to health authorities, even for suspected bird flu [30, 31]. Loop analysis does not consider these thresholds, as it assumes a continuous interaction between variables. Working at a national level, i.e. aggregating all the Cambodian farmers' surveillance intensities, we can assume that the directed interaction always exists. However, at an individual level, this assumption is not borne out: below the farmers' reporting threshold the links I_{SI}-I_{Si} and I_{SI}-I_{Sn} do not exist. A new model should therefore be built eliminating these links for the individual level.

Finally, loop analysis does not allow for the weighting of the strength of interactions between the variables. Nonetheless, these strengths can be assumed to vary within our model for Cambodia. For example, the link I_{SI}–I_{CI} is stronger than I_{SI}-I_{Si}. Indeed, all farmers attempt to put in place their own controls when they observe dead or diseased animals, while only some report the cases to veterinary authorities, among other reasons because their trust in these institutions is limited [32]. However, the weighting of interactions can be integrated into our model by the use of the Fuzzy Cognitive Maps method, a semi-quantitative modeling tool for dynamic systems which would allow the refining of the models proposed by loop analysis [33].

The progression from the 3I model to a more precise model was carried out by specifying the variables and interactions laid out by the minimal model. This method requires only few data and is therefore easily adapted to the context of developing countries. We applied it to the case of surveillance and control of HPAI H5N1 in Cambodia, which led us to elaborate a model with six variables. The construction of the model improved our understanding of the system, especially because this model highlights the link between surveillance and control components, which is usually unclear in traditional representations of this system (hierarchical organigrammes). We feel that this breakdown method can be applied to any system for the surveillance and control of an endemic disease. However, the single example of Cambodia is not sufficient to demonstrate the method's generality, i.e. the possibility to apply it to other mitigation systems. We are currently conducting further field work to this end in Thailand and Laos.

Interpretation of the mitigation system's behavior: the study of the system's efficiency

As the 3I model and the model for Cambodia are stable, we were able to study the responses of the systems to perturbations of their compartments. We propose here to attribute a financial, human and material cost to the intensity of surveillance and control. The hypothesis of a positive correlation between cost and intensity of surveillance and control seems acceptable from a qualitative point of view. We can then interpret a positive perturbation of a surveillance or control compartment as an investment made in this compartment in terms of financial, human and material resources deployed to achieve the surveillance or the control of the disease. Starting from there, we can study the effect of an investment in a certain compartment on the costs of surveillance or control and on the disease occurrence level. We considered here that the investments causing both a reduction of the surveillance and control associated costs and the disease occurrence level a priori optimize the system's efficiency, that is to say its effectiveness in reducing the occurrence of the disease by optimizing the financial, human and material resources deployed. However, we should keep in mind that optimizing the system's efficiency does not necessary mean reducing the surveillance and control activities costs to reduce the disease occurrence level. It could also consist in making better use of the resources available, at a constant cost level.

Nevertheless, we think that this qualitative modeling approach is an interesting first step to study the efficiency of surveillance and control systems in animal health. Indeed, it is very hard to quantify the effect of an investment, especially when we work with partial or imprecise data. Additionally, the application of policies for the surveillance and control of diseases relies on human behavior, which is also hard to measure or quantify [34]. The loop analysis being a qualitative method, we cannot, by definition, quantify the decrease or increase in efficiency. However, we can calculate its trend according to an investment choice. This is the result of the analysis of the perturbations, which allow us to rank various situations according to the perturbation's effect on the system's efficiency.

The results of the analysis by the 3I and Cambodia-specific models show a dynamic common to both systems: investing in control measures corresponds to a reduction in the occurrence of the disease without increasing the intensity of surveillance. On the contrary, an investment in surveillance requires an additional increase in control to achieve a reduction in the occurrence of the disease. Investing in control rather than surveillance a priori yields a better improvement in the system's efficiency.

In the case of the model for Cambodia, the model shows that investments in local surveillance (for example working to improve farmers' trust in veterinary services, and encouraging them to report observed cases of disease), may contribute to reduce the intensity of the disease (Figure 4A). This reduction requires increased intensity of both surveillance and control of all stakeholders within the system, meaning at national level and all local stakeholders in Cambodia. (There is some ambiguity regarding intermediate stakeholders, we are unable to conclude whether the intensity of their surveillance increases or decreases). Even if we cannot quantify the different costs associated with the rises in intensity, the model predicts that this situation will be characterized by an overall increase in the cost of surveillance and control relative to the former equilibrium state. Thus, results indicate that the occurrence of the disease would be reduced but with an elevated cost, so the efficiency of the system would decrease in this case.

Conversely, results suggest that investing in control measures carried out at the national level can reduce the intensity of surveillance (Figure 4B), thereby limiting the costs incurred by this surveillance. However, the effect on I_D is ambiguous in this scenario, and we cannot therefore draw any conclusions regarding the effectiveness of this type of investment to reduce the occurrence of the disease. This ambiguity can be explained by the negative influence of uncompensated slaughter on the reporting of cases by local and intermediate stakeholders. If we construct an identical system but eliminate the negative links I_{Cn}-I_{Sl} and I_{Cn}-I_{Si}, we notice that an investment in control at national level allows for the occurrence of the disease to be unambiguously reduced. Thus, by working on the acceptability of national control measures, the effectiveness of the system to control the disease can be improved [35]. The type of investment presented in Figure 4B therefore allows for the reduction of surveillance costs, but its effectiveness in reducing disease occurrence is ambiguous, which means the efficiency of the system is ambiguous in this case.

Finally, a positive perturbation on the intensity of local control (Figure 4C) reduces unambiguously the intensity of the disease, while reducing the intensity of the surveillance and control at national level. Local control has two distinct advantages: on the one hand it does not negatively influence the intensity of surveillance, and on the other hand it does not require laboratory confirmation of HPAI H5N1 cases. The case in Figure 4C therefore represents a situation which should in principle be less costly than the one in Figure 4A (by reduction in the intensity of surveillance and control), and more effective that the one in Figure 4B in reducing the occurrence of the disease. The scenario represented in Figure 4C therefore provides for the optimization of the system's efficiency.

We adopted here the perspective of a decision-maker, with a good overall visibility of the system, to ask ourselves what scenario would optimize its efficiency. However, it should be noted that depending on the approach, a single given system can be represented by one of several loop analysis models. Indeed, the validity of any particular model structure depends only on a compromise between different points of view, and the interpretation of impacts depends on the question at hand. Our qualitative approach does not allow the quantification of the investment necessary to optimize the functioning of the system. However, by understanding the system's dynamics, the model allows different situations to be ordered hierarchically, and the most favorable configuration for optimizing the system's efficiency to be identified. This analysis can therefore be considered to be a precursor to a large-scale quantitative project, and allows for the targeting of data to collect. Thus, in the case of the model for Cambodia, it could be interesting to evaluate quantitatively and precisely the costs of reinforcing the control of HPAI H5N1 by local stakeholders. Indeed, according to the results of the model, the scenario in Figure 4C allows for the best optimization of the system's efficiency. This underlines the importance of training local stakeholders, specifically to control avian diseases at their level, and also supports the conclusions of other studies that have shown the importance of involving local stakeholders in the improvement of the efficiency of systems of surveillance and control of HPAI H5N1 [36], and animal diseases more widely [37].

CONCLUSION

Loop analysis can be adapted to the context of animal disease mitigation systems. This only requires defining the system's components and the sign of their interactions, but not their value. This is the main strength of the method (it is easy to design), but also its main limit, as we can only say whether the variables increase or decrease after a perturbation, but not how much or how fast. That is why loop analysis should not be used as a single method but in combination with quantitative ones.

Starting from a simple and generic model, it is possible to develop a model suitable to a specific context by breaking down the variables into sub-variables. In this way we were able to obtain a model with six variables organized into three levels, describing the mitigation system of HPAI H5N1 in Cambodia. This improved our understanding of the system's organization, especially highlighting the link between surveillance and control. Loop analysis also allows for the impacts on the system's different compartments to be studied. We chose to interpret these impacts in terms of financial, human and material investments. The results show, among others, that investing in the compartment corresponding to local-level control represents the most favorable scenario for optimizing the efficiency of the system.

We think the loop analysis could also be used as a participatory epidemiological tool. For example, it would allow different stakeholders of the network to build a common representation of their system, and to study its dynamic through various perturbations. This type of approach is already applied in the field of ecology, for example in teaching the carbon cycle and the greenhouse effect [38]. The value of this method lies in its simplicity and flexibility, as these models are fast to build, analyze and modify, providing a good aid for study and debate. A systemic approach to a disease mitigation system can allow its complexity to be reduced and its dynamics to be better understood, thereby stimulating new ways of conceptualizing these systems. This could also raise local stakeholders' awareness of these dynamics, thereby giving them a better understanding of the consequences of their actions.

ACKNOWLEDGEMENTS

We extend our thanks to the NaVRI members as well as the veterinarians and village volunteers of Takeo and Kompomg Cham provinces in Cambodia, for their contribution to this study. We acknowledge the assistance of Nget Kiry, Sophie Baron and Kunthy Nguon in interviewing the Cambodian stakeholders, and Dr Mak Chanthol in commenting our models. We also thank the CIRAD GREASE Network for its support. Many thanks to Chris Dickson for the translation of this article into English. This study was funded by the CIRAD's REVASIA programme and the CIRAD AGIRs Unit.

REFERENCES

1. Dufour B, Hendrikx P: Epidemiological surveillance in animal health 2[nd]ed. FAO, OIE and AEEMA: CIRAD; 2009.
2. Thrusfield M: Surveillance. In: Veterinary Epidemiology. BlackWell Science; 2007, 168-186.
3. Häsler B, Howe K, Stärk K: Conceptualizing the technical relationship of animal disease surveillance to intervention and mitigation as a basis for economic analysis. *BMC Health Serv Res.* 2011, 11: 225. 10.1186/1472-6963-11-225
4. McDermott J, Coleman P: Comparing apples and oranges – model-based assessment of different tsetse-transmitted trypanosomosis control strategies. *Int J Parasitol.* 2001, 31: 603-609. 10.1016/S0020-7519(01)00148-5
5. Groenendaal H, Nielen M, Hesselink JW: Development of the Dutch Johne's disease control program supported by a simulation model. *Prev Vet Med.* 2003, 60: 69-90. 10.1016/S0167-5877(03)00083-7
6. Stansfield SK, Walsh J, Prata N, Evans T, *et al*: Information to improve decision making for health. Disease control priorities in developing countries. Edited by: Jamison DT. New York: Oxford University Press; 2006, 1017-30.
7. Hendrikx P, Gay E, Chazel M, Moutou F, Danan C, Richomme C, Boue F, Souillard R, Gauchard F, Dufour B: OASIS: an assessment tool of epidemiological surveillance systems in animal health and food safety. *Epidemiol Infect.* 2011, 139: 1486-1496. 10.1017/S0950268811000161
8. Drewe JA, Hoinville LJ, Cook AJ, Floyd T, Stärk KD: Evaluation of animal and public health surveillance systems: a systematic review. *Epidemiol Infect.* 2012, 140: 575-90. 10.1017/S0950268811002160
9. OIE (World Organization for Animal Health): OIE Tool for the evaluation of performance of veterinary services (OIE PVS tool). Paris: OIE; 2010.http://www.oie.int/fileadmin/Home/eng/Support_to_OIE_Members/pdf/A_PVS_Tool_Final_Edition_2013.pdf
10. Drewe J, Hoinville L, Cook A, Floyd T, Gunn G, Stärk K: SERVAL: a new framework for the evaluation of animal health surveillance. *Transbound Emerg Dis.* 2013, 10.1111/tbed.12063.
11. Goutard FL, Paul M, Tavornpanich S, Houisse I, Chanachai K, Thanapongtharm W, Cameron A, Stärk KDC, Roger F: Optimizing

early detection of avian influenza H5N1 in backyard and free-range poultry production systems in Thailand. *Prev Vet Med.* 2012, 10.1016/j.prevetmed.2011.12.020.

12. Vergne T, Grosbois V, Durand B, Goutard F, Bellet C, Holl D, Roger F, Dufour B: A capture–recapture analysis in a challenging environment: assessing the epidemiological situation of foot-and-mouth disease in Cambodia. *Prev Vet Med.* 2012, 105: 235-243. 10.1016/j.prevetmed.2011.12.008
13. Redmond L, Beck H: Economics in animal health policy making in Northern Ireland. The economics of animal health and production. Edited by: Rushton J. Cambridge, MA: CAB International; 2009, 269-273.
14. Puccia CJ, Levins R: Qualitative modeling of complex systems: an introduction to loop analysis and time averaging. Cambridge: Harvard University Press; 1985.
15. Arkoosh MR, Johnson L, Rossignol PA, Collier TK: Predicting the impact of perturbations on salmon (*Oncorhynchus* spp.) communities: implications for monitoring. *Can J Fish Aquat Sci.* 2004, 61: 1166-1175. 10.1139/f04-068. 10.1139/f04-068
16. Ortiz M, Wolff M: Application of loop analysis to benthic systems in northern Chile for the elaboration of sustainable management strategies. *Mar Ecol Prog Ser.* 2002, 242: 15-27.
17. Mohtashemi M, Higgs BW, Levins R: Infection and atherosclerosis: is there an association?. *Complex Systems-Champaign.* 2006, 16: 259-276.
18. Zavaleta JO, Rossignol PA: Community-level analysis of risk of vector-borne disease. *Trans R Soc Trop Med Hyg.* 2004, 98: 610-618. 10.1016/j.trstmh.2003.12.014
19. Dambacher JM, Luh HK, Li HW, Rossignol PA: Qualitative stability and ambiguity in model ecosystems. *Am Nat.* 2003, 161: 876-888. 10.1086/367590
20. Dambacher JM, Li HW, Rossignol PA: Qualitative predictions in model ecosystems. *Ecological Modeling.* 2003, 161: 79-93. 10.1016/S0304-3800(02)00295-8.
21. Howe KS, Häsler B, Stärk KD: Economic principles for resource allocation decisions at national level to mitigate the effects of disease in farm animal populations. *Epidemiol Infect.* 2012, 141: 91-101.

22. Diekmann O, Heesterbeek JAP: Mathematical epidemiology of infectious diseases: model building, analysis and interpretation. Chichester: Wiley; 2000, Volume 5.
23. Justus J: Loop analysis and qualitative modeling: limitations and merits. *Biol Philos.* 2006, 21: 647-666.
24. The Loop analysis website.http://www.ent.orst.edu/loop/
25. Sorn S, Sok T, Ly S, Rith S, Tung N, Viari A, Gavotte L, Holl D, Seng H, Asgari N, Richner B, Laurent D, Chea N, Duong V, Toyoda T, Yasuda CY, Kitsutani P, Zhou P, Bing S, Deubel V, Donis R, Frutos R, Buchy P: Dynamic of H5N1 virus in Cambodia and emergence of a novel endemic sub-clade. *Infect Genet Evol.* 2012.http://dx.doi.org/10.1016/j.meegid.2012.05.013
26. Desvaux S, Sorn S, Holl D, Chavernac D, Goutard F, Thonnat J, Porphyre V, Ménard C, Cardinale E, Roger F: HPAI surveillance programme in Cambodia: results and perspectives. *Dev Biol (Basel).* 2006, 124: 211-224.
27. Ear S, Burgos Cáceres S: Livelihoods and highly pathogenic avian influenza in Cambodia. *Worlds Poult Sci J.* 2009, 65: 633-10.1017/S0043933909000440.
28. Burgos S, Hinrichs J, Otte J, Pfeiffer D, Roland-Holst D, Schwabenbauer K, Thieme O: Poultry, HPAI and livelihoods in Cambodia–a review. *2008, Mekong Team Working Paper No.* 3, Food and Agriculture Organization, Rome: A collaborative research project funded by DFID, BMELV, FAO, ICB, ILRI, Mekong Team Working Paper No. 3.
29. Van Kerkhove MD, *et al*: Poultry movement and sustained HPAI risk in Cambodia. Health and animal agriculture in developing countries. Edited by: Zilberman D. New York: Springer; 2012, 233-263.
30. Elbers ARW, Fabri THF, De Vries TS, De Wit JJ, Pijpers A, Koch G: The highly pathogenic avian influenza a (H7N7) virus epidemic in the Netherlands in 2003—lessons learned from the first five outbreaks. *Avian Dis.* 2004, 48: 691-705. 10.1637/7149
31. Ly S, Van Kerkhove MD, Holl D, Froehlich Y, Vong S: Interaction between humans and poultry, rural Cambodia. *Emerg Infect Dis.* 2007, 13: 130-132. 10.3201/eid1301.061014
32. Palmer S, Fozdar F, Sully M: The effect of trust on West Australian farmers' responses to infectious livestock diseases. *Sociol Rural.* 2009, 49: 360-374. 10.1111/j.1467-9523.2009.00495.x.

33. Ferrarini A: Fuzzy cognitive maps outmatch loop analysis in dynamic modeling of ecological systems. *Computat Ecol Software.* 2011, 1: 55-59.

34. Gramig BM, Horan RD: Jointly determined livestock disease dynamics and decentralized economic behavior. *Aust J Agric Resour Econ.* 2011, 55: 393-410. 10.1111/j.1467-8489.2011.00543.x.

35. Cohen NE, Van Asseldonk MAPM, Stassen EN: Social-ethical issues concerning the control strategy of animal diseases in the European Union: a survey. *Agr Hum Values.* 2007, 24: 499-510. 10.1007/s10460-007-9086-9.

36. Alders RG, Bagnol B, Brum E, Lubis AS, Young MP: Continuing education in the prevention and control of HPAI: a case study on Indonesia. *Worlds Poult Sci J.* 2009, 65: 529-531. 10.1017/S0043933909000397.

37. Souza Monteiro DM, Carrasco LR, Moffitt LJ, Cook AJC: Robust surveillance of animal diseases: an application to the detection of bluetongue disease. *Prev Vet Med.* 2012, 105: 17-24. 10.1016/j.prevetmed.2012.01.011

38. Salles P, Bredeweg B: A case study of collaborative modeling: building qualitative models in ecology. Artificial intelligence in education: shaping the future of learning through intelligent technologies. Edited by: Hoppe U, Verdejo F, Kay J. Japan, Osaka: IOS-Press/Ohmsha; 2003, 245-252.

CHAPTER 4

Design and Descriptive EpiDemiology of the Infectious Diseases of East African LiveStock (IDEAL) Project, a LonGitudinal Calf Cohort Study in Western Kenya

Barend Mark de Clare Bronsvoort[1], Samuel Mwangi Thumbi[3], Elizabeth Jane Poole[2], Henry Kiara[2], Olga Tosas Auguet[3,6], Ian Graham Handel[1,] Amy Jennings[1], Ilana Conradie[5], Mary Ndila Mbole-Kariuki[2,4], Philip G Toye[2], Olivier Hanotte[2,4], JAW Coetzer[5] and Mark EJ Woolhouse[3]

[1]The Roslin Institute and Royal (Dick) School of Veterinary Studies, University of Edinburgh, Roslin, UK.

[2]ILRI, Nairobi 00100, Kenya.

Citation: de Clare Bronsvoort, B.M., Thumbi, S.M., Poole, E.J. et al. Design and descriptive epidemiology of the Infectious Diseases of East African Livestock (IDEAL) project, a longitudinal calf cohort study in western Kenya. BMC Vet Res 9, 171 (2013). https://doi.org/10.1186/1746-6148-9-171

[3]Centre for Immunology, Infection & Evolution, University of Edinburgh, Kings Buildings, Edinburgh, UK.

4University of Nottingham, University Park, Nottingham, NG7 2RD, UK.

[5]Department of Veterinary Tropical Diseases, Faculty of Veterinary Science, University of Pretoria, Private bag X04, Onderstepoort, South Africa.

[6]Centre for Clinical Infection and Diagnostics Research, Department of Infectious Diseases, King's College London and Guy's and St Thomas' NHS Foundation Trust, London, UK.

ABSTRACT

Background

There is a widely recognised lack of baseline epidemiological data on the dynamics and impacts of infectious cattle diseases in east Africa. The Infectious Diseases of East African Livestock (IDEAL) project is an epidemiological study of cattle health in western Kenya with the aim of providing baseline epidemiological data, investigating the impact of different infections on key responses such as growth, mortality and morbidity, the additive and/or multiplicative effects of co-infections, and the influence of management and genetic factors.

A longitudinal cohort study of newborn calves was conducted in western Kenya between 2007-2009. Calves were randomly selected from all those reported in a 2 stage clustered sampling strategy. Calves were recruited between 3 and 7 days old. A team of veterinarians and animal health assistants carried out 5-weekly, clinical and postmortem visits. Blood and tissue samples were collected in association with all visits and screened using a range of laboratory based diagnostic methods for over 100 different pathogens or infectious exposures.

Results

The study followed the 548 calves over the first 51 weeks of life or until death and when they were reported clinically ill. The cohort experienced a high all cause mortality rate of 16% with at least 13% of these due to infectious diseases. Only 307 (6%) of routine visits were classified as clinical episodes, with a further 216 reported by farmers. 54% of calves reached one year without a reported clinical episode. Mortality was mainly to east coast fever, haemonchosis, and heartwater. Over 50 pathogens were detected in this population with exposure to a further 6 viruses and bacteria.

Conclusion

The IDEAL study has demonstrated that it is possible to mount population based longitudinal animal studies. The results quantify for the first time in an animal population the high diversity of pathogens a population may have to deal with and the levels of co-infections with key pathogens such as *Theileria parva*. This study highlights the need to develop new systems based approaches to study pathogens in their natural settings to understand the impacts of co-infections on clinical outcomes and to develop new evidence based interventions that are relevant.

Keywords: Cattle, Infectious disease, Kenya, Longitudinal study, Cohort, Epidemiology, Study design

BACKGROUND

It is estimated that by 2050 the global population will have risen to 9 billion with much of this growth predicted to occur in sub-saharan Africa [1]. There is therefore an urgent need to improve food production in these regions, and livestock production constitutes an essential part of this. In addition to providing food through milk and meat, livestock also provide hides, draught power, manure for fertiliser, building and fuel, capital reserves and cultural services and in many marginal regions are the only useful way of utilising poor quality grazing land. Livestock are key to poor peoples' livelihoods and offer an important route out of poverty.

Constraints on livestock production are varied and include nutrition, management, access to markets, natural catastrophes and importantly infectious disease (eg. [2]). Sub-Saharan Africa (SSA) harbours 12 of the 15 former World Organisation for Animal Health (O.I.E.) list A diseases considered most contagious including African swine fever, Rift Valley fever and African horse sickness. In addition, many less contagious but arguably more important diseases such as East Coast fever, trypanosomosis, brucellosis and leptospirosis are widespread. A systematic literature review on causes of morbidity and mortality among smallholder dairy farms in Eastern and South Africa identified tick-borne diseases, diarrhoea and trypanosomiasis as the most commonly documented causes of mortality [3]. This limits production directly but also ensures these regions are unable to trade animals and their products in international markets [4]. However, rinderpest is a clear example where a regional approach has produced a highly successful eradication programme and the world is now rinderpest free [5]. This points to the need for targeted research to understand the full spectrum of disease problems in a farming system and how an integrated control package might release the genetic potential of the existing livestock while maintaining genetic resilience to environmental or emerging disease threats.

Previous work in infectious disease epidemiology has focused on single disease studies eg. Zhang [6], Bronsvoort [7] and Gachohi [8] or a few closely related diseases eg. [9] but, in reality, organisms are normally infected with a number of more or less pathogenic organisms at any one time. There is increasing scientific interest in how pathogens interact, within both individuals [10, 11] and populations [12]. Examples include studies of viruses, bacteria, protozoa and helminth infections in both humans and livestock [11–20]. These interactions can be positive or negative and involve mechanisms such as: common risk factors and transmission routes (including shared vectors); non-specific immune responses; cross-reactive acquired immune responses; increased susceptibility of immuno-suppressed or immuno-compromised hosts; non-specific effects of genetic polymorphisms and nutritional deficiencies; the demographic and behavioural impacts of infectious diseases and of intervention measures. There may also be consequences of variations in the timing and ordering of exposure, infection, and disease caused by different pathogens, including responses to vaccinations [21, 22].

Animal health research in SSA has traditionally focussed on specific infections, particularly tick-borne and tsetse-borne diseases, not necessarily because they are the major diseases of cattle kept by the poor in these environments, but because they are known historically to be serious constraints to commercial systems using improved breeds. However, livestock in the tropics are routinely exposed to a wide variety of pathogens [23] whose direct and indirect impacts on animal health are unlikely to be independent of one another. Local breeds have been reared in these heavy disease challenge settings for many centuries which has resulted in selection for broad disease resistance likely at the expense of higher production [24]. Yet there have been no integrated studies of the co-distribution, co-incidence and overall impact of the major infectious diseases of livestock in the tropics. There is a need for detailed knowledge of the burden of infectious diseases impose on livestock as a prerequisite to informed decision making, resource allocation, prioritisation of research and selection of interventions. However, there is growing evidence that disease impacts cannot be fully understood by reference to single infections in isolation [25]. Instead, a holistic approach is required which considers both direct and indirect interactions between pathogens and the effects of these on the epidemiologies of infectious diseases of cattle and of the disease burdens they impose and, ultimately, of their impacts on human welfare [16, 26].

The Infectious Diseases of East African Livestock (IDEAL) project is a multi-disciplinary study which addresses two major issues: 1) the widely recognised lack of baseline epidemiological data on the dynamics and impacts of infectious diseases of cattle in the tropics; and 2) improving understanding of interactions between multiple infections and their sequelae by testing two specific hypotheses: i) that the negative impacts of different infections are not independent; ii) that 'positive' traits (e.g. resistance to infection, higher growth rates, low morbidity) cluster in certain individuals. In order to test these hypotheses we designed a longitudinal epidemiological field study to follow a random sample of newborn indigenous short horn zebu calves, with known genotype, through the first 12 months of life and to monitor them closely to identify when and what pathogens they were exposed to and the impact these had individually and in combination.

This paper describes the study design and reports the descriptive epidemiology of the IDEAL project. In particular we provide baseline data on the farm demographics and characterise the small holder African Shorthorn Zebu farming system of western Kenya which may be representative of the wider Lake Victoria basin. We also report the overall infectious disease related mortality rates and incidence of clinical episodes, the range of pathogens and exposures observed and the proportion of the cohort affected by each to provide a context for future papers on specific aspects of mortality and morbidity.

METHODS

Study setting

There has been intensive work to define the distribution of different agricultural production systems in East Africa (eg. [27, 28]). This study focused on a specific production system, sedentary mixed crop-livestock small holdings. This system encompasses >50% of poor people (defined as income below US<DOLLAR/>15 per month [29]) resident in East Africa [30], covers extensive areas of Kenya and beyond, and is of increasing importance as populations grow.

The study site was an area of western Kenya approximately 45 x 90km covering some or all of Busia (95.9%), Teso (96.3%), Siaya (55.5%), Butere/Mumias (26.9%) and Bungoma (20.4%) districts. Each district is further divided into sub-locations which are the smallest administrative unit in Kenya for which data was available on cattle numbers. A SL typically contains 60 to 90 households per km^2 and is 10-20km^2 in area. Land plots are typically 1-5 ha in size, with around 60% of households owning 2-3 breeding cattle grazed communally. The study site included 280 sub-locations (excluding 2 that were in Busia and Mumias towns) across 5 agro-ecological zones (AEZ). AEZ is a way to describe the type of land and its suitability for different crops and combines data on soil, topography, and climate. The areas of Kakamega, Vihiga, Lugari and Mt Elgon districts were not included as they were considered less representative of smallholder livestock farmers in East Africa (e.g. Mt Elgon slope, large-scale dairy farming more prominent) and due to logistic restrictions (i.e. the diagnostic laboratory was in Busia town, to which samples were transported daily).

Study Design and Recruitment

A stratified 2-stage random cluster sample of calves was drawn. The 1st stage cluster sample (by sub-location) was selected by random sampling sub-locations with replacement within each AEZ stratum. A total of 20 sub-locations were selected (Figure 1 and Table 1). A second stage sample size of 28 calves per sub-location was chosen at random to achieve the desired minimum sample size of 500 calves (based on logistical constraints and ability to detect a minimum relative risk of 3 with 80% power) and to allow for some losses (Table 2). A reporting system was established in each of the 20 selected sub-locations using a reporting pathway from Farmer → Sub-location-chief → Sub-chief → IDEAL Office. Each recruitment day the animal heath assistants (AHA) collated the eligible calf births for the sub-location and randomly selected 1-3 calves randomly from a hat each day. In order to be eligible the calf had to meet a set of specific selection criteria which were (1) the calf had to be between 3 and 7 days old at recruitment; (2) it was not as a result of artificial insemination; and (3) the dam was not managed under zero-grazing conditions. These criteria were set to give a reasonable window to capture calves being born without being too old and to avoid recruitment of exotic breeds rather than indigenous cattle. The sub-locations were visited on a rolling 5 week cycle to ensure there was an even distribution of calves across space and season. Calves were recruited over the 5 week cycle with 4/20 sub-locations being visited each week, taking 2 years to recruit the complete cohort. Only one calf per dam was recruited and a farmer could only have onc calf at a time in the study. Recruitment was conditional on the farmer allowing access to the calf and willingness to report clinical episodes to the project and not "self treat". A flat rate of compensation was agreed with the local veterinary office for this. Owners were asked to call the IDEAL team if a calf was observed to be ill between visits and one of the project veterinary surgeons would examine the calf and treat if considered to be seriously ill or a welfare issue. Calves were censored after any visit where a treatment was begun.

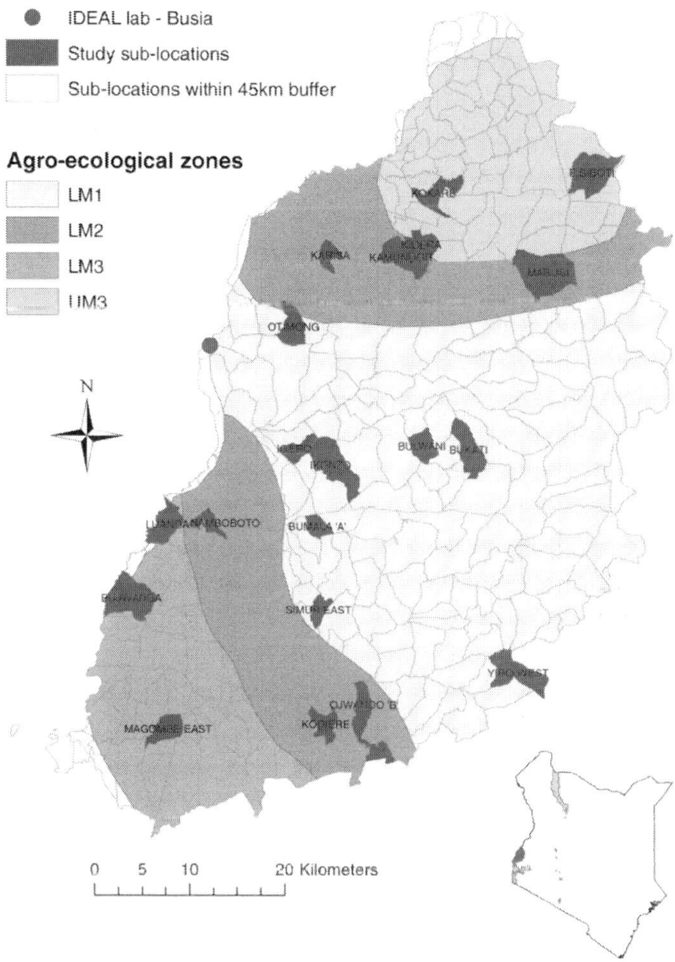

Figure 1. Map of western Kenya showing the study area, agro-ecological zones and sub-locations (selected sub-locations highlighted).

Table 1. Selected sub-locations with census/demographic characteristics (taken from the Human Population Census in Kenya 1999)

AEZ	Sub-location	No. households	Area(km²)	Cattle density per km²	Average herd size
UM3	East Siboti	1245	15.80	2439	3.4
	Kokare	325	8.29	937	6.1
	Kidera	314	7.36	728	4.8
LM1	Yiro West	1361	13.70	1187	3.9
	Simur East	415	4.32	425	3.8
	Igero	532	5.60	681	3.6
	Bumala A	724	4.38	222	2.3
	Ikonzo	1421	16.40	598	2.8
	Bulwani	478	6.87	578	3.2
	Bukati	993	11.20	1259	2.5
	Otimong	506	8.66	869	4.1
LM2 middle	Mabusi	1575	22.50	1575	3.1
	Kamunuoit	556	11.00	957	4.0
	Karisa	292	4.63	247	2.2
LM2 South	Ojwando B	832	12.60	1095	4.6
	Kodiere	630	6.38	849	4.7
	Namboboto	351	4.46	220	2.7
LM3	Luanda	726	9.76	730	4.7
	Bujwanga	1025	16.70	792	4.2
	Magombe East	578	7.67	852	5.4

Table 2. Distribution of sub-locations (SL) across agroecological zones (AEZs) in western Kenya and number selected for the IDEAL study

AEZ	No. SL/AEZ	Proportion/AEZ	No. SLs selected
LM1	114	0.40	8
LM2	86	0.30	6
LM3	28	0.10	3
LM4	4	0.01	0
UM3	53	0.19	3
Total	285		20

Upon recruitment a household questionnaire was completed by interview with the owner/head of the household. The questionnaire included questions about the farm size, crops, water sources, and other livestock. The dam was examined and a form completed and if it or the calf failed any of the eligibility criteria, the calf was excluded. The calf was then examined and a recruitment form and routine visit form completed. The calf was examined for congenital deformities and excluded if any were found. This is summarised in Figure 2.

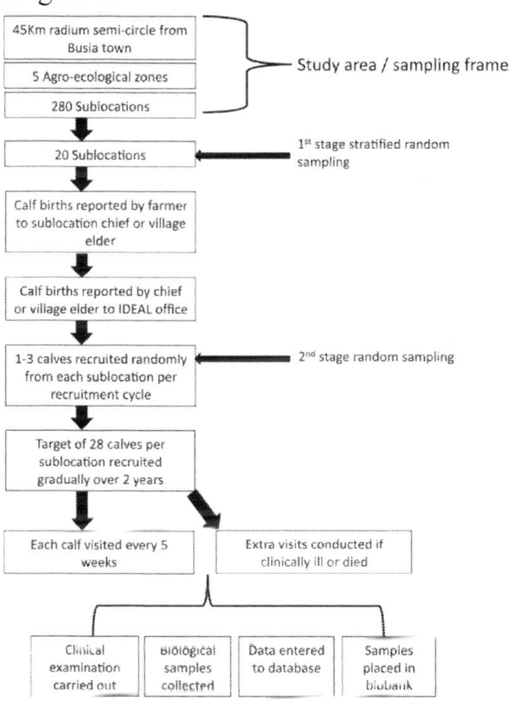

Figure 2. Schema showing the design and sampling used in the IDEAL project.

Data Collection and Training of Data Collectors

Data collection took place at the farm. A team comprising a veterinary surgeon/senior AHA and two AHAs went to each animal and followed a standard protocol for the physical examination and collection of compulsory samples. If the dam was also being visited there was an additional protocol for dam examinations. The AHAs were also trained in data collection and all questionnaires and data collection tools were piloted over about 9 months during the set-up phase of the project in western Kenya. Data were collected via a hand held Palm OS®; Personal-Digital Assistant (PDAs) and simultaneously on a paper questionnaire form. Barcodes were used to identify and link samples to individual animals. At the diagnostic field laboratory in Busia, data were downloaded from the hand held device to a database and cross-checked against the paper records and any discrepancies resolved with the AHA who collected the data.

Routine Clinical Examination of Calves

The clinical examination consisted of a systematic physical examination of the calf. This included observation of the animal at rest, posture, alertness, rectal temperature, weight, girth, FAMACHA score [31], mucus membrane colour, skin elasticity, presence and species of ticks and other ectoparasites and full palpation of the body checking for lesions and discharges. In addition to the physical examination of the calf a short questionnaire was used to update other activities on the farm such as any animal purchases or sales, treatment of the other livestock or cases of illness in other livestock.

A standard set of samples were collected at recruitment (7D), 5 weekly (5W), and 51 weeks (Y) visits as detailed below. A marginal ear vein sample was used to make a thick and a thin blood smear to screen for haemoparasites and for manual differential cell counts following shipment to Pretoria University. A jugular vein sample was collected into plain tubes for total serum protein estimation using a refractometer (model RHC-200ATC, Westover Scientific) and storage for antibody screening for a range of haemoparasites, bacteria and viruses and 0.5ml was added to RNAlater®; (Ambion®;) and stored at 4°C. An EDTA sample with 'magic buffer' was collected for genomic analysis (7D only). An EDTA sample for: (a) DNA extraction for pathogens; (b) direct microscopy on thick and thin smears for haemoparasities and (d) routine haematology including WBC, RBC, PCV, MCV, HGB,

MCH, MCHC using a Sysmex pocH-100iV Diff automated blood analyser (Sysmex®; Europe GMBH) was also collected. A further EDTA sample was stored at -80°C until DNA extraction and shipping to Pretoria University for screening (Y or last visit before death) for a large range of blood borne parasites using the reverse line blot (RLB) [32]. A heparinised blood sample was collected for *Mycobacterium bovis* screening using the "Bovigam" ELISA (Prionics®;, Celtic Diagnostics Ltd., Ireland) (Y only). In addition samples were collected for white blood cell stimulation, however, this was discontinued early in the study because of logistical constraints.

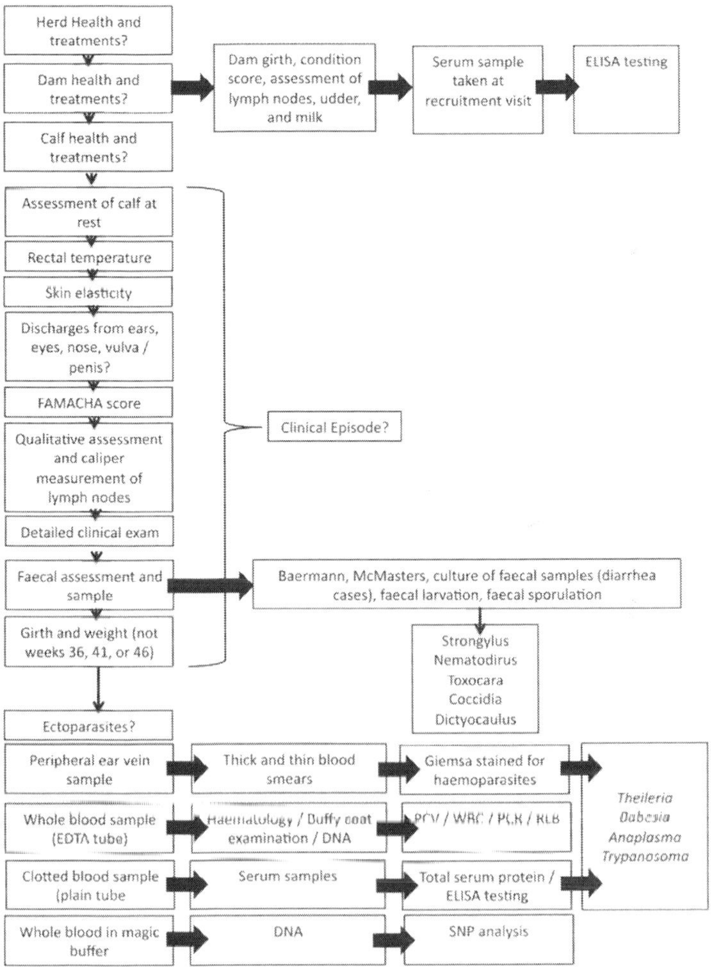

Figure 3. Schema showing the types and timings of clinical examination of calves and the types of sample collection for the IDEAL project.

Faecal samples were collected via rectal palpation for screening for helminths using standard techniques [33]. Samples were divided and one part put in a plastic bag and stored overnight at 4°C for screening by McMasters technique for strongyle eggs, by the direct Baermans technique for *Dictyocaulus vivperia* larvae, by Ziehl-Neelsen stained smear for *Cryptosporidium spp.* and *M. avium paratuberculosis* and by sedimentation for fluke species eggs. The second part was stored in a pot at room temperature overnight and then prepared for larval culture to speciate strongle eggs. Samples with >2000 coccidia oocyts were also cultured to type the species of coccidia present. Three superficial skin snips were taken from the ventral abdomen and incubated directly in RPMI-1640 (Sigma-Aldrich®;) to screen for *Onchocerca spp.* microfilaria [34]. Results from diagnostic tests done in the field laboratory in Busia were entered directly in a separate laboratory database. In addition at the final visit to a calf a standard set of measurements of height at wither, nose to tail length and phenotypical measures such as coat colour and hump and dewlap were recorded. This is summarised in Figure 3.

Clinical episodes and post mortem examinations in calves

In addition to routine clinical examinations and in order to capture as many clinical episodes as possible local AHAs working for the Kenyan Department of Veterinary Services in the sub-locations made weekly visits to each calf. These weekly visits involved a limited clinical examination focusing on identifying any acute disease and in particular any pyrexic or traumatic episodes. In the event that they identified pyrexia, enlarged lymph nodes or respiratory distress, they contacted an IDEAL project veterinary surgeon and an extra non-routine visit was made. The main triggers for a visit were a temperature of >40.5°C, generalised lymphadenopathy, anorexia, diarrhoea, generalised skin conditions, non-weight bearing lameness, coughing or respiratory distress. However each case report was considered and was visited depending on history and if there was believed to be a compromise in welfare. A full clinical examination was carried out and additional samples were collected based on the clinical syndrome observed. These included swabs of any discharges for bacteriological culture and typing, viral swabs and heparin blood samples for virological culture, and needle aspirates from enlarged lymph nodes for microscopy. If calves were in a severely diseased state the project veterinarian used their professional judgement and a set of criteria agreed with the ethics committee at University of Edinburgh/ International Livestock Research Institute and the animal was euthanised if necessary.

In the event that an animal died or was euthanised a full gross post mortem examination was carried out following standard veterinary approaches working through the body systems. A standard set of tissues was collected from each animal, including lung, liver, duodenum, ileum and lymph nodes, with additional samples specific to the suspected aetiology where appropriate. In the event of a history of sudden death a marginal ear vein blood smear was made and stained with methylene blue and checked for the presence of anthrax bacilli prior to further examination. In the event of a positive smear no post mortem was performed and the carcass buried. If there were neurological signs and/or a history consistent with rabies the head was removed and sent for testing at the Central Veterinary Laboratories at Kabete, Kenya and the remainder of the carcass incinerated. For those animals with neurological signs and no history of possible bites, a brain smear was prepared using the standard approach for identification of *E. ruminantium* the cause of heartwater disease.

Examination of the Dams

In addition to the above the calf's dam was examined at each visit. At recruitment a full clinical examination was done (including manual palpation of the udder for evidence of mastitis), the girth measured and the animal was condition scored using a standard 10 point score [35]. In addition phenotypic measurements of coat colour and pattern, horn length and shape, ear shape, size of hump and dewlap were recorded. Two plain and 3 EDTA vacuutainers of blood were collected for possible use later. At each 5 weekly visit up to the visit after the calf was weaned the dam was re-examined, the girth was measured, the animal was condition scored and the udder examined.

In the initial phase of the study we attempted to collect milk samples from dams at each visit. These are low production animals and have very small udders and teats compared to a holstein for example. In the majority of cases we were unable to collect samples as the calf would have suckled before we arrived and/or the owner had milked the dam. Similarly the AHAs were initially trained to use the California milk test [36] but again it proved very difficult to get enough milk to test. Both these activities were suspended after the first 3 months in December 2007.

Laboratory Analysis

A full list of pathogens that the project attempted to identify that we believed likely to be present in this setting is given in Table 3 and includes 100 different pathogens. The various techniques used and the time points at which they were done are also provided for reference. In some cases there is overlap as some techniques will only differentiate to genus level while others will allow species specific identification.

Table 3. Pathogens screened for during the study

Pathogen	Test	Visits tested	Pathogen	Test	Visits tested
Actinomyces sp.	RB	CE	*Hepatozoon spp.* catch-all	RLB	Y
Actinomycetes	RB	CE	*Hyalomma spp.*	CL	7D, 5W, Y
Amblyomma variegatum	CL	7D, 5W, Y	*Hypoderma bovis*	CL	7D, 5W, Y
Anaplasma bovis	RLB	Y	*Klebsiella ozaenae*	RB	CE
Anaplasma centrale	RLB	Y	*Klebsiella pneu-moniae*	RB	CE
Anaplasma marginale	RLB	Y	*Listeria spp.*	RB	CE
Anaplasma ovis	RLB	Y			
Anaplasma phagocytophi-lum	RLB	Y	Lumpy skin disease	PCR	CE
Arcanobacte-rium pyogenes	RB	CE	*Micrococcus spp.*	RB	CE
Babesia bicor-nis	RLB	Y	*Moniezia spp.*	FM, FC	7D, 5W, Y
Babesia bige-mina	RLB	Y	*M. avium paratuber-culosis*	ZN	Y
Babesia bovis	RLB	Y	*Nematodirus spp.*	FM, FC	7D, 5W, Y
Babesia caballi	RLB	Y	*Non-pathogenic Staphylococci*	RB	CE
Babesia canis	RLB	Y	*Oesophagostomum radratium*	FM, FC	7D, 5W, Y
Babesia diver-gens	RLB	Y	*Onchocerca spp.*	SNP, MIC	Y

Babesia felis	RLB	Y	*Ostertagia ostertagi*	FM+FC	7D, 5W, Y
Babesia gibsoni Japan	RLB	Y	*Pasteurella multocida*	RB	CE
Babesia microti	RLB	Y	*Rickettsia spp.* catch-all	RLB	Y
Babesia motasi	RLB	Y	*Rickettsia spp. (DnS14) raoultii*	RLB	Y
Babesia odocoilei	RLB	Y	*Riphicephalus appendiculatus*	CL	7D, 5W, Y
Babesia ovis	RLB	Y	Rotavirus	ELISA	CE
Babesia rossi	RLB	Y	*Salmonella spp.*	RB	CE
Babesia vogeli	RLB	Y	*Sarcocystis spp.*	HIS	PM
Bacillus anthracis	RB	PM	*Staphylococcus aureus*	RB	CE
Bluetongue virus	PCR	Y, CE	*Staphylococcus epidermicus*	RB	CE
Bacillus spp.	RB	CE	*Staphylococcus epidermidis*	RB	CE
Boophilus spp.	CL	7D, 5W, Y	*Staphylococcus spp.*	RB	CE
Borrelia afzelii	RLB	Y	*Streptococcus bovis*	RB	CE
Borrelia burgdorferi s. lato	RLB	Y	*Streptococcus spp.*	RB	CE
Borrelia burgdorferi s. stricto	RLB	Y	*Theileria annae*	RLB	Y
Borrelia garinii	RLB	Y	*Theileria annulata*	RLB	Y
Borrelia valaisiana	RLB	Y	*Theileria bicornis*	RLB	Y
Bunostomum trigonocephalum	FM	7D, 5W, Y	*Theileria buffeli*	RLB	Y
Bovine Viral Diarrrhoea Virus	ELISA - ag	Y	*Theileria cervi*	RLB	Y
Calicophoron spp.	FM,FC	7D, 5W, Y	*Theileria equi*	RLB	Y
Chabertia ovina	FM,FC	7D, 5W, Y	*Theileria equi*-like	RLB	Y
Clostridium spp.	RB	CE	*Theileria lestoquardi*	RLB	Y
Coccidia spp.	FM,FC	7D, 5W, Y	*Theileria mutans*	RLB	Y
Coccobacillary	RB	CE	*Theileria orientalis* 1	RLB	Y
Cooperia spp	FM,FC	7D, 5W, Y	*Theileria parva*	RLB, PCR	Y
Corynebacterium spp.	RB	CE	*Theileria spp.* (buffalo)	RLB	Y

Cryptosporidium spp.	ZN,MIC	7D, 5W, Y	*Theileria spp.* (duiker)	RLB	Y
Dermatophilus congolensis	RB	CE	*Theileria spp.* (kudu)	RLB	Y
Dictyocaulus viviparus (L1)	FB	7D, 5W, Y	*Theileria spp.* (sable)	RLB	Y
E.coli	RB	CE	*Theileria spp.*	MIC, (RLB)	7D, 5W, Y, CE
Ehrlichia chaffeensis	RLB	Y	*Theileria taurotragi*	RLB	Y
Ehrlichia ruminantium	RLB,MIC, PCR	Y, CE	*Theileria velifera*	RLB	Y
Ehrlichia spp. (Omatjenne)	RLB	Y	*Toxocara vitulorum*	FM,FC	7D, 5W, Y
Eimeria alabamensis	FM,MIC	7D, 5W, Y	*Trichophyton spp.*	MIC	CE
Eimeria auburnensis	FM,MIC	7D, 5W, Y	*Trichostrongylus axei*	FM,FC	7D, 5W, Y
Eimeria bovis	FM,MIC	7D, 5W, Y	*Trichuris spp.*	FM,FC	7D, 5W, Y
Eimeria cylindrica	FM,MIC	7D, 5W, Y	*Trypanosoma brucei*	HCT, DG, PCR	7D, 5W, Y
Eimeria ellipsoidalis	FM,MIC	7D, 5W, Y	*Trypanosoma congolense*	HCT, DG, PCR	7D, 5W, Y
Eimeria subspherica	FM,MIC	7D, 5W, Y	*Trypanosoma spp.*	HCT, DG, PCR	7D, 5W, Y
Eimeria zuernii	FM,MIC	7D, 5W, Y	*Trypanosoma theileri*	HCT, DG, PCR	7D, 5W, Y
Epizootic haemorrhagic disease	PCR	Y, CE	*Trypanosoma vivax*	HCT, DG, PCR	7D, 5W, Y
Fasciola spp.	FS,MIC	7D, 5W, Y	*Weksella zoohelcum*	RB	CE
Haemonchus placei	FM,FC	7D, 5W, Y			

RB routine bacteriology, *CE* clinical episode, *CL* clinical examination, *RLB* reverse line blot, *7D* recruitment visit, *5W* routine 5 weekly visit, *Y* final visit at 51 weeks, *FM* faecal examination by McMaster's technique, *FC* faecal culture, *MIC* routine microscopy, *SNP* skin snip and culture, *ZN* Ziehl–Neelsen stain, *DG* dark ground microscopy, *HCT* haematocrit, *PCR* polymerase chain reaction.

In addition, the project screened stored sera from calves at 51 weeks or from their last visit prior to death for evidence of exposure to a number of other diseases believed likely to be important in this region. Further, plasma and DNA were analysed at a number of external laboratories (Table 4).

Table 4. Serological screening tests to pathogens

Pathogen	Ab/Ag based	Test name	Manufac-turer	Visits tested
M. bovis	Ab	Bovigam ELISA	Prionics	Y
RespiratorySyncitialvirus	Ab	ELISA	Svanova	Y
Bluetonguevirus	Ab	ELISA	PI	Y
T. parva	Ab	ELISA	ILRI in house	7D, 5W, Y
T. mutans	Ab	ELISA	ILRI in house	7D, 5W, Y
A. marginale	Ab	ELISA	ILRI in house	7D, 5W, Y
B. bigemina	Ab	ELISA	ILRI in house	7D, 5W, Y
Parainfluenza3 virus	Ab	ELISA	Svanova	Y
Bovine ViralDiarrhoea virus	Ab	ELISA	Svanova	Y
Bovine ViralDiarrhoea virus	Ag	ELISA	Svanova	Y
EpizooticHaemor-rhagicdisease virus	Ab	ELISA	PI in house	Y
Akabanediseasevirus	Ab	ELISA	PU in house	Y
Palyam group	Ab	ELISA	PU in house	Y
InfectiousBovineRhino-tracheitisvirus	Ab	ELISA	Svanova	Y
Neospora caninum	Ab	ELISA	Svanova	Y
Brucella spp.	Ab	ELISA	IDEXX	Dam 7D
Leptospira hardjo	Ab	ELISA	Linnodee	Dam 7D

PI is the Pirbright Institute (formerly the Institute for Animal Health). *Ab* antibody, *Ag* antigen.

Whole blood samples in EDTA were stored in "magic buffer" (Biogen Diagnostica, Spain) and were genotyped using the Illumina 50K bovine SNP chip (Illumina Inc.®;).

Database and Sample Tracking

The project managed data in a set of linked Access databases (Microsoft Corp.). All reports of calf births and recruitment visits were managed in the reporting database. After animals were recruited the main household questionnaire and the routine clinical visits, clinical episodes and post mortems were recorded using palm pilots running Satellite Forms (SatelliteForms.net). These were connected to the field database and daily downloaded. Every animal was tagged with a bar coded ear tag and visit sheets for each individual were kept. At every visit, the bar code was scanned to minimise recording errors. The field database generated a list of samples and then tests that were to be carried out on them in the local Busia laboratory and this was synchronised each evening so the laboratory staff knew what testing to do each day. The laboratory database linked all the barcoded samples in the field database to the respective calf, to the test results, to where the samples and any daughter samples generated from the original field sample were stored and when they were moved to the ILRI lab in Nairobi or to other laboratories outside Kenya. At the end of the field work the field and laboratory databases were merged and moved to a multiuser MySQL database that could be accessed and updated remotely giving all staff access to the data for analysis. All samples eventually were moved to ILRI Nairobi and were appended to the ILRI laboratory information management system for sample management and tracking. Samples where possible were stored in duplicate and only one of the duplicates moved at a time to reduce the risk of losing complete sample sets. At ILRI duplicates are stored in separate buildings in either -20°C or -80°freezers or in vapour phase in large liquid nitrogen biobank chambers as appropriate.

Tropical Livestock Units

Tropical Livestock Unit (TLU) is a standardising measure used to quantify different types and sizes of livestock. It gives a reference unit that captures the total number of livestock units present in a farm, with 1 TLU being the equivalent to an animal of 250 kg liveweight. One TLU is equivalent to 1 cow, 10 goats or sheep, 5 pigs, 100 chickens, and 0.7 camels [37, 38]. This unit has been used for different purposes, including calculating insurable livestock units in the index-based livestock insurance programmes in northern arid areas of Kenya. The different species and sizes of livestock kept in the farms were converted in to TLU's to serve as a proxy indicator for livestock wealth of each household. The conversion factors used here are those reported by Njuki *et al.*[39].

Analysis

The R software version 2.9.1 (http://cran.r-project.org/) was used to generate the descriptive statistics and graphics of the farm characteristics and frequencies of pathogens. All statistical tests were interpreted at the 5% level of significance.

Survival time for each calf was defined as the age at which the study calf died due to infectious causes. Animals that died for reasons other than infectious causes, or that were lost or removed from the study before one year for non-compliance were censored. These contributed "at-risk" time only up to the censoring point. All survivors to one year were censored at the time of leaving the study. Kaplan-Meier estimates of the survival function were used to determine the overall mortality rates [40].

RESULTS

Cohort Characteristics

A total of 548 calves were recruited and followed for up to 51 weeks or until they died over the 3 year period of the field work. The spatial distribution of the selected sub-locations is given in Figure 1 and the number of calves recruited as a proportion of the breeding dams in each sub-location is given in Table 2. The cattle densities in each sub-location ranged from 220/km^2 to 2439/km^2 and the sub-locations ranged in size from 4.38 km^2 to 22.5 km^2. The average herd size across all sub-locations ranged from 2.2 breeding cows in Karisa a more hilly area compared to 6.2 animals in Kokare. The life line for each calf is illustrated in Figure 4 and highlights the drop out of calves from death and euthanasia and the pattern of clinical episodes. In addition there were 2 periods where sampling and particularly recruitment were suspended. The first was following the political unrest in 2008 and work in the field was suspended for 6 weeks. This resulted in a small number of calves missing visits for one or two 5 weekly visits. The second was over an extended holiday period in 2009/2010.

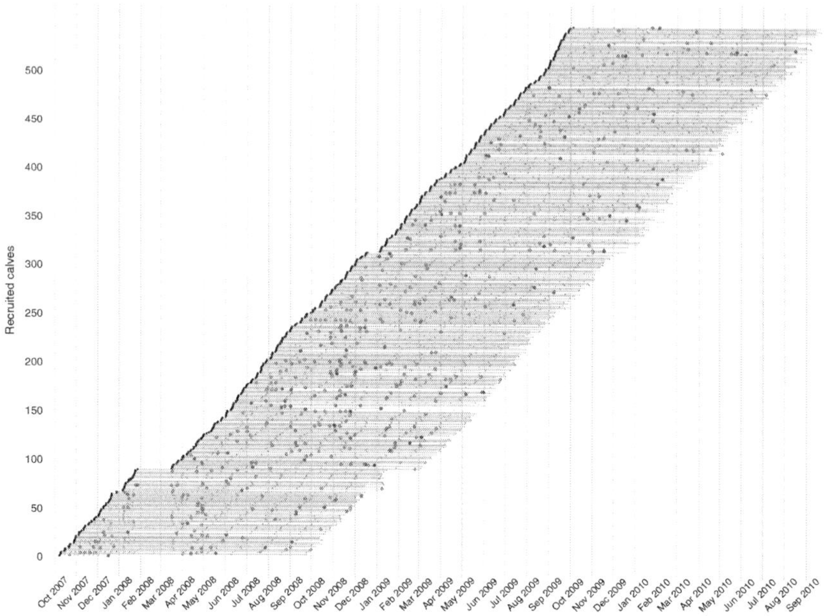

Figure 4. Life lines for each calf showing the time of recruitment, routine examinations and clinical episodes or deaths over the 3 years of the IDEAL project. Black dot=recruitment date, grey bar=weekly visit, grey circle=clinical episode; red dot= died and blue dot = euthanised.

Farm Characteristics

A total of 548 owners/household heads were interviewed. Data on the owner's age, gender, education and training level attained, and main occupation are summarised in Figure 5 and table 5. Of the 548 owners, 69% were men and 31% women. The mean age in years for male owners was 50.7 (range 22 - 85) and that for females 49.0 (range 20 - 78). Differences in ages between male and female farmers were statistically insignificant (p=0.1679, df= 352, 2-sample t-test). Approximately 15% of the farmers had no formal education, and none had attained university education. A small percentage (21%) had gained technical skills allowing them to work in the informal markets with the common ones being masonry, tailoring and carpentry. The majority (86.2%) of the interviewed owners reported farming as their only source of income, with the rest reporting teaching, civil service, pension and business as their main sources of income with farming offering supplementary income.

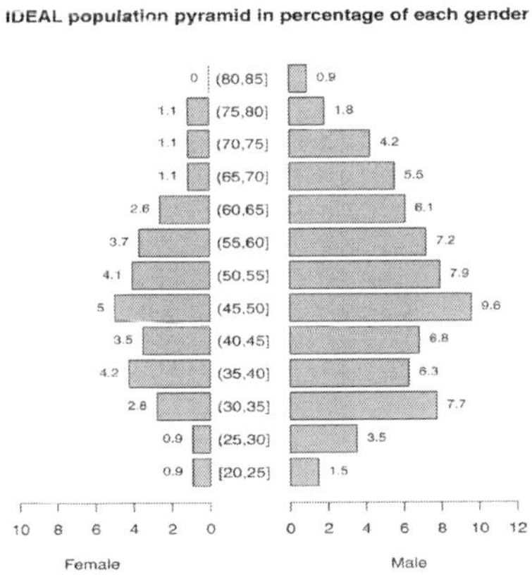

Figure 5. Population pyramid showing the age structure for male and female household heads. Each bar value represents the percent number of farmers in that age group.

Table 5. Descriptive statistics for farmer's demographic variables

	N*	Frequency	Percent**
Sex of house head	548		
Male		370	69
Female		178	31
Education level of house head	544		
None		81	14.9
Primary education		337	61.9
Secondary education		126	23.2
University education		0	0
Technical training	541		
No		415	76.7
Yes		126	23.3
Main occupation	544		
Farmer		469	86.2
Teacher		6	1.1
Civil servant		11	2
Business		22	4.1
Retired with pension		14	2.6
Other		22	4

*Not all the farmers responded to the questions in the questionnaires and N notes the number of respondents to the particular question.

**The proportions are calculated using the number of respondents to the question.

The average farm size was only 1.98 ± (0.1 SE) hectares (range 0.1 to 23.1 ha), with majority (96.1%) being owned. Such land is continuously sub-divided, to give adult sons an inheritance and ownership rights. This practice results in families owning small pieces of land that are sometimes not economically viable for agriculture. The rest (3.9%) rented the land they farmed on. All the farms selected for the study kept cattle and also planted food crops, with each farm having a median 5 (range 1 to 131) cattle. The indigenous short horn zebu cattle were the predominant breed kept, with only a small percentage (3.1%) also keeping zebu crosses. Farmers kept more than one species of livestock; an attribute identified as a strategy for spreading risk of losses [41, 42]. Different livestock species serve different purposes within the farm enterprise. The general herd structure is given in Table 6, with adult females comprising 41.4% of all cattle kept, and adult males 9.8%.

Table 6. Land sizes, livestock species kept and the herd structure

	N	**Per-cent**	**Mean**	**me-dian**	**s.d.**	**Min.**	**Max.**
Land size owned (hectares)							
	517	94.3	1.98	1.37	2.28	0.1	23.1
Livestock numbers							
All cattle	548	100	6.5	5	7.6	1	131
Indigenous cattle	548	100	6.5	5	7.6	1	131
Cross breds	17	3.1	1.4	1	1	1	5
Goats	209	38.1	3.5	3	3.8	1	33
Sheep	112	20.4	3.9	2.5	5.3	1	48
Pigs	150	27.3	2.2	1	2.2	1	13
Chickens	485	88.5	14.3	10	12.7	1	120
Dogs	297	54.2	2.04	2	1.4	1	9

Tropical livestock units	546	99.6	5.8	4.1	6.71	0.48	114.3
Herd structure (indigenous)	548	Frequency	Mean/farm	Percent			
Adult females		1463	2.7	41.4			
Adult males		345	0.6	9.8			
Female calves		465	0.8	13.2			
Male calves		446	0.8	12.6			
Weaning females		399	0.7	11.3			
Weaning males		417	0.8	11.8			
Total		3535	6.5	100.0			

Husbandry and Management Practices

Almost 60% of the farms provided housing for livestock. This was usually in the form of an open yard/kraal surrounded by a fence made of untreated wood or bushes with no roof. The remaining 40% of farms provided no housing and the animals were left free or tethered within the homestead during the night. Among those providing housing, 83.1% housed calves separate from the dams/bulls.

Calves were not allowed to graze with adults (in 94.4% of the farms) until after weaning. This was mainly to prevent calves suckling dams while out in the field. Calves were allowed to suckle as the farmer milked, with some farmers reporting that milk-let-down in their short horn zebus only happened when stimulated by calves. Other farmers obtained their share first and left the rest for the calf to suckle.

During the dry season, 49.1% of the farms reported providing drinking water for the cattle within the homestead. The rest drove their animals to a water source. These proportions did not differ significantly between the dry and the wet seasons. Distances to the watering points were below 1 km for 73.8% and 75.8% of the farms in the dry and wet seasons respectively, with the rest travelling more than 1 km to access drinking water. Table 7 shows data on the housing, distances to watering points, frequency of watering, and quality of water both in the dry and wet seasons.

Table 7. Description of housing, and watering practices in the dry and wet seasons

	N	Dryseason		Wet season	
Housing	545	Freq	Percent	Freq	Percent
Kraal/yard		321	58.9	322.0	59.3
None		224	41.1	223.0	40.7
Access to water	547				
Distance to furthest					
watering point					
At homestead		91	16.6	100.0	18.3
<1 km		313	57.2	314.0	57.5
1–5 km		141	25.8	131.0	24
6–10 km		2	0.4	1.0	0.2
Frequency of					
watering					
Freely available		11	2	13.0	2.4
Once a day		149	27.2	446.0	81.5
Twice a day		367	67.1	87.0	15.9
Thrice a day		20	3.7	1.0	0.2
Water quality					
Good, clear		533	97.4	508.0	92.9
Muddy		14	2.6	39.0	7.1

Cattle Trading and Breeding Practices

Almost all the cattle purchases and sales (98.9%) were done through cattle markets (Table 8). The rest (1.1%) of the farms reported trading animals directly with neighbouring farms. A total of 24 different cattle markets were reported serving the 20 sub-locations, spanning four administrative districts. However, a quarter (6/24) of these markets served 71.2% of all the farmers in the study, an indication that farmers preferred trading in big markets, where they are likely to get more competitive prices.

Table 8. Location of trading markets and sources of breeding bulls

	N	Freq	Percent
Location of	504		
purchasing point			
Within sublocation		75	14.9
Neighbouring sublocation		396	78.6
Other		33	6.5
Purchasing point	539		
Market		533	98.9
Neighbouring farm		6	1.1
Breeding practices	548		
Own bull (bred)		63	11.5
Own bull (bought)		45	8.2
Bull donated		2	0.4
Bull borrowed		422	77
Communal area bull		19	3.5
Other		1	0.1

There were no reports of organised breeding programmes, and farmers did not keep any written breeding records. The choice of breeding bulls was mostly based on availability of a bull, and if more than one then the farmer decided on personal preferences. Only 11.4% and 8.2% of the farms kept own-bred or purchased breeding bulls respectively (see table 8). Most farmers (76.2%) borrowed breeding bulls whenever their cows needed service. Based on this, only a few bulls are available to serve animals, raising the chances of widespread inbreeding. A few farmers (3.4%) indicated they did not make any direct breeding decisions and depended on their cows being served while grazing in the same communal areas or at watering points. This number is likely to be much higher than reported as animals mix freely and frequently at watering points and communal grazing fields.

Access to Veterinary Services

During the farmer interview at the recruitment visit, most farmers (84.7%) reported accessing some form of veterinary services, mainly provided by private animal health workers, and to a lesser extent by government animal health workers, and veterinary drug suppliers (see Table 9). A few farmers indicated they did not use the services of an animal health worker, and instead treated their sick animals themselves. Approximately 90% of farmers reported using some form of tick control of which most (89.9%) reported using whole body spraying with acaricides at the farm. Only a few farmers reported accessing communal cattle dips. Most of the cattle dips in the study sub-locations are abandoned and not in use. Only just over 50% of farmers reported using any form of anthelminthic treatment and only around 18% reported using any form of tsetse control. A moderate proportion of farmers reported using vaccination (52%) although most of those reporting using vaccines did not know what vaccine they had given their animals or what they were protected against (76.7%) and their use seems to be largely driven by need rather than a regular programme of control.

Table 9. Description of access to veterinary services and disease control practices in the farm as reported during the calf recruitment visit

		Frequency	**Percent**
Access to veterinary services	544		
Yes		461	84.7
No		83	15.3
Type of Veterinary support	461		
Private animal health worker		264	57.3
Government animal health worker		176	38.2
Veterinary drug supplier		23	5.0
Farmer		10	2.2
Tick-control	548		
Yes		498	90.9
No		50	9.1
Application method	498		
Spraying whole body		462	92.7
Spraying legs only		9	1.8

Pour on		6	1.2
Hand dressing		25	5.0
Dipping		8	1.6
Other (traditional,manual removal)		10	2
Worm control	548		
Yes		309	56.4
No		239	43.6
Application method	309		
Drench		265	85.8
Bollet		47	15.2
Others (injectables/unknown)		2	0.6
Traditional		5	1.6
Trypanosome control	548		
Yes		98	17.9
No		450	82.1
Method used	98		
Spraying whole body		51	52.0
Chemotherapy		32	32.7
Pour-on		10	10.2
Other (dipping/head dressing/			
unknown)		8	8.2
Use of Vaccines	546		
Yes		284	52.0
No		262	48.0
Frequency of use	277		
Routinely		9	2.9
When need arises		269	97.1
Vaccine type used	284		
Unknown		230	81.0
Anthrax		8	2.8
Black quarter		11	3.9
Contagious Bovine Pleural Pneumonia		1	0.4
Foot and mouth disease		25	8.8
Lumpy skin disease		18	6.3
Other		6	2.1

There was a notable difference between the proportion of farmers who reported carrying out disease control measures such as tick control and worming during the initial visit, and the actual proportion of farmers who reported using any preventive measures during the one year follow up period. This suggests that farmers are answering what they think they should be doing or maybe have done but a significant proportion actually then appeared to not carry out these measures over the course of our observations. Interestingly with tsetse control stated and observed activities seem to align well possibly reflecting the recent inputs form NGOs in this area. In contrast vaccination use was much higher than stated and this is not clear why such a discrepancy should arise. This highlights the need for caution in interpreting responses especially from cross-sectional data (see Table 10).

Table 10. Table comparing the proportion of farms reporting using each disease control measure at initial visit alongside actual proportion of farms that carried out the measures during the follow up period (n = 548)

Type of control	Initial visit %	Actual practice %
Tick control*		
Yes	90.9	69.9
No	9.1	30.1
Worm control*		
Yes	56.4	26.8
No	43.6	73.2
Tsetse and trypanosome control		
Yes	17.9	14.1
No	82.1	85.9
Vaccine use*		
Yes	52.0	96.4
No	48.0	3.6

*significantly different at the 5% level using a McNemar's chi-squared test.

Morbidity and Mortality

The 548 recruited calves contributed a total of 175,732 calf days of life to the study. Figure 4 shows the temporal pattern of deaths and clinical episodes over the 3 years of the study. A total of 88 calves died before reaching 51 weeks of age giving an crude mortality rate of 16.4 (13.2-19.5) per 100 calves in their first year of life (Table 11).

Table 11. Counts of primary cause of deaths attributed by expert committee

Cause of death	No. calves
East coast fever	32
Unknown	20
Haemonchosis	9
Heartwater	6
Trauma	3
Actiomyces pyogenes	1
Babesiosis	1
Bacterialpneumonia	1
Black Quarter	1
Cassava	1
Foreign body	1
Mis-mothering	1
Rabies	1
Salmonellosis	1
Trypanosomiasis	1
Turning sickness	1
Viral pneumonia	1
No post mortem carried out	6
Total	88

NB 2 additional calves were considered to have died with ECF as a secondary contributing cause; one with heartwater and another with black quarter.

Fifteen calves were euthanised and were considered to have died from the primary pathology reported on *post mortem*. The distribution of times of deaths by AEZ is given in the Kaplan-Meier plot (Figure 6) showing that AEZ5 which is UM3 in Figure 1 and includes Magombe East, followed by AEZ1 (LM1) which include Bumala A had much higher death rates than other AEZs. The reasons are not yet clear and are the subject of ongoing analyses. Deaths were also attributed to a secondary or contributing cause of death when this was appropriate.

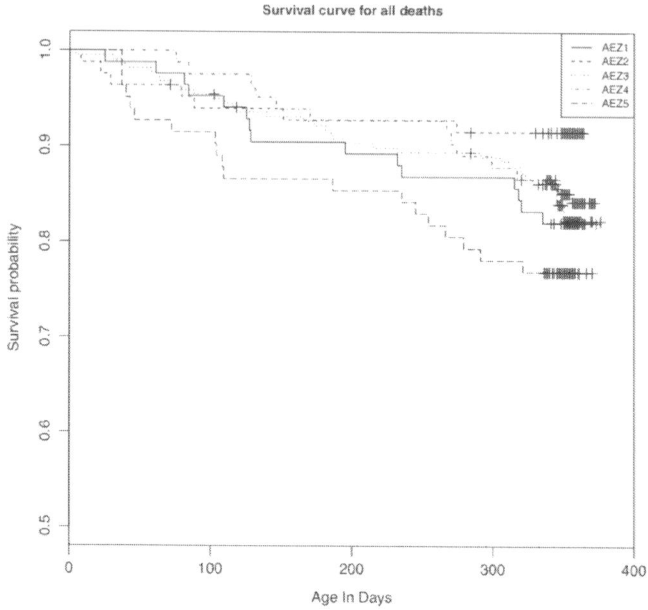

Figure 6. Kaplan-Meier survival curves for deaths due to all causes. + mark censoring for reasons other than death and are mainly at 51 weeks when visits stopped. The dashed lines give the 95% confidence intervals on the survival probability.

Unfortunately due to logistical reasons *post mortems* were not carried out on 6 of these calves so their cause of death remained unknown. Of the remaining 82 deaths all received a post-mortem examination. A further 4 of this 82 were of unknown cause (a total of 10 calves that died of a completely unknown cause). Seven died from a known non-infectious cause (cassava poisoning (1), foreign body pneumonia (1), mismothering (1), starvation (1), trauma (3)) and 1 died from an unidentified non-infectious cause. Eleven calves had clinical signs indicative of an infectious agent but the definitive cause remained unidentified and 59 died of an infectious cause that was diagnosed by post-mortem examination, appropriate testing and clinical history (East Coast fever (33), turning sickness (1), haemonchosis (10), heartwater (6), babesiosis (1), rabies (1), salmonellosis (1), trypanosomiasis (1), black quarter (1), viral pneumonia (1), multifocal abcessation due to *Actinomyces pyogenes* (1), and *Arcanobacterium* infection (1). This gives a minimum of 70 deaths attributable to infectious diseases and a minimum mortality rate due to infectious causes of 13.3% (10.4-16.2) per 100 calves in the first year of life.

Of the 32 cases of East Coast Fever 8 had a contributing cause of helminthiasis, 5 of which were due to haemonchosis and 2 of trypanosomiasis. Of the 10 cases of haemonchosis 2 had a contributing cause of Theileriosis, and 1 of lung worm (*Dictyocaulus viviparous*). Of the 6 heartwater cases 1 had a contributing cause of East Coast fever as did the case of black quarter. It is interesting to note that in an area generally considered to have high tsetse challenge there seemed to be little clinical trypanosomiasis.

A further 307 clinical episodes were observed by the AHAs on their routine 5 weekly visits and 216 clinical episodes were reported during non routine visits in response to reported illness. The details of all the clinical signs and patterns is currently under analysis but the overall distribution of clinical episodes by age is given in Figure 7. This suggests a bimodal pattern with a large peak around 16 weeks at the time when maternal antibodies might be expected to be waning. There is a second smaller peak later around 41 weeks when many calves are weaned.

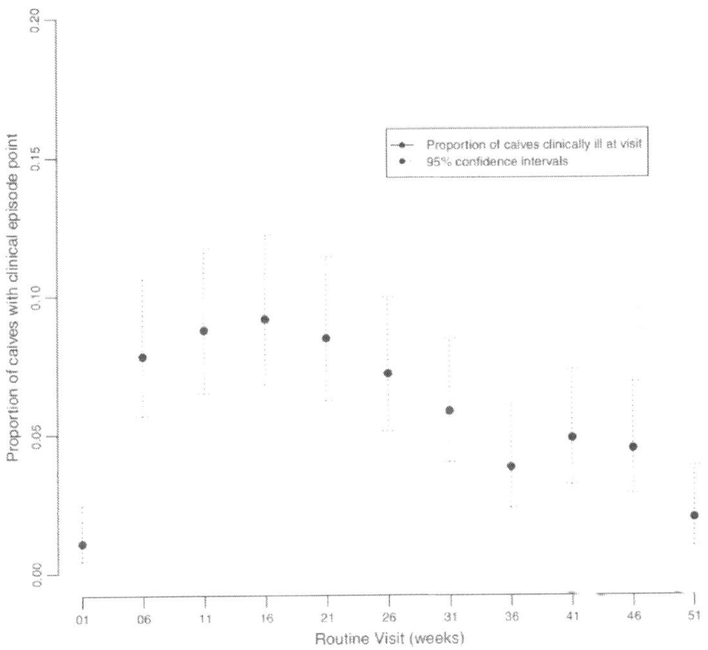

Figure 7. The distribution of the proportion of calves classed as having a clinical episode stratified by visit number over the 51 weeks of observations for each calf in the IDEAL project.

Pathogens and Exposures

Figure 8 shows the list of pathogen/test combinations experienced by the calf by the time of publication crudely stratified into endoparasites, haemoparasites, bacteria and viruses. Some of the common pathogens such as *Theileria spp.* appear several times as a number of techniques were used to identify them. In addition, some assays, such as microscopy, do not distinguish between species. More detailed analysis of these co-infections is on going. What this Figure shows very clearly is that this population of calves is infected with over 50 different pathogens and has been exposed to at least a further 6 bacteria and viruses. However, relatively few pathogens were found in the majority of calves, and the main pathogens were helminths and protozoan haemoparasites. What is of particular interest is that, given such high incidences of these key pathogens such as *T. parva*, *A. marginale*, *B. bigemina* and *H. placei*, why more of these calves did not die. One of the main objectives of the continuing analyses of this dataset is to unravel the coinfections and relate these to the calf genotype and key outcomes such as growth rate, morbidity and mortality. It is also interesting that there are very few bacterial diagnoses and these appear to have only sporadic occurrence and rarely contributed to death. We plan to look in more detail at the dam serology, but of the 2 bacterial pathogen exposures already measured in the dams, *Brucella spp.* and *Leptospira hardjo* the seroprevalences were extremely low, 0.036 (0.022-0.050 adjusted 95% CI) and 0.068 (0.035-0.101 95% adjusted CI) respectively. Also there was little clinical evidence of some of the major viral diseases such as foot-and-mouth disease.

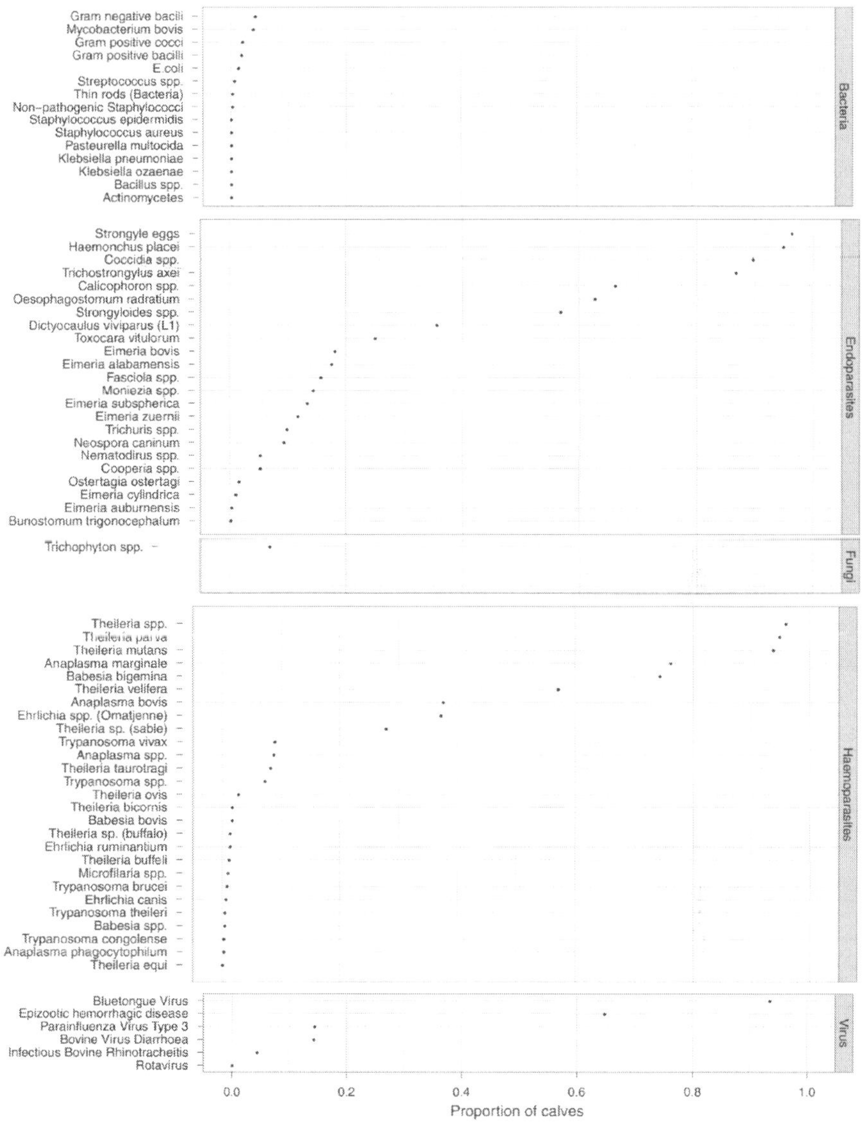

Figure 8. The proportion of animals positive for a given pathogen/test combination at any time through the course of the 51 weeks of observation on each calf in the IDEAL project.

DISCUSSION

The IDEAL project is the first attempt to describe the entire disease burden of any naturally occurring population. Funding was only available to follow calves for the first 12 months of life. The use of a longitudinal design, though enormously logistically challenging in this environment, allowed us to generate a unique dataset to study the effects of co-infections in the SHZ breed in this small holder setting. This may be applicable across a large sector of the Great Lakes basin where very similar breeds and husbandry are in operation.

When designing the project a number of different approaches were considered. They included stratification by management system, wealth/herd size, livestock distribution, location, ethnicity, etc. However, the lack of available data on several of these factors led to the decision to stratify by agro-ecological zone only. Random cluster sampling will have ensured that reasonable representation was provided for the various levels of each of the un-stratified factors, i.e. the total sample size will include farmers with varying herd sizes and management systems. The proportion of sub-locations sampled in each AEZ is in proportion of each AEZ in the total survey area (based on numbers of sub-locations). The study was constrained by logistics to an area of 45 km radius from Busia town in order to make repeated visits possible. Initially other options were reviewed but following piloting of sampling in the field it became clear that given the road conditions and number of animals that would have to be sampled per day at the peak of sampling in year 2 this was the most practical approach.

Owners were paid a retainer for the year to allow access to the animals and therefore compliance was very high. There were a small number of instances of animals being stolen and of owners treating the calves with anthelminthics without consulting the project vet. Where these were identified animals were censored and their data from the visit following treatment discarded.

The descriptive analysis from the recruitment interview indicate that livestock production in this system is characterised by low-input, with as few as 30% of the farms carrying out any form of disease control during the follow-up time. Even for those farms that reported carrying out disease control measures, the frequency of these per year was below what would be effective. This level of management would likely be insufficient to support the use of improved "exotic" breeds which are kept in the region but which we intentionally excluded from this study. The Western Province of Kenya accounts for only 4% of Kenya's total exotic dairy herd [43]. This is despite major breed improvements programs instituted to support smallholder farmers in the region through increased livestock productivity [43, 44].

Livestock disease and vector control are required for increased livestock productivity, and prevention of losses through disease-related morbidity, mortality, and loss of markets for livestock products. The observed lack of disease control has implications on some of the strategies envisaged to rapidly improve livestock-dependent livelihoods. It also highlights the need to provide support not just for the imported exotic breeds but also for the indigenous breeds in order to minimise the losses and maximise productivity. The consistent use of disease control practices has contributed to the relative success of the smallholder dairy sector in the Kenyan highlands [45]. The benefits of such controls, carried out at community level, have also been demonstrated in other settings [46]. Failure to consider these disease issues is recognised as a factor that could seriously reduce rural growth [47].

Many countries in sub-Saharan Africa have had to make structural adjustments to their veterinary infrastructure and the services they provide which leaves farmers and herdsmen without the support needed to introduce exotic genetic stock. Further, Rege *et al.*[48] argue that breeding strategies in the context of smallholder farms should be based on improving food security, income and overall livelihoods of the livestock keepers and should not be based on genetic improvement of livestock. Focus should be on providing the most appropriate genotypes in a local context. However, identifying these appropriate genotypes is itself complex. Mwachara *et al.*[49] identify the need to involve the livestock keepers in designing the breeding programmes to take into account the full array of contributions to livelihoods that these animals make and so identify genetic characteristics related to these functions. Whereas most programs have concentrated on cross-breeding, there exists a lot of potential and advantages for improvements based on within-breed selection.

The mortality rates in this indigenous calf population were higher than anticipated at the design stage. There are few reports that we could find from similar systems but other reports from the region suggest a range of mortalities. Barnett [50] reported a mortality rate of 29% from a study based in Western Province Kenya. In a Tanzanian smallholder dairy system mortality rates of 35% were reported [51] within the first year with 42% reported as of unknown cause and 19% due to redwater (babesiosis). Swai *et al.*[52] reported mortality rates of 12% in small holder dairy systems in

Zimbabwe with 56% ascribed to tick borne disease particularly east coast fever. Gitau *et al.*[53] reported 7% mortality in calves up to 6 months of age from the same area of Western. A more recent large study of calf mortality in Mali [54] reported an overall calf mortality of 17% but when this was broken down by system the more intensive systems had high mortality rates of 19% and 25% compared to 10% in the traditional pastoralist systems. Interestingly they report gastrointestinal disorders as causing 28% of their overall mortality followed by perinatal problems (16%) and accidents (14%). Direct comparisons are very difficult to make with many of these studies as the design, breeds, environment etc are not the same. However, it is useful to get an overall impression of how these animals are performing in this system. The mortality rate in the IDEAL cohort appears high given it is an indigenous breed that might be expected to have had time to adapt to the conditions. There are likely to be many contributing causes including possible inexperience in raising cattle compared to traditional cattle owning groups such as the Maasai or Fulani and the co-infection combinations present in the region.

The identification of pathogens at all time points in the study is on going. We adopted a very pragmatic approach using the best field techniques available as the method of diagnosis but for many pathogens this is not sufficient. For example speciation of *Theileria* parasites requires more detailed analysis such as RLB [55]. It must be noted that detection of pathogens is limited by the sensitivity of the assay, the presence of the pathogen at the time of sampling and its location in the tissue which is sampled. This presents many challenges in trying to produce a definitive list of pathogens at every time point for each calf. For this preliminary presentation of the pathogens we have simply summed across all visits to estimate the proportion of calves with each pathogen (or pathogen/test combination). This ignores the dynamics of the order of exposure but this is to be reported in a number of other papers. The list of pathogens is extensive but there are actually only a few very high prevalence pathogens. Theses are mainly gut helminths and tick borne haemoparasites, in particular *T. parva*.

The IDEAL project provides unique data on total livestock disease burden in the region, which will allow for ranking of infectious diseases in order of importance. Such data are important for prioritising interventions, the absence of which up to this point has led to a lack of metrics to assess the impact of livestock diseases leads and therefore inefficient resource allocation [2]. In addition, the project will provide data on the within breed variation of key traits such as growth rates, clinical tolerance and resistance, and survival. This provides a basis for identifying desirable traits that may be taken up while designing within-breed improvement programs. Within-breed selection may not achieve increased productivity per animal as rapidly as cross-breeding methods, but offers the opportunity to retain the adaptive characteristics already present in indigenous breeds and which may offer opportunity for adaptation to changing climates. The findings of positive associations between knowledge of diseases and access to veterinary support with whether farmers carry out disease control practices supports the idea that increased extension services would have significant positive effect on livestock productivity.

CONCLUSIONS

This population of calves is the first to have a comprehensive investigation of the pathogen burden and exposures of any animal population. The analyses of the biobanked samples will continue and it is expected that there will be further pathogens added to the list. This preliminary report on the IDEAL project design and data collection offers an overview of the farming system in Western, Kenya, and of the infectious disease challenges experienced by the calves of the region. It provides a detailed description of the methods used to collect this detailed longitudinal dataset. This provides more information for those reading analytical papers from the project, and acts as a supporting document to the extensive biobank held at ILRI, Nairobi. It gives preliminary results and offers an overview of more detailed analyses that result from the IDEAL project.

ACKNOWLEDGEMENTS

The authors would like to thank the Wellcome Trust (grant No. 079445) for financial support for the project. We would like to thank the Kenyan Department of Veterinary Services for their enormous support in setting up the study and working with our teams in the field. We would also like to thank the many local animal health assistants and farmers who participated in the running of the project. Finally we would like to recognise the enormous contribution of Dr. Magai Kaare who worked as the IDEAL project manager until his untimely death following a car accident in 2008. He will be greatly missed by all who knew him.

REFERENCES

1. United Nations Department of Economic and Social Affairs Population Division: World population prospects: The 2008 revision, highlights, working paper no. ESA/P/WP,. 2009, 210,
2. Perry B, Grace D: The impacts of livestock diseases and their control on growth and development processes that are pro-poor. Philos Trans R Soc Lond B Biol Sci. 2009, 364 (1530): 2643-2655. 10.1098/rstb.2009.0097.
3. Phiri BJ, Benschop J, French NP: Systematic review of causes and factors associated with morbidity and mortality on smallholder dairy farms in Eastern and Southern Africa. Prev Vet Med. 2010, 94 (1–2): 1-8.
4. Perry BD, Rich KM: Poverty impacts of foot-and-mouth disease and the poverty reduction implications of its control. Vet Rec. 2007, 160 (7): 238-+. 10.1136/vr.160.7.238. [Times Cited: 0]
5. Morens DM, Holmes EC, Davis AS, Taubenberger JK: Global rinderpest eradication: lessons learned and why humans should celebrate too. J Infect Dis. 2011, 204: 502-505. 10.1093/infdis/jir327.
6. Zhang XS, Woolhouse MEJ: Escherichia coli O157 infection on Scottish cattle farms: dynamics and control. J R Soc Interface. 2011, 8 (60): 1051-1058. 10.1098/rsif.2010.0470.
7. Bronsvoort B, Radford A, Tanya V, Kitching RP, Morgan K: The molecular epidemiology of foot-and-mouth disease viruses in the Adamawa Province of Cameroon. J Clin Microbiol. 2004, 42 (5): 2186-2196. 10.1128/JCM.42.5.2186-2196.2004.
8. Gachohi JM, Kitala PM, Ngumi PN, Skilton RA: Environment and farm factors associated with exposure to Theileria parva infection in cattle under traditional mixed farming system in Mbeere District, Kenya. Trop Anim Health Prod. 2011, 43: 271-277. 10.1007/s11250-010-9688-x.
9. Geiger SM: Immuno-epidemiology of Schistosoma mansoni infections in endemic populations co-infected with soil-transmitted helminths: Present knowledge, challenges, and the need for further studies. Acta Tropica. 2008, 108 (2–3): 118-123.
10. Woolhouse M, Taylor L, Haydon D: Population biology of multihost pathogens. Science. 2001, 292 (5519): 1109-1112. 10.1126/science.1059026.

11. Woolhouse M, Webster J, Domingo E, Charlesworth B, Levin B: Biological and biomedical implications of the co-evolution of pathogens and their hosts. Nat Genet. 2002, 32 (4): 569-577. 10.1038/ng1202-569.
12. Rohani P, Green C, Mantilla-Beniers N, Grenfell B: Ecological interference between fatal diseases. Nature. 2003, 422 (6934): 885-888. 10.1038/nature01542.
13. Morand S, Manning S, Woolhouse M: Parasite-host coevolution and geographic patterns of parasite infectivity and host susceptibility. Proc R Soc London Ser B-Biol Sci. 1996, 263 (1366): 119-128. 10.1098/rspb.1996.0019.
14. Donoghue H, Marcsik A, Matheson C, Vernon K, Nuorala E, Molto J, Greenblatt C, Spigelman M: Co-infection of Mycobacterium tuberculosis and Mycobacterium leprae in human archaeological samples: a possible explanation for the historical decline of leprosy. Proc R Soc B-Biol Sci. 2005, 272 (1561): 389-394. 10.1098/rspb.2004.2966.
15. Fall A, Diack A, Diaite A, Seye M, d'Ieteren G: Tsetse challenge, trypanosome and helminth infection in relation to productivity of village Ndama cattle in Senegal. Vet Parasitol. 1999, 81 (3): 235-247. 10.1016/S0304-4017(98)00213-1.
16. Zinsstag J, Ankers P, Ndao M, Bonfoh B, Pfister K: Multiparasitism, production and economics in domestic animals in sub-Saharan West Africa. Parasitol Today. 1998, 14 (2): 46-49. 10.1016/S0169-4758(97)01183-6.
17. Kaufmann J, Dwinger R, Hallebeek A, Vandijk B, Pfister K: The interaction of Trypanosoma congolense and Haemonchus contortus infections in trypanotolerant ndama 520 cattle. Vet Parasitol. 1992, 43 (3–4): 157-170.
18. Flynn RJ, Mulcahy G, Welsh M, Cassidy JP, Corbett D, Milligan C, Andersen P, Strain S, McNair J: Co-Infection of cattle with Fasciola hepatica and Mycobacterium bovis - immunological consequences. Transbound Emerg Dis. 2009, 56 (6–7): 269-274.
19. Carslake D, Grant W, Green LE, Cave J, Greaves J, Keeling M, McEldowney J, Weldegebriel H, Medley GF: Endemic cattle diseases: comparative epidemiology and governance. Philos Trans R Soc Lond B Biol Sci. 2011, 366 (1573): 1975-1986. 10.1098/rstb.2010.0396.

20. Rafi W, Ribeiro-Rodrigues R, Ellner JJ, Salgame P: 'Coinfection-helminthes and tuberculosis'. Curr Opin HIV AIDS. 2012, 7 (3): 239-244. 10.1097/COH.0b013e3283524dc5.
21. Cooper P, Chico M, Losonsky G, Sandoval C, Espinel I, Sridhara R, Aguilar M, Guevara A, Guderian R, Levine M, Griffin G, Nutman T: Albendazole treatment of children with ascariasis enhances the vibriocidal antibody response to the live attenuated oral cholera vaccine CVD 103-HgR. J Infect Dis. 2000, 182 (4): 1199-1206. 10.1086/315837.
22. Yazdanbakhsh M, Kremsner P, van Ree: Immunology - Allergy, parasites, and the hygiene hypothesis. Science. 2002, 296 (5567): 490-494. 10.1126/science.296.5567.490.
23. Cleaveland S, Laurenson MK, Taylor LH: Diseases of humans and their domestic mammals: pathogen characteristics, host range and the risk of emergence. Philos Trans R Soc Lond B Biol Sci. 2001, 356 (1411): 991-999. 10.1098/rstb.2001.0889.
24. Hanotte O, Dessie T, Kemp S: Time to tap Africa's livestock genomes. Science. 2010, 328 (5986): 1640-1641. 10.1126/science.1186254.
25. Claridge J, Diggle P, McCann CM, Mulcahy G, Flynn R, McNair J, Strain S, Welsh M, Baylis M, Williams DJ: Fasciola hepatica is associated with the failure to detect bovine tuberculosis in dairy cattle. Nat Commun. 2012, 3 (5): 853-
26. Zinsstag J, Schelling E, Roth F, Bonfoh B, de Savigny D, Tanner M: Human benefits of animal interventions for zoonosis control. Emerg Infect Dis. 2007, 13 (4): 527-531. 10.3201/eid1304.060381.
27. Jaeztold R, Schmidt H: Farm Management Handbook of Kenya, Part IIA, West Kenya. Nyanza and Western Provinces. 1983, Kenya: Kenyan Ministry of Agriculture, p.387
28. Thornton P, Kruska R, Henninger N, Kristjanson P, Reid R, Atieno F, Odero A, Ndegwa T: Mapping Poverty and Livestock in the Developing World. ILRI (International Livestock Research Institute). 2002, Nairobi: ILRI
29. GOK: Geographical Dimensions of Well-Being in Kenya. Where the Poor Live. Ministry of Planning and National Development. 2003, Nairobi: Regal Press
30. IFAD: Assessment of Rural Poverty: Eastern and Southern Africa. 2002, Rome: International Fund for Agricultural Development

31. Reynecke DP, Van Wyk Ja, Gummow B, Dorny P, Boomker J: Application of ROC curve analysis to FAMACHA(□) evaluation of haemonchosis on two sheep farms in South. Vet Parasitol. 2011, 177 (3–4): 224-230.

32. Gubbels JM, De Vos AP, Van der Weide M, Viseras J, Schouls LM, De Vries E, Jongejan F: Simultaneous detection of bovine Theileria and Babesia species by reverse line blot hybridization. J Clin Microbiol. 1999, 37 (6): 1782-1789.

33. Taylor MA, Coop RL, Wall R: Veterinary Parasitology. 2007, Oxford, UK: Blackwell Publishing Ltd, ISBN 1405119640

34. Renz A, Trees AJ, Achukwi D, Edwards G, Wahl G: Evaluation of Suramin, Ivermectin and Cgp-20376 in a New Macrofilaricidal drug screen, Onchocerca-Ochengi in African cattle. Trop Med Parasitol. 1995, 46: 31-37. [QQ463 TROP MED PARASITOL]

35. Nicholson M, Butterworth M: A Guide to Condition Scoring of Zebu Cattle. 1986, Addis Ababa: International Livestock Centre for Africa

36. OW Schalm DN: Experiments and observations leading to development of the California mastitis test. J Am Vet Med Assoc. 1957, 130: 199-204.

37. Jahnke HE: Livestock Production Systems and Livestock Development in Tropical Africa. 1982, Kiel: Kieler Wissenschaftsverlag Vauk

38. ILCA: Handbook of African livestock statistics. Tech. rep., Addis Ababa: International Livestock Centre for Africa; 1993

39. Njuki J, Poole N, Johnson J, Baltenweck I, Pali P: Gender, livestock and livelihood indicators. Tech. rep., International Livestock Research Institute (ILRI) 2011. http://results.waterandfood.org/handle/10568/3036,

40. Kaplan EL, Meier P: Nonparametric estimation from incomplete observations. J Am Stat Assoc. 1958, 53 (282): 457-481. 10.1080/01621459.1958.10501452.

41. Kristjanson P, Krishna A, Radeny M, Nindo W: Pathways out of Poverty in Western Kenya and the Role of Livestock. 2004, FAO, http://hdl.handle.net/10568/1212,

42. Zaibet L, Traore S, Ayantunde A, Marshall K, Johnson N, Siegmund-Schultze M: Livelihood strategies in endemic livestock production systems in sub-humid zone of West Africa: trends, trade-offs and implications. Env Dev Sustainability. 2011, 13: 87-105. 10.1007/s10668-010-9250-z.

43. Mudavadi PO, Otieno K, Wanambacha JW, Odenya JO: Smallholder dairy production and marketing in Western Kenya: a review of literature. Tech. rep., MOARD/KARI/ILRI 2001. http://hdl.handle.net/10568/1781,
44. Waithaka M, Nyangaga J, Staal S, Wokabi A, Njubi D, Muriuki K, Njoroge L, Wanjohi P: Characterization of dairy systems in the western Kenya region. Tech. rep., Smallholder Dairy (R&D) Project MALDM-KARI-ILRI-DFID, Nairobi, Kenya 2002. http://dspacetest.cgiar.org/handle/10568/1859,
45. Omore A, Muriuki H, Kenyanjui M, Owango M, Staal S: The Kenya dairy sub-sector: a rapid appraisal. Tech. rep., ILRI 1999. http://dspacetest.cgiar.org/handle/10568/2054,
46. Hüttner K: The effect of a community-based animal health service programme on livestock mortality, off-take and selected husbandry applications A field study in northern Malawi. Livest Prod Sci. 2001, 72 (3): 263-278. 10.1016/S0301-6226(01)00187-7. http://linkinghub.elsevier.com/retrieve/pii/S0301622601001877,
47. FAO: The state of food and agriculture, livestock in the balance. Rome, 2009. ISBN 9789251062159
48. Rege J, Marshall K, Notenbaert A, Ojango J, Okeyo A: Pro-poor animal improvement and breeding — What can science do?. Livest Sci. 2011, 136: 15-28. 10.1016/j.livsci.2010.09.003. http://linkinghub.elsevier.com/retrieve/pii/S1871141310004774,
49. Mwacharo JM, Drucker AG: Production objectives and management strategies of livestock keepers in South-East Kenya: implications for a breeding programme. Trop Anim Health Prod. 2005, 37 (8): 635-652. 10.1007/s11250-005-4253-8. http://www.springerlink.com/index/10.1007/s11250-005-4253-8,
50. Barnett S: Theileriasis control. Bull Epiz Dis Afr. 1957, 5: 343-357.
51. French N, Tyrer J, Hirst W: Smallholder dairy farming in the Chikwaka communal land, Zimbabwe: birth, death and demographic trends. Prev Vet Med. 2001, 48: 101-112. 10.1016/S0167-5877(00)00191-4.
52. Swai E, Karimuribo ED, Kambarage DM, Moshy WE: A longitudinal study on morbidity and mortality in youngstock smallholder dairy cattle with special reference to tick borne infections in Tanga region, Tanzania. Vet Parasitol. 2009, 160: 34-42. 10.1016/j.vetpar.2008.10.101.

53. Gitau G, Perry B, McDermott J: The incidence, calf morbidity and mortality due to Theileria parva infections in smallholder dairy farms in Murang'a District, Kenya. Prev Vet Med. 1999, 39: 65-79. 10.1016/S0167-5877(98)00137-8.
54. Wymann MN, Bonfoh B, Schelling E, Bengaly S, Tembely S, Tanner M, Zinsstag J: Calf mortality rate and causes of death under different herd management systems in peri-urban Bamako, Mali. Livest Sci. 2006, 100: 169-178. 10.1016/j.livprodsci.2005.08.010.
55. Oura CAL, Bishop RP, Wampande EM, Lubega GW, Tait A: Application of a reverse line blot assay to the study of haemoparasites in cattle in Uganda. Int J Parasitol. 2004, 34 (5): 603-613. 10.1016/j.ijpara.2003.12.012. [815TG INT J PARASITOL]

CHAPTER 5

Unmanned Aircraft Systems for Studying Spatial Abundance of Ungulates: Relevance to Spatial Epidemiology

Jose´ A. Barasona[1], Margarita Mulero-Pa´ zma´ny[2], Pelayo Acevedo1, Juan J. Negro[2], Marı´a J. Torres[3], Christian Gorta´ zar[1], Joaquı´n Vicente[1]

[1]SaBio IREC, National Wildlife Research Institute (CSIC-UCLM-JCCM), Ciudad Real, Spain,

[2]Department of Evolutionary Ecology, Don˜ana Biological Station, CSIC, Seville, Spain,

[3]Department of Microbiology, Universidad de Sevilla, Seville, Spain

Citation: Barasona JA, Mulero-Pázmány M, Acevedo P, Negro JJ, Torres MJ, Gortázar C, et al. (2014) Unmanned Aircraft Systems for Studying Spatial Abundance of Ungulates: Relevance to Spatial Epidemiology. PLoS ONE 9(12): e115608. https://doi.org/10.1371/journal.pone.0115608

ABSTRACT

Complex ecological and epidemiological systems require multidisciplinary and innovative research. Low cost unmanned aircraft systems (UAS) can provide information on the spatial pattern of hosts' distribution and abundance, which is crucial as regards modelling the determinants of disease transmission and persistence on a fine spatial scale. In this context we have studied the spatial epidemiology of tuberculosis (TB) in the ungulate community of Doñana National Park (South-western Spain) by modelling species host (red deer, fallow deer and cattle) abundance at fine spatial scale. The use of UAS high-resolution images has allowed us to collect data to model the environmental determinants of host abundance, and in a further step to evaluate their relationships with the spatial risk of TB throughout the ungulate community. We discuss the ecological, epidemiological and logistic conditions under which UAS may contribute to study the wildlife/livestock sanitary interface, where the spatial aggregation of hosts becomes crucial. These findings are relevant for planning and implementing research, fundamentally when managing disease in multi-host systems, and focusing on risky areas. Therefore, managers should prioritize the implementation of control strategies to reduce disease of conservation, economic and social relevance.

INTRODUCTION

Understanding the spatial distribution of risks for multi-host diseases in complex ecological scenarios is a key issue for disease management in wildlife [1] and also at the domestic/wildlife interface [2]. Population demography, resource allocation and behaviour determine the spatiotemporal structure of interactions among hosts. These characteristics may therefore play an important role in pathogen transmission [3], [4], [5]. The distribution range and abundance of domestic and wild animals can be affected by natural (biotic and abiotic) and human-mediated factors [6]. In an epidemiological context these factors are able to drive intra and inter-specific disease transmission rates [7]. For instance, areas commonly grazed by both cattle and wildlife - including artificial feeding and watering areas - have been proved to increase the risk of pathogen transmission between species [8], [9]. But risk factor analyses are highly dependent on the spatial scale, as the relationships between factors and the transmission rates may change with the selection of different areal units [10].

Determining the patterns of host abundance at a finer resolution for large areas requires an impressive sampling effort and/or the use of predictive modelling (e.g. [11], [12]). The use of expensive methods, such as manned aircrafts, may apply to wildlife monitoring [13]. Low cost unmanned aircraft systems (hereafter UAS) have recently emerged as an efficient alternative for wildlife monitoring and ecological research [13], [14]. These systems offer the ability to monitor biological processes of large areas remotely and rapidly. Thus, UAS may become an affordable, safe and accurate option for a wide variety of studies [15], [16], [17]. UAS equipped with on board cameras have been used to obtain high-resolution images of wildlife occurrences in quasi real time in highly dynamic landscapes (e.g. [16], [18]), but their possible efficacy in predicting distribution and abundance of animals has not been proven. UAS therefore have a high potential as regards collecting precise information on hosts, which is highly prized by epidemiologists since it is useful to determine the spatial risk of disease transmission (e.g. [19]).

Most multi-host diseases (e.g. foot and mouth disease, rabies, anthrax, brucellosis and tuberculosis) often emerge and/or persist in complex ecological communities involving several domestic and wild hosts [20], [21]. Of special relevance are diseases involving wild ungulates in their epidemiological cycles, as these species are undergoing a generalised expansion (e.g. [22]). This promotes epidemiological scenarios in which disease transmission often results in the maintenance or amplification of diseases that affect not only wild ungulates, but also livestock production, public health and endangered wildlife species (e.g. [23], [24]). This research proposes an epidemiological approach to evaluate applications of UAS as feasible tool for monitoring wildlife distribution and abundance.

Tuberculosis (TB), which is caused by the *Mycobacterium tuberculosis* complex, is an important re-emerging zoonotic disease that is shared between cattle and wildlife, and the existence of wildlife reservoirs limits the control of this disease [21], [25]. The TB infection generally develops into chronic infections, with long-term persistence in populations, and is able to induce a period of infectiousness during which direct and indirect contact favouring disease transmission between individuals (intra and inter-specific) occurs. In South-central Spain, complex epidemiological connections between TB prevalence in cattle and wild ungulates (wild boar *Sus scrofa*, red deer *Cervus elaphus* and fallow deer *Dama dama*) have been established [23], [26], thus highlighting the role of wildlife as TB maintenance hosts in this complex system (see also [8], [23], [27]).

Since environmental features (e.g. natural water sources, water holes, marsh-shrub ecotone and/or grazing areas) frequently used by hosts could act as important sources of TB and/or favour closer contact of individuals both between and within the species, we hypothesise that the pattern of TB frequency can be explained by the spatial patterns of the host abundance. Our aim when using UAS high-resolution images was specifically to: i) model the spatial pattern of abundance for each potential host species in a Mediterranean scenario, and ii) evaluate the explanatory capacity of the predicted abundance of hosts with the spatial risk of TB throughout the ungulate community as a way to validate the predictions of host abundance obtained from UAS data.

MATERIALS AND METHODS

Ethics Statement

This study utilized samples from wild ungulates captured (shot by the Park's rangers) and necropsied in the context of the Doñana National Park health-monitoring programme. Sampling followed a protocol approved by the Animal Experiment Committee of Castilla-La Mancha University and by the Spanish Ethics Committee, and designed and developed by scientists (B and C animal experimentation categories) in accordance with EC Directive 86/609/EEC for animal handling and experiments. The work was conducted complying with the current Spanish legislation involving aviation safety and field technicians had the required licenses to operate in the frequencies they used. Doñana National Park authorities (P43-2010), MINECO (AGL2010-20730-C02-01) and Aeromab Project (P07-RNM-03246) field permits approved these methods.

Study Area

The study was carried out in Doñana National Park, DNP (37°0′ N, 6°30′ W, 54,000 ha), a nature reserve located on the Atlantic coast of South-western Spain. DNP has the highest level of environmental protection in Spain and is one of the most important natural reserves in Europe in terms of biodiversity. The region has a dry sub-humid Mediterranean climate with marked seasons. In the wet season (winter and spring), the marshlands are flooded and ungulates graze in the more elevated shrublands. The hardest season for ungulates in this area is summer (from July to September), when herbaceous vegetation, wetlands and water bodies (i.e., there is no water flow during this season) dry up in most habitats, and only a few meadows remain green on the boundary between the upper area of shrubs and the low dry marsh (ecotone). During the dry season, a north-south oriented longitudinal humid ecotone – dominated by wetland plants including *Scirpus maritimus* and *Galio palustris* with *Juncus maritimus* associations – can be identified between the shrublands and the edge of the dry marshlands (Fig. 1).

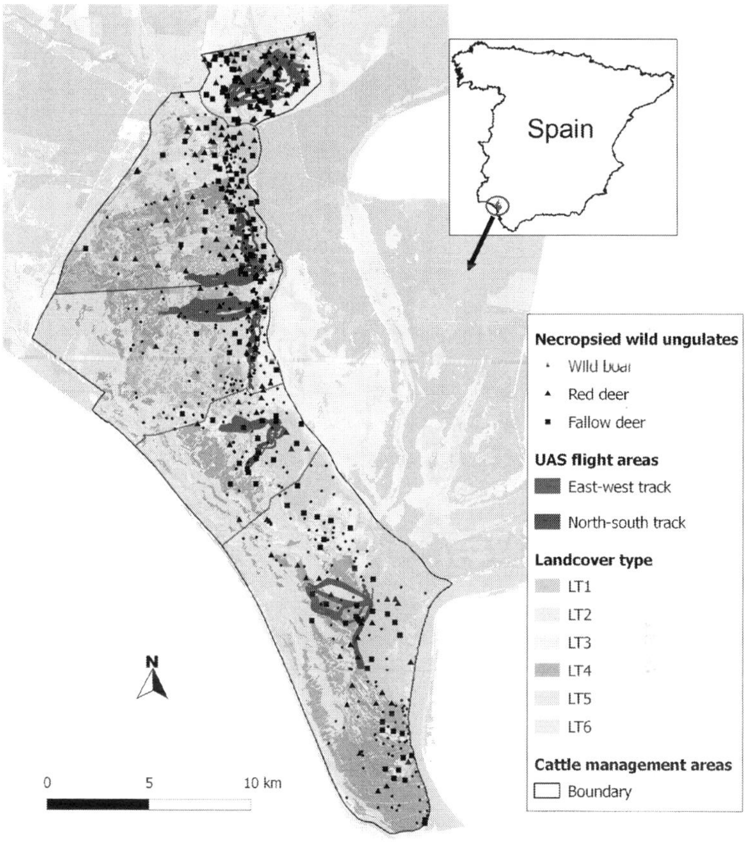

Figure 1. Map of the study area, Doñana National Park in Southern Spain, obtained from [36]. Six main habitats can be differentiated: dense shrubland (LT1), low-clear shrubland (LT2), herbaceous grassland (LT3), woodland (LT4), bare land (LT5), watercourse vegetation and water body (LT6). Locations of the necropsied wild ungulates and of the UAS tracks at the five cattle management areas are shown.

https://doi.org/10.1371/journal.pone.0115608.g001

A traditional breed of cattle (locally called "marismeña") is present in DNP, in which 5 different cattle management areas (MA) can be identified from north to south (Fig. 1). The southern MA, Marismillas (MA1, n=318 cattle) has a density of 3.1 cattle/km^2. The central area comprises 3 MA (MA2, n=152 cattle; MA3, n=168 cattle and MA4, n=350 cattle) and an average of density of 4.2 cattle/km^2. There are no cattle in the northern MA, which is called Coto del Rey (MA5). Wild ungulates in the area include: wild boar, red deer and fallow deer. Agriculture and hunting are prohibited in DNP, and artificial feeding practices are not conducted. However, cattle TB reactor rates are still high in DNP despite compulsory testing and the culling of infected animals [23], [28]. Previous studies have revealed that *M. bovis* infection prevalence is spatially structured, leading to among the highest rates of TB reported in wildlife worldwide [23], [27]. In addition to behavioural factors, it is probable that environmental features may explain variations in TB rates in wildlife since they favour direct and/or indirect contact between individuals and species during the dry season [8], [29].

Unmanned Aircraft Systems (UAS) methodology

We performed UAS fieldwork during the summer (August and September) of 2011, the season when water resources are more limited for wildlife, the availability of food is severely reduced [30] and the aggregation of individuals around water resources is therefore expected to be at its maximum [31]. The UAS platform was built using the foam fuselage of a radio controlled model Easy Fly plane (St-models, China) propelled by a brushless electrical engine. The embarked systems are: an on-board video camera used for First Person View Flight (FPV), a GPS (10 Hz, Mediatek, model FGPMMOPA6B), an Ikarus autopilot (Electronica RC, Spain) which provides flight stabilization, On Screen Display (OSD), a Panasonic Lumix LX-3 digital photo camera 11MP (Osaka, Japan), a three dimension waypoint following capability and an "emergency return home" function. The ground station is composed of a case containing a monitor, a DVD recorder, a video receiver and the control signal transmitter with the corresponding antennas. It also includes a Laptop with which to program the autopilot, store the pictures and data logs, and decode in-flight telemetry, thus allowing the position of the UAS to be tracked in real time on a Microsoft map (Redmond, WA, USA). For further details please see [17] and [18].

We conducted a total of 60 aerial tracks (each of which was ~4 km in length and 0.1 km wide: total surveyed length 240 km, covering 10.1% of the study area) between 17.30 h and 21.00 h local time, the time when the animals are most active as the temperature begins to decrease [32]. Two types of tracks – east-west and north-south oriented transects (Fig. 1) – were surveyed in each MA and replicated six times. The UAS were programmed to fly at an altitude of 100 m and at an average speed of 40 km/h. Details of the image processing can be found in [18]. The information obtained was divided into patches of 1 ha (our territorial unit for modelling). The patch size was adjusted to the width (0.1 km) of aerial tracks which is also consistent with a previous study on deer spatial ecology in the study area [33]. For each territorial unit and species, the abundance index (our response variables for modelling) was estimated as the number of individuals seen per ha.

Environmental Predictors for Modelling Host Abundance

Environmental variables were chosen on the basis of their availability for the study area and their potential predictive power (e.g. [14], [31], [34]). The predictors calculated for each territorial unit were: straight-line distance (km) to nearest artificial water hole (DW); straight-line distance (km) to nearest marsh–shrub ecotone (DE); exacted grid area (ha) surveyed by UAS (GA); number of water points per km^2 (WDN); proportion of dense scrub (LT1); proportion of low-clear shrubland (LT2); proportion of herbaceous grassland (LT3); proportion of woodland (LT4); proportion of bare land (LT5); and proportion of watercourse vegetation (LT6; summarized in Table 1). Landcover data was obtained from Andalusia Environmental Information [35].

Table 1. Environmental predictors used for spatial modelling of both host species abundance and *Mycobacterium tuberculosis* **complex transmission risk factors.**

Codes	Descriptions	Mean	SD
DW	Distance to nearest water point (km)	**0.45**/0.97	**0.36**/0.71
DE	Distance to nearest marsh-shrub ecotone (km)	**1.01**/2.33	**0.89**/2.15
GA	UAS grid area (ha)	**1.23**/1	**0.80**/0.00
WDN	Water point density (n/km^2)	**0.82**/0.51	**1.18**/0.35
LT1	Dense shrub (%)	**20.94**/11.39	**34.47**/26.51
LT2	Low-clear shrub (%)	**23.72**/30.56	**33.08**/39.41
LT3	Herbaceous grassland (%)	**24.91**/11.55	**36.09**/30.78
LT4	Woodland (%)	**14.66**/17.69	**31.43**/34.04
LT5	Bare land (%)	**7.12**/10.64	**21.83**/26.11
LT6	Water course vegetation (%)	**5.87**/10.92	**19.25**/27.84
MA	Cattle management area (factor, 5 categories)	–	–

For continuous predictors, mean values and standard deviations (SD) are reported for both the surveyed area (UAS track grids: in bold) and the whole study area (surveyed/study area).

doi:10.1371/journal.pone.0115608.t001

https://doi.org/10.1371/journal.pone.0115608.t001

Canopy cover reduces visibility and thus the detection of ungulates in UAS images. We therefore calculated detection coefficients in order to take this fact into consideration. Detection coefficients were calculated for both types of vegetation cover, dense shrub land (LT1) and woodland (LT4), in order to correct the reduction of visibility when ungulates were found below these canopy covers. We analysed the probability of detection of 100 random circle points (1 m^2 size) per habitat image (surface=1 ha), considering any point above vegetation cover as "not detected" and any point without vegetation cover as "detected". This was repeated in ten different images of each cattle management area and landcover type (n=100 images). For each landcover type, the detection coefficient is calculated as the proportion of points that were detected per image. The detection coefficients used in the statistical analyses were 0.538 and 0.359 as thresholds of visibility in LT1 and LT4 cover types, respectively. This means that we could only detect 53.8% of the ungulates located in LT1 habitat and 35.9% of the ungulates located in LT4. Collinear variables were excluded using a variance inflation factor (VIF) coefficient>3 threshold cut-off value [36].

Modelling Ungulate Abundance

Since information on ungulate distribution for the whole DNP was incomplete, we used spatially explicit modelling for predicting abundance values for overall study area. Habitat-species relationships can be parameterised with this feasible and reliable tool from the sampling obtained with UAS. The statistical model obtained could then be used to forecast species abundance in areas where information about the target species was not available [12]. When calibrating the abundance model, we only used the east-west UAS track data because we found very low habitat variation in the north-south UAS tracks (mainly overlapping with ecotone habitat), whereas the east-west tracks randomly sample the different habitats present in DNP. Previously to modelling, the dataset from the east-west UAS was randomly split by using a subset to parameterise the models (70%), whereas the rest (30%) was reserved for model validation on independent data. The north-south UAS track data were, however, also used to validate predictions from modelling using another set of independent data.

For each species (red deer, fallow deer and cattle - wild boar were excluded owing to low diurnal detection of the species; see [37]), the response variable was the abundance index for the territorial units. Then, the values of each environmental predictor were summarized for each sampling unit using 'zonal statistic tool' with Quantum GIS version 1.8.0 Lisboa [38].

The local abundance of each species per territorial unit was modelled using a generalised linear model, with a negative binomial distribution and a logarithmic link function [39]. We opted for the negative binomial distribution owing to high levels of overdispersion in the data when the models were fitted with Poisson distributions. The final models were obtained using a forward-backward stepwise procedure based on Akaike Information Criteria (AIC; [40]).

After modelling, predicted abundances were quantitatively compared with the data observed in the validation datasets by using Pearson's correlations (significant threshold; $p<0.05$) on both the east-west UAS independent data and the north-south UAS data. All statistics were carried out using R 2.15.2 [41].

Sampling and Tuberculosis Diagnosis

From 2006 to 2012, 949 wild ungulates comprising wild boar (n=570), red deer (n=190) and fallow deer (n=189) were randomly captured, necropsied and sampled in the context of the DNP health-monitoring programme (Fig. 1). The exact shooting location of each animal sampled was geo-referenced with a portable GPS (GarminLtd., Cayman Islands). Species, sampling year, gender and age were recorded on each of the animals sampled using similar criteria to previous studies (see [23]). Briefly, age classes were typified as juveniles (<24 months old) or adults (>24 months old) [8]. Necropsies were performed by qualified wildlife veterinarians with a wide experience in the diagnostics of macroscopic TB compatible lesions via the detailed inspection of lymph nodes and abdominal and thoracic organs (see [42]). Cultures were performed in order to confirm TB, using Lowenstein-Jensen piruvate medium. In the four MA in which cattle were present, veterinarian authorities used skin tests on the cattle populations and positive reactors were slaughtered during same period, as part of the National TB control Programme.

Determining the risk of TB at spatial scale

We used a Bayesian multivariable logistic regression approach [43] to evaluate the association among the occurrence of TB at individual level (response variable) and the predicted species abundances, while adjusting for the other risk factors hypothesised to influence disease status in DNP [26]. Predicted abundances of each modelled species (obtained in the UAS approach at 1 ha resolution; Fig. 2) were assigned to each TB-tested animal by using 'point sampling tool' with Quantum GIS. In addition, the covariates related to aggregation risky points (i.e., DW and WDN, see Table 1), species, sex, age class and sampling year of TB-tested animals were also included in the mode. In Bayesian inference, uncertainty in estimating predictor is fully reflected in the estimates and credible intervals (CIs) of all other variables, which is particularly important when there are a number of explanatory variables and smooth terms available [44]. This Bayesian framework was run using Integrated Nested Laplace Approximations (R-INLA package; [45]) to avoid the convergence and mixing algorithms of the Markov Chain Monte Carlo based sampling methods [46]. Deviance information criterion (DIC) was used for Bayesian model selection by following a backward stepwise procedure and the model with the lowest value of DIC was retained [47]. The statistical significance of each explanatory variable was assessed by 95% CIs of the probability distribution.

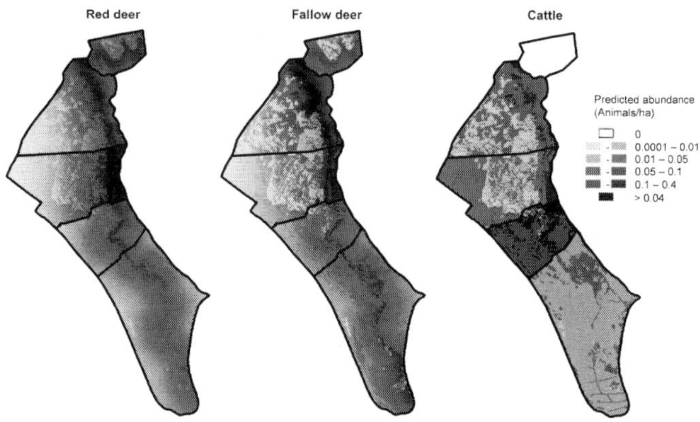

Figure 2. Predicted patterns of species relative abundance (animals/ha) in Doñana National Park obtained by modelling abundance data obtained from Unmanned Aircraft Systems (see Table 2).

https://doi.org/10.1371/journal.pone.0115608.g002

RESULTS

Spatial Distribution of Ungulates Abundance

The aerial tracks allowed a total of 3,149 ungulates to be located, identified and recorded from the UAS images. This included 51.44% of red deer,

33.18% of fallow deer, 11.36% of cattle, 2.89% of horses and only 0.22% of wild boar, while 0.89% was recorded as unknown species. The results of the best-fitting abundance models and their coefficients for the environmental predictors are summarized in Table 2. The common landscape predictors affecting the abundance indices explained by the best-fitting models for all species were: LT1 (negatively related to dense shrub), LT3 (positively related to herbaceous grassland) and MA (Table 2).

Table 2. Results of the generalised linear models (negative binomial error distribution and logarithmic link function) used to predict red deer, fallow deer and cattle abundance on a spatial scale in Doñana National Park.

Response	Final model
Red deer abundance	~ −0.001·DE −2.96·[1]LT1+0.78·LT3+0.44·[2]MA2+2.15·MA3+1.1·MA4+0.4·MA5
Fallow deer abundance	~ −2.17·DE −2.25·[1]LT1+2.9·LT3+7.22·[1]LT4 −1.42·[2]MA2+2.73·MA3+3.68·MA4 −1.34·MA5
Cattle abundance	~ −1.19·[1]LT1+2.93·LT3+3.36·[2]MA2+6.6·MA3+9.17·MA4 −3.41·MA5

Coefficients are shown for the most parsimonious models according to AIC. Measures for model support and statistical parameters (test and p-values) for the variables selected in the final models can be found in S1 Table. Variable codes are described in Table 1.
[1]LT1 and LT4 were corrected by detection coefficients, 0.538 and 0.359, respectively. [2]Reference value of the parameter estimator was 0 for "cattle management area 1 (MA1)". MA from 2 to 5 codify for the location of each sampling unit in each management area (1 when present and 0 when absence).

doi:10.1371/journal.pone.0115608.t002

https://doi.org/10.1371/journal.pone.0115608.t002

Validation on independent datasets for both east-west and north-south UAS track data showed significant correlations between observed and predicted abundance values (Table 3). After being assessed using independent data, the final models were used to predict species abundance in the whole of DNP (Fig. 2). Common distribution patterns can be observed in the specific predictions of species abundance throughout DNP. It is remarkable that all ungulate species tended to spatially aggregate at the ecotone. Differences between MA were more evident for cattle.

Table 3. Assessment of the predictive performance of the abundance models in independent datasets by using Pearson correlations ($p<0.001$ in all cases).

Abundance models		
Model	East-west UAS track data (30%)	North-south UAS track data
	Pearson's r, n=501 grids	*Pearson's r, n=1013 grids*
Red deer	0.32	0.26
Fallow deer	0.36	0.23
Cattle	0.19	0.34

doi:10.1371/journal.pone.0115608.t003

https://doi.org/10.1371/journal.pone.0115608.t003

Spatial risk factors for TB infection

Infection was detected in 55.69% (SE=4.12%; n=570) of wild boar, 35.79% (SE=6.86%; n=190) of red deer and 17.99% (SE=5.49%; n=189) of fallow deer. From 2006 to 2013, official skin testing of the 1,139 cattle in the DNP study area revealed a maintained mean incidence of 9.04% (SE=4.91%) TB reactors.

Variables in the best-fitting Bayesian model included the following predictors: species sampled, WDN, predicted abundance of red deer and cattle (Table 4; the distribution probability for each parameter.). The Bayesian inference revealed that the increasing of TB risk was significantly associated with the high predicted abundances for red deer (estimated as 0.21; SD=0.05) and cattle (estimated as 0.12; SD=0.07). The infection risk was higher in wild boar than red deer, but higher in red deer than in fallow deer. In general, areas with low values of WDN were mostly related to wild ungulates infections.

Table 4. Results of the best-fitting (i.e. the lowest DIC value) Bayesian multivariable logistic regression model used to determine the most relevant factors explaining species positivity as regards tuberculosis.

Variable		Coefficient	SD	95% CI	
				2.5%	97.5%
Intercept		−8.17	0.19	−8.57	−7.79
Species	Red deer	0.71*	0.21	0.30	1.13
	Wild boar	1.31*	0.18	0.95	1.68
WDN		−1.50*	0.30	−2.11	−0.89
PRD		0.21*	0.05	0.09	0.31
PCT		0.12*	0.07	−0.04	0.26

Explanatory categorical variable was the TB-tested species: fallow deer, red deer and wild boar. Quantitative variables were water point density (WDN) and predicted species abundance (PRD in red deer and PCT in cattle) obtained from the models summarised in Table 2 and Fig. 2. Coefficients, standard deviations (SD) and 95% credible intervals (CI) are shown. Coefficients of species categories are relative to the fallow deer. Significant variables are marked with (*).

doi:10.1371/journal.pone.0115608.t004

https://doi.org/10.1371/journal.pone.0115608.t004

DISCUSSION

This research provides a multidisciplinary approach to study complex ecological and epidemiological systems, which may benefit from innovative technologies previously implemented in other fields, such as UAS. Although other study designs are possible, this research could be considered as a case study where UAS technology became a valuable tool to address host community abundance at finer spatial scale. Thus, UAS is a promising alternative in studies that require the estimation of spatial patterns of different species in large areas –like epidemiological ones- where traditional methods (e.g. line transects, drive counts, etc.) are not feasible due to logistical reasons. Our study identified points of high host abundance, which were associated with a high risk of TB infection, probably because these areas could act as important sources of TB and/or favour effective intra- and inter-specific contacts. Results allowed to identify the cattle/wildlife interface in which control strategies for TB – and other diseases that are modulated by similar risk factors – should be prioritised, which is particularly crucial for the development of disease control policies in the context of shared diseases.

On the Methodological Approach

Remote observation technologies based on low cost UAS have the potential to be a reliable and efficient alternative to study the distribution of ungulate abundance in open landscapes where multiple wild species can be differentiated by high quality imagery. Therefore, UAS constituted an excellent tool for wildlife monitoring in general and spatial epidemiology in particular. The valuable advantages of UAS respect to other methodologies are its low impact on wildlife behaviour and its capacity to provide high spatio-temporal accurate information on the surveyed populations at open landscapes. For instance, high-resolution images obtained from UAS allowed us to record, for each species, the local number of individuals and to avoid the animals' reactive responses to observer presence [16]. In DNP, a relatively open study area, we successfully identified the species in 99.1% of the animals detected. This result contrasts with previous studies using UAS pictures with rapid naked-eye image analysis, that were not able to distinguish, at species level, herbivores smaller than elephants, also in open habitats [16]. Differences between studies could be related with a combination of methodological factors such as: the height of the UAS (100–750 m [16] versus 100 m in this study) and the cruise speed (an average of 80 km/h [16] versus 40 km/h in this study).

The high precision obtained with UAS and the cost-effectiveness of the method suppose a notable advantage of this system when monitoring wildlife distribution in large territories, because the effort required to obtain accurate information for large areas – as demanded by epidemiologists – is unaffordable and/or unapproachable with the use of traditional methodologies (e.g. [11]). Other cost-effective indirect indicators of species presence and/or abundance, such as droppings (e.g. [48]) and browsing indices (e.g. [49]), may not be able to distinguish between species, and therefore are not useful to study multihost systems. However, accuracy and variation in animal detection and identification from UAS images may depend, among other things, on habitat visual permeability and the species' behavioural traits (e.g. [50]). On the one hand, habitat detectability can be taken into account to estimate corrected abundances in closed habitats, as it was done in this study. On the other, daytime elusive cryptic species such as wild boar [37] are under-represented in the surveys (compared with independent data, Natural Processes Monitoring Team at EBD, http://www-rbd.ebd.csic.es/Seguimiento/mediobiologico.htm). No information with which to model wild boar abundance was consequently obtained in this study. Future improvements to the UAS must address this by for instance incorporating systems with thermal sensors that are able to detect animals at night [17], [51].

Spatial Patterns of Ungulate Abundance and TB Infection

We have modelled the spatial distribution of wild ungulates and cattle on a very fine-scale throughout DNP. In general, all the epidemiologically-connected species selected the ecotone. This habitat offers key resources, namely food and shelter (see also [31]), which are more relevant during the summer when drought severely reduces the availability of resources [30]. We found differences in the spatial patterns predicted for each species among the MA for all the species studied. In the case of wild ungulates, this probably resembles differences in the species' habitat selection, as they have different ecological requirements on a local scale (e.g. [52]). In particular, fallow deer had a higher predicted abundance than red deer and cattle in the southernmost MA. Stone pine *(Pinus pinea)* reforestations perform especially well in the southernmost MA of DNP, and favour good pasture, which attracts fallow deer, which is probably the European deer species that is most prone to grazing (e.g. [22]). Differences among MA were quite significant for cattle as a consequence of the DNP cattle farming plan which compulsorily establishes the maximum cattle stocks in each MA.

The host spatial patterns were recorded in summer 2011, when the maximum aggregation of hosts is expected. However, TB data in sampled animals were collected during 2006–2012. Therefore, to study the relationship between host abundance and disease, we assumed that the spatial patterns did not change from year to year or seasons. Our design is realistic due to TB is a chronic disease and seasonal changes are not marked. Published observations on ungulate distribution confirm these seasonal patterns are consistent every year [31]. In addition, the sampling years of the sampled animals were included in the risk model, which controlled the effect of the changes in the spatial pattern. Nonetheless future studies should address how yearly changes (or due to the precipitation regime and water distribution in summer) in spatial distribution affect the TB epidemiology and its spatial pattern.

By joining abundance, environmental and disease data in a unique Bayesian modelling approach we have shown that spatial variation in the TB risk throughout DNP could relate to: (i) TB-tested species, (iii) environmental features, and/or (iii) predicted abundance of ungulates from UAS data. TB risk was higher in wild boar than red and fallow deer which is consistent with the infection pattern founded in this shared disease at South-central Spain, as has previously been demonstrated by both field and molecular epidemiology [23], [27]. The Bayesian inference revealed that the high TB risk for wild ungulates was negatively associated to water point densities. As the density of water points increases, lower levels of aggregation are expected at these points, and the risk for (direct and/or indirect) disease transmission should subsequently decrease [8], [53]. Water points and irrigated and cultivated fields have been identified as high-risk areas for disease transmission between feral pigs (*Sus scrofa*) and cattle in Mediterranean areas from USA [5]. These complex epidemiological scenarios have also been described in dry areas in Africa, in which cattle share water points and diseases with wildlife [54], [55]. In South Spain, a recent study has evidenced that wildlife-livestock interactions occur much more often at water resources than would be expected to occur by chance, and has argued that water points should be considered as potential hotspots for TB transmission between wildlife and cattle on extensive farms [53]. As a step forward, an experimental study in South Spain evidenced that effective segregation strategies of wildlife and cattle at water points under dry Mediterranean conditions have the potential to reduce inter-specific contacts and TB transmission at the wildlife/cattle interface [29]. Further research should therefore focus on the environmental persistence of *M. bovis* at watering sites in ecosystems with marked dry seasons, particularly where water becomes limited and leads to high animal aggregations.

TB risk was also positively associated with the high-predicted abundance of ungulates and consequently served as an actual way to evaluate the explanatory capacity of the predictions obtained by the UAS approach. The effect of local host abundance on the inter-specific transmission is probably environmentally mediated (indirect transmission) in Mediterranean areas [53], whereas direct interactions, especially at intra social group level, may be more relevant as regards determining direct rates for intra-species transmission [27]. Blanchong et al. [56] evidenced that direct effective contacts within white-tailed deer *Odocoileus virginianus* family groups were a significant mechanism for disease transmission. Apart from water points, other scattered environmental resources have the potential for disease transmission, such as grasslands, especially at the ecotone, where ungulates aggregate in order to forage during the summer. Future studies, which may benefit from the continued use of UAS for wildlife monitoring, should take into account the seasonality of resources in a highly variable ecosystem like DNP in order to assess both intra and inter-annual differences in the species habitat selection and thus in the risk factors that drive disease transmission

REFERENCES

1. Osnas EE, Heisey DM, Rolley RE, Samuel MD (2009) Spatial and temporal patterns of chronic wasting disease: fine-scale mapping of a wildlife epidemic in Wisconsin. Ecol Applic 19:1311–1322.
2. De Garine-Wichatitsky M, Miguel E, Mukamuri B, Garine-Wichatitsky E, Wencelius J, et al. (2013) Coexisting with wildlife in transfrontier conservation areas in Zimbabwe: Cattle owners' awareness of disease risks and perceptions of the role played by wildlife. Comp Immunol Microbiol Infect Dis 36:321–332.
3. Grenfell BT, Dobson AP (1995) Ecology of infectious diseases in natural populations. New York, USA: Cambridge University Press.
4. Dobson A (2004) Population dynamics of pathogens with multiple host species. Am Nat 164:64–78.
5. Cooper SM, Scott HM, de la Garza GR, Deck AL, Cathey JC (2010) Distribution and interspecies contact of feral swine and cattle on rangeland in south Texas: implications for disease transmission. J Wildl Dis 46:152–164.
6. Soberón J (2007) Grinnellian and Eltonian niches and geographic distributions of species. Ecol Lett 10:1115–1123.
7. Tuyttens FAM, Delahay RJ, Macdonald DW, Cheeseman CL, Long B, et al. (2000) Spatial perturbation caused by a badger (*Meles meles*) culling operation: implications for the function of territoriality and the control of bovine tuberculosis (*Mycobacterium bovis*). J Anim Ecol 69:815–828.
8. Vicente J, Höfle U, Garrido JM, Acevedo P, Juste R, et al. (2007) Risk factors associated with the prevalence of tuberculosis-like lesions in fenced wild boar and red deer in south central Spain. Vet Res 38:451–464.
9. Palmer MV, Thacker TC, Waters W, Gortázar C, Corner LA (2012) *Mycobacterium bovis*: a model pathogen at the interface of livestock, wildlife, and humans. Vet Med Int 2012:e236205.
10. Meliker JR, Sloan CD (2011) Spatio-temporal epidemiology: principles and opportunities. Spat Spatiotemporal Epidemiol 2:1–9.
11. Etherington TR, Ward AI, Smith GC, Pietravalle S, Wilson GJ (2009) Using the Mahalanobis distance statistic with unplanned presence-only survey data for biogeographical models of species distribution and abundance: a case study of badger setts. J Biogeogr 36:845–853.

12. Acevedo P, Quirós-Fernández F, Casal J, Vicente J (2014) Spatial distribution of wild boar population abundance: Basic information for spatial epidemiology and wildlife management. Ecol Indic 36:594–600.
13. Watts AC, Bowman WS, Abd-Elrahman AH, Mohamed A, Wilkinson BE, et al. (2008) Unmanned Aircraft Systems (UASs) for ecological research and natural-resource monitoring (Florida). Ecol Res 26:13–14.
14. Jones GP, Pearlstine LG, Percival HF (2006) An assessment of small unmanned aerial vehicles for wildlife research. Wildl Soc Bull 34:750–758.
15. Getzin S, Wiegand K, Schöning I (2012) Assessing biodiversity in forests using very high-resolution images and unmanned aerial vehicles. Methods Ecol Evol 3:397–404.
16. Vermeulen C, Lejeune P, Lisein J, Sawadogo P, Bouché P (2013) Unmanned Aerial Survey of Elephants. PLoS One 8:e54700.
17. Mulero-Pázmány M, Stolper R, van Essen LD, Negro JJ, Sassen T (2014) Remotely Piloted Aircraft Systems as a Rhinoceros Anti-Poaching Tool in Africa. PLoS One 9:e83873.
18. Rodríguez A, Negro JJ, Mulero M, Rodríguez C, Hernández-Pliego, et al. (2012) The Eye in the Sky: Combined Use of Unmanned Aerial Systems and GPS Data Loggers for Ecological Research and Conservation of Small Birds. PLoS One 7:e50336.
19. Ostfeld RS, Glass GE, Keesing F (2005) Spatial epidemiology: an emerging (or re-emerging) discipline. Trends Ecol Evol 20:328–336.
20. Frölich K, Thiede S, Kozikowski T, Jakob W (2002) A review of mutual transmission of important infectious diseases between livestock and wildlife in Europe. Ann N Y Acad Sci 969:4–13.
21. Gortázar C, Ferroglio E, Höfle U, Frölich K, Vicente J (2007) Diseases shared between wildlife and livestock: a European perspective. Eur J Wildl Res 53:241–256.
22. Apollonio M, Andersen R, Putman R (2010) European ungulates and their management in the 21st century. New York: Cambridge University Press. 618 p.
23. Gortázar C, Torres MJ, Vicente J, Acevedo P, Reglero M, et al. (2008) Bovine tuberculosis in Doñana Biosphere Reserve: the role of wild ungulates as disease reservoirs in the last Iberian lynx strongholds. PLoS One 3:e2776.

24. Martínez-López B, Barasona JA, Gortázar C, Rodríguez-Prieto V, Sánchez-Vizcaíno JM, et al. (2013) Farm-level risk factors for the occurrence, new infection or persistence of tuberculosis in cattle herds from South-Central Spain. Prev Vet Med, in press.
25. Phillips CJC, Foster CRW, Morris PA, Teverson R (2003) The transmission of *Mycobacterium bovis* infection to cattle. Res Vet Sci 74:1–15.
26. Rodríguez-Prieto V, Martínez-López B, Barasona JA, Acevedo P, Romero B, et al. (2012) A Bayesian approach to study the risk variables for tuberculosis occurrence in domestic and wild ungulates in South Central Spain. BMC Vet Res 8:148.
27. Gortazar C, Torres MJ, Acevedo P, Aznar J, Negro JJ, et al. (2011) Fine-tuning the space, time, and host distribution of mycobacteria in wildlife. BMC Microbiol 11:27.
28. Boadella M, Vicente J, Ruiz Fons F, de la Fuente J, Gortázar C (2012) Effects of culling Eurasian wild boar on the prevalence of *Mycobacterium bovis* and Aujeszky's disease virus. Prev Vet Med 107:214–221.
29. Barasona JA, VerCauteren KC, Saklou N, Gortazar C, Vicente J (2013) Effectiveness of cattle operated bump gates and exclusion fences in preventing ungulate multi-host sanitary interaction. Prev Vet Med 111:42–50.
30. Bugalho MN, Milne JA (2003) The composition of the diet of red deer (*Cervus elaphus*) in a Mediterranean environment: a case of summer nutritional constraint? For Ecol Manage 181:23–29.
31. Braza F, Alvarez F (1987) Habitat use by red deer and fallow deer in Doñana National Park. Miscellània Zoològica 11:363–367.
32. Carranza J, Hidalgo de Trucios SJ, Medina R, Valencia J, Delgado J (1991) Space use by red deer in a Mediterranean ecosystem as determined by radio-tracking. Appl Anim Behav Sci 30:363–371.
33. Carranza J, Alvarez F, Redondo T (1990) Territoriality as a mating strategy in red deer. Anim Behav 40:79–88.
34. Vicente J, Barasona JA, Acevedo P, Ruiz-Fons JF, Boadella M, et al. (2013) Temporal Trend of Tuberculosis in Wild Ungulates from Mediterranean Spain. Transbound Emerg Dis 60:92–103.
35. REDIAM (2013) Red de información ambiental de Andalucia. Consejería de Medio Ambiente y Ordenación del Territorio. Available: http://www. juntadeandalucia. es/medioambiente/site/rediam.

36. Zuur AF, Ieno EN, Elphick CS (2010) A protocol for data exploration to avoid common statistical problems. Methods Ecol Evol 1:3–14.
37. Russo L, Massei G, Genov PV (1997) Daily home range and activity of wild boar in a Mediterranean area free from hunting. Ethol Ecol Evol 9:287–294.
38. QGIS Development Team (2012) Quantum GIS Geographic Information System. Open Source Geospatial Foundation Project. Available: http://qgis. osgeo. org.
39. Cameron AC, Trivedi P (2013) Regression analysis of count data (Vol. 53). Cambridge University Press.
40. Akaike H (1974) A new look at the statistical model identification. IEEE Transactions on Automatic Control 19:716–723.
41. R Development Core Team (2013) R: A language and environment for statistical computing. R Foundation for Statistical Computing, Vienna, Austria. Available: http://www. R-project. org.
42. Vicente J, Höfle U, Garrido JM, Fernández-De-Mera IG, Juste R, et al. (2006) Wild boar and red deer display high prevalences of tuberculosis-like lesions in Spain. Vet Res 37:107–119.
43. Besag J, York J, Mollié A (1991) Bayesian image restoration, with two applications in spatial statistics. Ann Inst Stat Math 43(1):1–20.
44. Brezger A, Lang S (2006) Generalized structured additive regression based on Bayesian P-splines. Comput Stat Data Anal 50(4):967–991.
45. Schrödle B, Held L (2011) Spatio-temporal disease mapping using INLA. Environmetrics 22:725–34.
46. Bisanzio D, Giacobini M, Bertolotti L, Mosca A, Balbo L, et al. (2011) Spatio-temporal patterns of distribution of West Nile virus vectors in eastern Piedmont region, Italy. Parasit Vectors 4:230.
47. Held L, Schrödle B, Rue H (2010) Posterior and cross-validatory predictive checks: a comparison of MCMC and INLA. In: Kneib T, Tutz G, editors. Statistical Modelling and Regression Structures. 91–110.
48. Acevedo P, Ferreres J, Jaroso R, Durán M, Escudero MA, et al. (2010) Estimating roe deer abundance from pellet group counts in Spain: An assessment of methods suitable for Mediterranean woodlands. Ecol Indic 10:1226–1230.
49. Morellet N, Gaillard JM, Hewison AJ, Ballon P, Boscardin Y, et al. (2007) Indicators of ecological change: new tools for managing populations of large herbivores. J Appl Ecol 44:634–643.

50. Anderson K, Gaston KJ (2013) Lightweight unmanned aerial vehicles will revolutionize spatial ecology. Front Ecol Environ 11:138–146.
51. Focardi S, de Marinis AM, Rizzotto M, Pucci A (2001) Comparative evaluation of thermal infrared imaging and spotlighting to survey wildlife. Wildl Soc Bull 29:133–139.
52. Acevedo P, Farfán MA, Márquez AL, Delibes-Mateos M, Real R, et al. (2011) Past, present and future of wild ungulates in relation to changes in land use. Landsc Ecol 26:19–31.
53. Kukielka E, Barasona JA, Cowie CE, Drewe JA, Gortazar C, et al. (2013) Spatial and temporal interactions between livestock and wildlife in South Central Spain assessed by camera traps. Prev Vet Med 112:213–221.
54. Renwick A, White P, Bengis R (2007) Bovine tuberculosis in southern African wildlife: a multi-species host-pathogen system. Epidemiol Infect 135:529–540.
55. Munyeme M, Muma JB, Skjerve E, Nambota AM, Phiri IGK, et al. (2008) Risk factors associated with bovine tuberculosis in traditional cattle of the livestock/wildlife interface areas in the Kafue basin of Zambia. Prev Vet Med 85:317–328.
56. Blanchong JA, Scribner KT, Kravchenko AN, Winterstein SR (2007) TB-infected deer are more closely related than non-infected deer. Biol lett 3:104–106.

SECTION 2:
VIRUS

CHAPTER 6

The Contribution of Molecular Epidemiology to the Understanding and Control of Viral Diseases of Salmonid Aquaculture

Michael Snow

Marine Scotland Science, 375 Victoria Road, Aberdeen, AB11 9DB, Scotland, UK

ABSTRACT

Molecular epidemiology is a science which utilizes molecular biology to define the distribution of disease in a population (descriptive epidemiology) and relies heavily on integration of traditional (or analytical) epidemiological approaches to identify the etiological determinants of this distribution. The study of viral pathogens of aquaculture has provided many exciting

Citation: Snow, M. The contribution of molecular epidemiology to the understanding and control of viral diseases of salmonid aquaculture. Vet Res 42, 56 (2011). https://doi.org/10.1186/1297-9716-42-56

opportunities to apply such tools. This review considers the extent to which molecular epidemiological studies have contributed to better understanding and control of disease in aquaculture, drawing on examples of viral diseases of salmonid fish of commercial significance including viral haemorrhagic septicaemia virus (VHSV), salmonid alphavirus (SAV) and infectious salmon anaemia virus (ISAV). Significant outcomes of molecular epidemiological studies include:

- Improved taxonomic classification of viruses
- A better understanding of the natural distribution of viruses
- An improved understanding of the origins of viral pathogens in aquaculture
- An improved understanding of the risks of translocation of pathogens outwith their natural host range
- An increased ability to trace the source of new disease outbreaks
- Development of a basis for ensuring development of appropriate diagnostic tools
- An ability to classify isolates and thus target future research aimed at better understanding biological function

While molecular epidemiological studies have no doubt already made a significant contribution in these areas, the advent of new technologies such as pyrosequencing heralds a quantum leap in the ability to generate descriptive molecular sequence data. The ability of molecular epidemiology to fulfil its potential to translate complex disease pathways into relevant fish health policy is thus unlikely to be limited by the generation of descriptive molecular markers. More likely, full realisation of the potential to better explain viral transmission pathways will be dependent on the ability to assimilate and analyse knowledge from a range of more traditional information sources. The development of methods to systematically record and share such epidemiologically important information thus represents a major challenge for fish health professionals in making the best future use of molecular data in supporting fish health policy and disease control.

Keywords:- Rainbow Trout, Molecular Epidemiology, Molecular Epidemiological Study, Infectious Salmon Anaemia Virus, Venezuelan Equine Encephalitis Virus

INTRODUCTION

What is molecular epidemiology?

Epidemiology, or the study of factors affecting the health of populations is as old as science itself, with Hippocrates (460-377 BC) being the first to examine the relationship between disease occurrence and environmental issues. More recent technological developments have facilitated the identification and exploitation of molecular biomarkers, whose use alongside traditional epidemiological approaches has led to a better understanding of the underlying mechanisms of disease transmission in populations. This young science of "molecular epidemiology" has emerged since the 1970s when the term was first coined in relation to the study of influenza virus [1]. Molecular epidemiology utilises molecular biology to define the distribution of disease in a population (descriptive epidemiology) but still relies heavily on integration of traditional (or analytical) epidemiological approaches to identify the etiological determinants of such relationships (Figure 1).

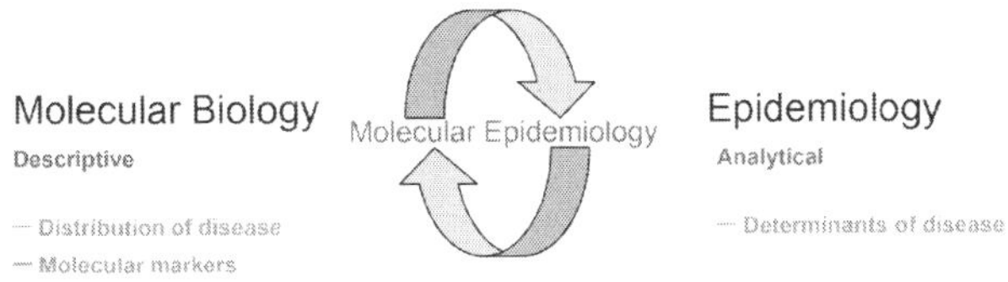

Figure 1. Definition of molecular epidemiology, spanning the disciplines of descriptive and analytical epidemiology.

Descriptive molecular epidemiology most often involves an attempt at establishing the evolutionary history of a given viral species called its phylogeny. A phylogenetic tree is a graphical summary of this inferred evolutionary relationship between isolates from which the pattern and in some cases timing of events that occurred as viruses diversified can be hypothesised. Since all life is related by common ancestry, viruses (or other organisms) in current circulation display genetic diversity which reflects their evolutionary history due to the accumulation and inheritance of mutations when genetic material is copied. In the absence of historical information or isolates, which are often unavailable, the evolutionary history of viruses in current circulation can be hypothesised based on the sampling of current genetic markers and by working backwards to infer the most likely series of events which best explain the observed relationships. Fundamental to reconstructing such an evolutionary history is the comparison of homologous characters (often nucleotide or amino acid sequence positions); i.e. those which descend from a common ancestor. In practice, this involves creating an alignment of sequences which represents a hypothesis of positional homology and provides the basis for reconstructing evolutionary history based on a mathematical model of evolution. This review aims to focus on the contribution that the application of such descriptive tools has made to the practical understanding and control of viruses of salmonid aquaculture, rather than the techniques themselves, which have been extensively reviewed by others (e.g. [2]).

Analytical epidemiology is based on observations of disease trends and incidence in different populations that turn into testable hypotheses, and entails rigorous collection of data for all aspects of study [3]. As noted by Porta et al. [4], the lack of careful application of the science of epidemiology can limit the amount of useful information obtained from studies of molecular epidemiology. This remains a significant challenge in the field of fish virology where the application of descriptive molecular epidemiology is gathering momentum, yet data required for adequate interpretation is often lacking or at best collected haphazardly. Despite this, molecular epidemiology has made a significant contribution to the understanding and control of viral diseases of fish. This review aims to illustrate some of these contributions, whilst highlighting some future challenges faced in this field to ensure that molecular epidemiology fulfils its potential of converting complex viral transmission pathways into relevant fish virus control policies.

1.2 Setting the scene: fish viruses, aquaculture and molecular epidemiology

Fish, unlike many other domestically reared farmed animals, are most often reared in open systems (cages) where they are exposed to a wide diversity of naturally occurring environmental pathogens. Recent estimates suggest the presence of a staggering 10^8 viruses/mL in coastal seawater [5]. Although the majority of these viruses appear to be infective [6], most are pathogens of planktonic organisms and have a narrow range of potential host species [5]. Whilst the importance of the role of viruses as agents of microbial mortality and regulators of global carbon and nutrient cycling in the oceans is being realised [7], some viruses are also, clearly, progenitors of disease in higher organisms such as fish. Given the extent of the viral genetic diversity present in both marine and freshwaters [8], it is perhaps not surprising that viruses of fish are represented in most of the major established viral genera (for review see [9]). Indeed, culturing fish in an open seawater (or freshwater) system rich in viral diversity clearly presents opportunity for the introduction of many potential pathogens into aquaculture systems. The good news is that the vast majority of these potential progenitors of disease do not cause a problem in healthy wild and farmed fish populations. In natural systems, where viruses and their piscine hosts have apparently existed for millenia, this would appear to be a consequence of co-evolution, where killing of the host might otherwise have led to viral extinction. In support of this, examples exist of naturally occurring marine reservoirs of fish viruses recovered from asymptomatic hosts with no demonstrated deleterious effect on fish populations [10–12]. However virulence can still be an adaptive response of pathogens [13], and disease can be a significant factor in regulating natural fish populations [14, 15].

The advent of intensive fish husbandry and associated international movements of fish has fundamentally altered the natural equilibrium by exposing animals to new environments and their viruses to new host species [16]. As a consequence, increased rates of evolution of fish viruses are apparent in aquaculture as compared to those in natural systems [17]. In intensive culture systems, higher rates of evolution result from increased selective pressures from factors associated with modern aquaculture including the use of novel host species, continuous addition of susceptible fish, high rearing densities, differing temperatures and the presence of immunised fish [18]. The effects of such selection pressures are further amplified by virtue of the fact that many significant viral pathogens of fish have an RNA genome. Viruses with RNA genomes are characterised by the presence of an error-prone RNA polymerase and high mutation rates. This results in the generation of "quasispecies" populations comprised of a spectrum of mutant genomes which are continually generated [19]. This constitutes the raw material for evolution, with such mutants potentially having a greater biological fitness, thus providing a mechanism for adaptation to changing conditions.

The accumulation of relatively high levels of genetic variation in viruses associated with aquaculture provides excellent and exciting opportunities for understanding the relative contribution of molecular epidemiology to their understanding and control. In order to identify the extent to which this capacity has been exploited in this field, this review focuses on three of the most significant and well studied RNA pathogens of salmonid aquaculture, viral haemorrhagic septicaemia virus (VHSV), infectious salmon anaemia virus (ISAV) and salmonid alphavirus (SAV). In this area the application of molecular epidemiology has provided important new knowledge of fundamental importance to disease management and control in the following key areas:

- Viral taxonomy and classification
- Understanding origins of diseases of aquaculture
- Viral phylogeography and risks associated with viral translocation
- Outbreak tracing and disease control
- Genetic diversity and viral surveillance

CASE STUDIES HIGHLIGHTING THE CONTRIBUTION OF MOLECULAR EPIDEMIOLOGY TO THE UNDERSTANDING AND CONTROL OF VIRAL DISEASE OF SALMONID AQUACULTURE

Viral Haemorrhagic Septicaemia Virus

Overview

Viral haemorrhagic septicaemia (VHS) was, until the late 1980s thought to be a disease exclusive to the freshwater rainbow trout industry of Continental Europe, which is characterised by extensive degeneration and necrosis of the internal tissues [20]. The disease can result in mortalities of up to 90% [21]. Multiple occurrences of disease were later associated with European [22, 23] and Asian [24] marine aquaculture and a widespread global marine distribution of VHS virus in an ever increasing variety of largely aysmptomatic marine host species is now acknowledged. A growing body of evidence suggests the original import of VHSV into freshwater systems from marine sources, with the most recent example of this being the apparent introduction of VHSV into the Great Lakes system of North America where extensive mortality in wild fish has resulted [25, 26].

Molecular epidemiology facilitates the taxonomic classification of VHSV and provides a basis for understanding the relationship between isolates

VHSV is one of several important fish viral species that were originally identified as belonging to the family *Rhabdoviridae* based on the distinctive morphology of viral particles in electron micrographs. Until relatively recently, insufficient data existed to assign any of the fish rhabdoviruses to novel genera with tentative groupings of "unassigned" and "vesiculovirus-like" viruses being proposed based on protein electrophoretic data and antigenic relatedness [27]. Sequence determination and phylogenetic analysis of the glycoprotein gene subsequently indicated that the lyssavirus-like viruses, including VHSV, infectious haematopoietic necrosis virus (IHNV) and hirame rhabdovirus (HIRRV) warranted classification within a new genus [28]. This distinction has since been supported by the comparison of genetic variation at the G-L gene junction which represents a major point of distinction between the different established genera of rhabdoviruses [29]. Indeed, the presence of an additional and characteristic NV gene at this junction in the fish viruses VHSV, IHNV and HIRRV supported their designation within a new genus (genus *novirhabdovirus*) [30]. Taxonomic classification of fish viruses is important in identifying their potential biological properties based on comparison with their often better studied counterparts (e.g. rabies virus). Such studies have also represented the first steps towards a deeper understanding of the classification of fish rhabdoviruses such as VHSV which have led to significant and tangible outcomes of practical relevance to disease control.

To date, four main genetic groups of VHSV are recognised worldwide [17, 31], as depicted in Figure 2. In the case of VHSV, these genotypes reflect a largely geographic rather than a species-specific distribution and in the absence of recombination in negative strand viruses have been shown to be robust and independent of the preferred gene of analysis [32]. Genotype I includes a wide range of viruses originating from freshwater rainbow trout farms in continental Europe [17, 33], and a large number of isolates originating from marine species in the Baltic Sea/Skagerrak/Kattegat/English channel [17, 31, 34]. These isolates fall within 2 of 5 proposed subgroups within Genotype I (subgroups 1a and 1b respectively) [17]. Marine isolates from subgroup 1b differ markedly from other Genotype I isolates since it has been experimentally shown that they are of low pathogenicity to rainbow trout [35]. The remaining subgroups identified within Genotype I include isolates recovered from Danish freshwater rainbow trout farms in the 1980s (subgroup Ic), marine-farmed rainbow trout in Scandinavia (subgroup Id) [17, 36] and isolates recovered from turbot and rainbow trout in Georgia and the Black Sea region (subgroup Ie) [37]. Genotype II includes a limited number of VHSV isolates recovered from the Baltic sea [31], while Genotype

III includes isolates originating from outbreaks of VHS in turbot farms in the British Isles along with a number of isolates recovered from a variety of marine species caught in Scottish and North Atlantic waters [31].

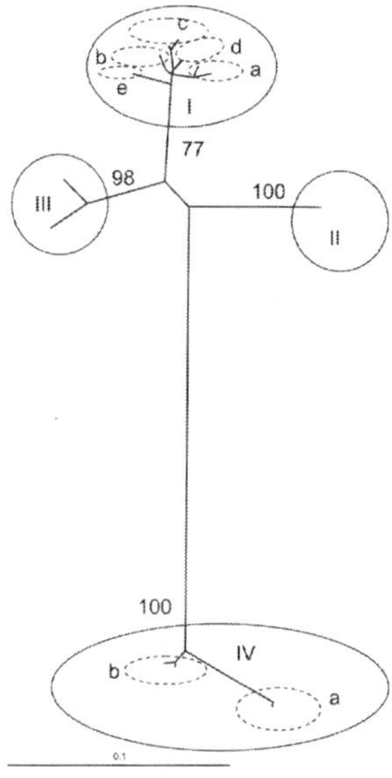

Figure 2. Overview of the genetic relationships between the main recognized genotypes of VHSV based on a dataset comprised of partial G gene data from 2 representatives of each of the established major genetic subgroups. Publicly available sequences derived from up to 2 isolates from each genotype recognized to date were imported into Bioedit version 7.0.5.3 [76] and a multiple alignment performed using Clustal X [77]. The final alignment consisted of 17 unique sequences spanning a partial region of the VHSV G Gene (649nt). The phylogenetic relationship among VHSV isolates was inferred using a maximum likelihood based approach implemented within PAUP* (version 4.0; [78]) and using the PaupUP interface v1.0.3.1 [79]. The jModeltest 0.1.1 program [80]) was used to identify the model that best fits the sequence data from 56 models using the Akaike Information Criterion (AIC; [81]). The optimal unrooted maximum likelihood tree was identified using a heuristic search implemented in PAUP* and evaluated using 100 bootstrap iterations [82]. Significant bootstrap values for the major clades were transferred to the unrooted tree derived from the original data.

The final genetic group of VHSV isolates identified to date comprises a range of isolates recovered from wild marine fish from the Pacific Northwest [24, 38, 39] and eastern coastal areas of Canada and the USA, isolates from Japanese flounder [24] and isolates originating from a recent series of epidemics in the Great Lakes of North America [25, 26]. Establishing the genetic relationships between currently circulating isolates of VHSV through molecular epidemiological studies has led to new understanding in a number of areas of critical importance to disease control and the development of sustainable aquaculture.

Understanding genetic relationships provides an insight into the origins of VHS in aquaculture

VHS was traditionally thought to be a disease exclusive to freshwater rainbow trout farming, where it has historically been responsible for significant economic loss. The first and subsequent identification of marine isolates of VHSV virus of differing genetic type in Europe [40] and North America [41] questioned this dogma. The fact that these genotypes were indicated to have become separated a long time before fish farming was established on both continents [28], coupled to the fact that VHS is not a disease problem in rainbow trout in North America, indicated existence of long standing natural marine reservoirs of VHS virus. The existence of naturally occurring marine reservoirs of VHS virus is now widely recognised and a long list of species which may be infected with the virus has been established (an updated list of natural and experimental hosts for VHSV can be found via the International Database on Aquatic Animal Disease [42]). The demonstration that some of these naturally occurring marine VHS viruses (Genotype 1b; Figure 2) bear a close genetic relationship to those responsible for disease in freshwater rainbow trout farms (Genotype 1a; Figure 2) suggested that such isolates represent the most likely origins of introduction of VHS into rainbow trout aquaculture in Europe (for a review see Skall et al. [43]). Since the use of unpasteurised "trash" marine fish was widely used as a source of aquaculture feed in the early days of the industry, this may be a mechanism via which the pathogen was introduced. Whatever the mechanism, since naturally occurring marine genotype 1b viruses have been shown to be non-pathogenic for rainbow trout [35], it is clear that the virus must have adapted a new pathogenic life-strategy within rainbow trout aquaculture. This may be a consequence of the high selective pressures and continuous availability of susceptible hosts associated with aquaculture. Such conditions may favour development of an alternate "pathogenic" life-strategy to that in the natural environment where death of the host may represent an evolutionary dead-end. The introduction of VHSV from the marine environment into rainbow trout aquaculture appears to have occurred on more than one occasion. Indeed, the identification of Genotypes 1c, 1d and 1e [17] has led to the interpretation that host adaptation from marine environment/species to rainbow trout has occurred 3 or 4 times in freshwater farms in Denmark and more recently in fish reared at the coast of Finland and Sweden [17].

Until recently, only the adaptation of marine Genotype I strains of VHSV in rainbow trout had been shown to occur. Much like the Genotype Ib marine isolates, other naturally occurring genotypes of VHSV have been experimentally shown to be of low virulence to rainbow trout [35]. The recent demonstration that VHS disease in a marine rainbow trout farm in Norway was caused by a Genotype III virus [44], suggests that the potential for adaptation to this species is not an exclusive property of Genotype I viruses. Indeed, given the potential for continuous adaptation of RNA viruses, it is perhaps not surprising to find that other viruses are capable of dramatic shifts in host specificity given the appropriate opportunity. Such opportunity in an aquaculture context can include factors such as the use of continuous production systems without synchronous fallowing which can allow pathogens to persist for long periods resulting in increased potential for adaptation to occur.

While the above examples assume the input and subsequent adaptation of naturally non-pathogenic viruses in aquaculture systems, naturally occurring viruses can in some cases also cause disease without requirement for change. Following outbreaks of disease in marine farmed turbot in the British Isles caused by VHSV [22], experimental trials with a range of naturally occurring GIII isolates indicated that turbot were susceptible to these viruses [45]. The susceptibility of turbot to naturally occurring VHSV is thus a significant risk to the sustainable production of turbot within the natural range of such isolates. The susceptibility of turbot to naturally occurring VHSV is at odds with the conventional wisdom that pathogens evolve toward non-pathogenic lifestyles in natural systems. This may be a consequence of the lack of exposure to the pathogen in nature with commercial aquaculture resulting in the exposure of previously naïve species to new pathogens. Another example of a pathogen which can cause disease without adaptation is that of VHS disease epidemics which occur periodically in wild Pacific herring and which result from a naturally occuring and widespread reservoir of virus [15]. In this instance, evolution towards a virulent state appears to have been an adaptive response of VHSV in this species.

Understanding the range of naturally occurring pathogens highlights the risk associated with their translocation

Understanding the phylogenetic relationships between VHSV isolates highlights the existence of major viral genotypes confined largely to different geographical areas as discussed above. The physical separation of these areas has resulted in limited geneflow and independent evolution of the viruses circulating in these regions. It stands to reason that VHS viruses in these regions have not evolved independently but have rather co-evolved with their hosts to ensure continued survival. Since viruses rely on the availabilty of hosts to avoid extinction it is perhaps not surprising to find many examples where viruses within their natural range do not cause disease in their hosts. Anthropogenic factors, including those associated with international aquaculture, have in general lead to the increased potential for VHSV viruses to be translocated outwith their naturally occurring range. An extreme example of the potential consequence of such occurrence is that of the recent series of VHS epidemics in the Great Lakes region of North America. This series of epidemics, which spread rapidly through the region and resulted in dramatic fish kills, was shown to have been caused by a virus whose natural range appeared to be marine species inhabiting the eastern coastal areas of the USA or Canada [46]. The introduction of this Genotype 1b virus into the freshwater system of the Great Lakes is thought to have originated from one of several possible sources including ballast water, baitfish translocation or by anadromous or catadromous species that can enter the Great Lakes via the St. Lawrence river [46]. The Great Lakes system is one of the largest freshwater systems in the world and is noted for its rich biodiversity [47]. The fact that VHSV virus spread rapidly through the Great Lakes system causing large scale mortality in a wide range of susceptible species, has highlighted the potential risk of translocation of viruses and exposure to previously naïve species.

Identification of genetic relationships is used to trace the spread of pathogens in aquaculture

Molecular epidemiology is a powerful tool, which alongside more traditional epidemiological tools has potential for tracing the origins and spread of new disease outbreaks. Such an approach relies not only on the availability of large genetic datasets relating to phylogenetically informative and defined regions of the genome but also on availability of data relating to potential epidemiological contact. The latter is often more challenging to obtain. An example of the application and potential of such tools is the recent occurrence of VHS disease in rainbow trout in the British Isles, which until this point had a history of freedom from VHS. The disease was first identified and contained on a farm in North Yorkshire, England in May 2006 and initial investigations into the likely origin of introduction proved inconclusive. Subsequent molecular epidemiological analysis of the causative virus identified a Genotype Ia virus which was very similar to isolates from Denmark and Germany circulating between 2004 and 2006 and suggested likely introduction from this region [48]. However, the lack of identification of an exact matching sequence and lack of epidemiological contact information hindered identification of a more exact source.

A sound knowledge of molecular epidemiology supports strategies for surveillance and further knowledge

Understanding the natural sequence variation and divergence among isolates is of fundamental importance to ensuring adoption of an appropriate surveillance regime for VHSV and in interpretation of its results. Molecular detection methods such as real-time PCR are increasing being employed in this field due to benefits including sensitivity, specificity, high throughput, ease of interpretation and ability to include appropriate controls. Molecular methods rely on the specificity of primers and probes to ensure detection of their intended targets. Targetting such methods to the required well characterised and conserved regions of the genome thus requires a thorough knowledge of molecular variation of the species. Even with this knowledge, the risk that highly specific single-plex assays such as probe-based real-time PCR could fail to detect new emerging variants of pathogens should be recognised. Such methods as developed for VHSV have to date, however, proven to be robust in detecting all known variants of the virus in known circulation [49].

Since phenotypic properties of viruses may in some cases be consistent with genetic origin, selecting isolates representative of the different genotypes can be a sensible strategy for further research into establishing their biological properties. Previous work on VHSV has investigated the consistency of pathogenic properties of different genotypes for different species [45, 50]. Selection of representative strains for selection on this basis also ensures that the widest knowledge of the properties of the species as a whole can be obtained. Alternatively, other studies have specifically selected virus isolates of known genetic similarity that were expected to display different pathogenic properties based on their history of origin. In this way, very similar isolates with only 5 amino acid substitutions were identified, which displayed differing virulence for rainbow trout [51]. Such work provides an important basis for resolving those epitopes responsible for conferring the virulent phenotype, which in turn can lead to improved methods for disease control.

Salmonid Alphavirus (SAV)

Overview

Salmonid alphaviruses are responsible for salmon pancreas disease (SPD) and sleeping disease (SD) conditions, primarily of farmed Atlantic salmon and rainbow trout, respectively. Clinical signs associated with SPD include abnormal swimming behaviour and lack of appetite, while characteristic histopathological signs include severe degeneration of exocrine pancreas, cardiomyopathy and skeletal myopathy [52]. SD involves similar lesions but often manifests itself as fish lying on their side or "sleeping" as a consequence of extensive necrosis of skeletal red muscle [53]. Until recently, the presence of SAV has not been reported outside of salmonid aquaculture, where it is responsible for significant economic loss, principally in Western Europe.

Molecular epidemiology facilitates the taxonomic classification of SAV and provides a basis for understanding the relationship between isolates

Molecular analyses of salmonid alphaviruses have recently demonstrated that they represent a group of viruses within the genus *alphaviridae* in the family *Togaviridae*, and differ markedly from the previously established New World viruses of Venezuelan equine encephalitis virus (VEEV) and eastern equine encephalitis virus and the Old World viruses of Aura and Sindbis virus [54]. Salmon pancreas disease virus (SPDV) has been officially recognised as a species by the International Committee for the Taxonomy of Viruses (ICTV) which includes isolates causing both SD and SPD in rainbow trout and Atlantic salmon [55].

Classification in this way can give clues as to the potential biological properties and functions of salmonid alphaviruses based on the established properties of their better studied relatives. Interestingly, most alphavirus lifecycles involve an obligate, commonly arthropod intermediate host.

Such an intermediate host does not appear to be a requirement of salmonid alphavirus transmission since these viruses have been shown to efficiently transmit horizontally via direct water-borne transmission [56]. Whether arthropods (e.g. sealice) have a potential role on the epizootiology of salmonid alphaviruses remains an interesting research question.

The most comprehensive molecular epidemiological analysis of the genetic relationships within the salmonid alphaviruses was recently conducted by Fringuelli et al. [57]. This relationship is illustrated in Figure 3.

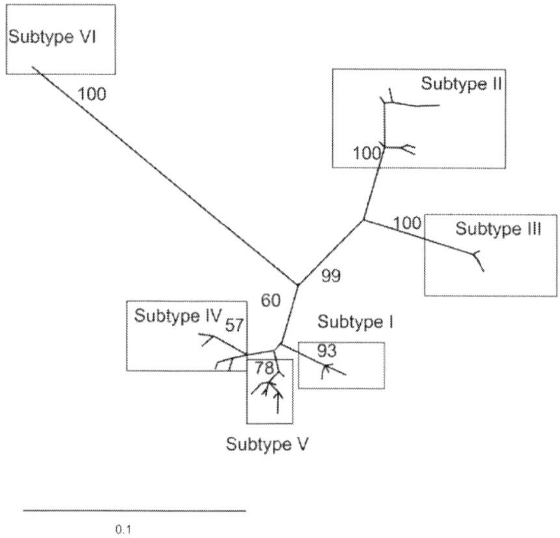

Figure 3. Overview of the genetic relationships between the main recognized genotypes of SAV based on a dataset of partial E2 sequence data (298nt). Publically available sequences derived from isolates from each subtype recognized to date were imported into Bioedit version 7.0.5.3 [76] and a multiple alignment performed using Clustal X [77]. Duplicate sequences were removed and the final alignment consisted of 36 unique sequences representing a total of 83 isolates and spanned a region of 298nt of the open reading frame of SAV E2 gene. The phylogenetic relationship among SAV isolates was inferred using the method described in the legend of Figure 2.

The existence of a total of six discrete and identifiable subtypes has been suggested based on analysis of partial E2 and nsP3 sequences. Subtypes I, IV and V are genetically very closely related and include isolates from Atlantic salmon farms in the British Isles responsible for salmonid pancreas disease (SPD). Subtype VI included a single isolate from Ireland. The disease in Norway has to date been exclusively associated with a different genetic group of viruses belonging to subtype III. Subtype II isolates include those which have caused sleeping disease (SD) in rainbow trout farms from Europe. While viruses responsible for SPD in Atlantic salmon and SD in rainbow trout generally fall within different genetic subgroups there are exceptions. This fact, coupled to their close genetic similarity supports the taxonomic designation of the salmonid alphaviruses as a single species.

Understanding genetic relationships provides an insight into the origins of SAV in aquaculture

The monophyletic nature of the freshwater rainbow trout subtype of SAV (Subtype II, Figure 3) has been interpreted to indicate a single introduction and subsequent dissemination of SAV within this industry. This, based on current knowledge might be expected to have occurred from a salmonid source. Recent information, however, derived from the screening of marine fish species by real-time PCR in areas remote from aquaculture, has indicated the presence of SAV RNA in marine species including common dab (*Limanda limanda*) and plaice (*Pleuronectes platessa*) [58]. The identification of subtype V SAV sequences associated with these detections suggests that, as in the case for VHSV, marine reservoirs of SAV might exist. Whilst the exact implications of this remain unclear, it is possible that the origins of SAV in freshwater rainbow trout may have been from a similar marine source or indeed, import via contact with infected Atlantic salmon. It is interesting to speculate as to whether in time a more extensive marine reservoir of SAV will become apparent, as has been shown for VHSV, and whether this might suggest independent import of SAV into Norwegian and UK aquaculture respectively.

Identification of genetic relationships is used to trace the spread of pathogens in aquaculture

Molecular epidemiology has been applied to the study of SAV and has provided evidence to support farm to farm transmission in both Scotland and Ireland where clusters of identical sequences from isolates from different farms located within the same bodies of water have been demonstrated [57]. Furthermore, the fact that sequences from one cluster of outbreaks in the West of Ireland was associated with a subtype of virus distinct to that prevalent in the region of origin of these fish supports horizontal rather than vertical transmission as the predominant source of infection. Horizontal transmission can occur directly through the water or via fomites, and SAV has been shown to have a mimimum half-life of 5.7 days in seawater at 10°C [59]. Having said this, understanding the epizootiology of SAV is complex for a number of reasons. Firstly, the disease is not notifiable in the UK, meaning than information on disease occurrence and characterisation of responsible agents is often lacking. Secondly, the virus appears to have a relatively low rate of evolution and identical viral sequences have been reported to be detected over extended periods of time within single populations, within farms and between different farms. This makes interpretation of epidemiological links difficult. A clearer explanation has, however, been identified for the import of sleeping disease into the UK, since evidence of French common ancestry of contemporary isolates circulating in both France and Scotland/England supports epidemiological evidence that import was associated with the processing of fish imported from France in 2002 [57, 60, 61].

A sound knowledge of molecular epidemiology supports strategies for surveillance and further knowledge

Knowledge of molecular epidemiology and genetic diversity of SAV has led to the development of sensitive and specific molecular diagnostic methods [62] capable of detecting the range of reported SAV subtypes. Further, these methods are capable of subtype discrimination and might prove useful in rapid screening of isolates to highlight translocation of new subtypes to new geographic areas. Genotypic differentiation has again been used to identify isolates for further detailed research, in particular to understand the clinical presentation of disease following infection with different genotypes [63]. Recent work developing reverse genetics approaches to study SAV [64] offers great potential in realising the biological significance of the genetic differences between strains. Such work has important implications for disease surveillance programmes in addition to contributing to a deeper understanding of the biology of the causative agent and thus how to apply practical disease mitigation control strategy.

Infectious Salmon Anaemia Virus (ISAV)

Overview

Infectious salmon anaemia (ISA) is a disease of farmed Atlantic salmon that has been responsible for extensive losses in all major regions where this species is farmed including, Norway, Canada, Scotland, the Faroe Islands and Chile. The disease is characterised by severe anaemia, ascites and haemorrhagic liver necrosis and congestion [65]. While continued disease problems occur in Norway, Canada and Chile, the disease has successfully been eradicated from both Scotland and the Faroes. In the latter case, however, this was only achieved at the expense of collapse of the entire industry.

Molecular epidemiology facilitates the taxonomic classification of ISAV and provides a basis for understanding the relationship between isolates

Molecular epidemiological study has made a significant contribution to our understanding of ISAV. Following identification and initial characterisation, the virus was recognised as an orthomyxovirus [66], and ultimately assigned to its own genus.

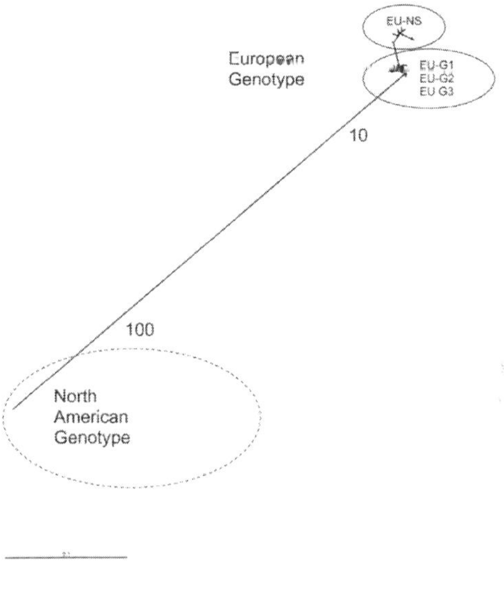

Figure 4. Overview of the genetic relationships between the two main identified genotypes of ISAV. Publically available ISAV sequences were imported into Bioedit version 7.0.5.3 [76]and a multiple alignment performed using Clustal X [77]. Identical sequences were identified and only a single representative of each sequence type retained in the dataset to reduce subsequent analytical bias. The final alignment consisted of 38 unique sequences spanning a region of 987 (positions 25-1012 with respect to the open reading frame of ISAV segment 6). The phylogenetic relationship among ISAV isolates was inferred using the method described in the legend of Figure 2.

Orthomyxoviruses such as influenza viruses are characterised by a high potential for adaptation via a repertoire of mechanisms for genetic change including mutation, reassortment and recombination. Indeed, these properties have all been demonstrated for ISAV within aquaculture [67]. Relationships among ISAV isolates have been explored most extensively on the basis of their surface protein coding genes, namely the haemagglutinin esterase (HE) and fusion (F) protein genes [68, 69]. Though some complicated subdivisions of ISAV subtypes have been proposed, these analyses have revealed the same general pattern of existence of single major European and North American genotypes of ISAV (Figure 4). Within the European genotype of ISAV, different subtypes have been proposed to classify isolates from Europe (EUG1, EUG2 and EUG3), and some isolates from North America (EU-NS) which appear to share a more recent common ancestor with European ISAV than with the true North American ISAV genotype.

Understanding genetic relationships provides an insight into the origins of ISAV in aquaculture

North American and European evolutionary lineages of ISAV appear to have been present and diverged long before the development of commercial Atlantic salmon aquaculture [70]. Interestingly, all pathogenic variants of ISAV associated with disease outbreaks are characterised by the occurrence of a sequence deletion within the so-called highly polymorphic region (HPR), with respect to a putative ancestral form of the HE gene [69]. This putative ancestral form is designated HPR0 and its presence has been detected in healthy farmed and wild fish in all the major regions of Atlantic salmon production worldwide including, Norway [12, 69], Scotland [71, 72], Canada [73], Chile [68] and the Faroe Islands (Dr Christiansen, personal communication). Molecular epidemiological evidence supports the suggestion that pathogenic ISAV which causes disease in fish farms likely originates from a reservoir of the ISAV-HPR0 variant since a wide variety of HPR0 sequences reflecting the broad diversity of virulent isolates have been identified and are distributed throughout the tree depicted in Figure 4. The mechanisms governing adaptation of the ISAV-HPR0 variant in aquaculture are, however, poorly understood. The fact that new cases of ISA which have emerged in countries previously free of the disease, however, does suggest that a lack of synchronous fallowing and thus provision of opportunity for pathogens to persist over long time periods may contribute to the risk of emergence of disease. In this instance a better understanding of the potential risk factors in emergence of ISA disease has contributed to practical fish health policy aimed at mitigating such risks. Such solutions include the creation of management areas, which amongst other measures aim to achieve periods of synchronous fallowing in order to limit opportunity for pathogens, including ISAV to reside within production systems.

Identification of genetic relationships is used to trace the spread of pathogens in aquaculture

Molecular epidemiogical study has been applied to understanding the appearance of ISA in Chile [68], which has caused a significant decline in production. Molecular evidence in this instance pointed strongly towards a Norwegian source, with vertical or trans-generational transmission being highlighted as a plausible explanation, based on the significant and historical import of fertilised eggs into Chile from this area. Such knowledge has raised concerns over the future risk of import of disease in association with ova and has lead to an application of biosecurity measures aimed at guarding against such import. This includes broodstock testing regimes based on molecular detection, accompanied by destruction of ova originating from positive-testing parents and the development of protocols for the disinfection of imported egg products. Whether true vertical transmission of ISAV (ie the internal transmission of virus within the gametes) occurs in ISAV remains somewhat controversial though it has been suggested [69]. It seems clear from evidence in Chile that transgenerational transmission is highly likely to have occurred. Whether this represents true vertical transmission or resulted from inadequate disinfection of eggs remains uncertain. In this case, the application of molecular epidemiological study has identified the probable origin of import of ISAV into Chile, thus highlighting a potential risk that can be mitigated based on the implementation and maintenance of appropriate biosecurity measures. Since horizontal transmission remains the main mechanism for dissemination of ISAV where it is present, the major challenge now faced, is to apply molecular epidemiological tools to understand and control the spread of ISA within Chile.

A sound knowledge of molecular epidemiology supports strategies for surveillance and further knowledge

Molecular detection methodologies have been developed to cover the broad range of ISAV viruses in global circulation [74], providing confidence that the pathogen can be accurately detected. Information on the existence of the ISAV-HPR0 variant has been invaluable in interpreting the results of molecular diagnostic tests which currently do not differentiate between this and other variants of the virus. The fact that the ISAV-HPR0 variant has been shown to be present, even in territories with freedom of ISA disease, suggests a clear distinction in risk associated with the presence of ISAV-HPR0 and other disease causing variants. Differentiation of viruses with different biological properties such as HPR0, also provides the basis for developing understanding of the genetic determinants of biological phenotype. Such knowledge is fundamental to managing pathogen-associated risk in open aquaculture systems.

FUTURE PERSPECTIVES AND CHALLENGES

The application of molecular epidemiology to the study of fish viruses has to date largely focussed on the use of descriptive techniques and interpretation of genetic relationships has often been hampered by a lack of associated analytical epidemiological information. The development of next generation sequencing technology promises a revolution in the ability to generate sequence data and thus information of potential epidemiological relevance. Descriptive molecular data alone is, however, inadequate in tracing pathogen spread, especially when variation is limited and evolution does not occur in a clock-like or regular fashion. The ability of molecular epidemiology to fulfil its potential to translate complex disease pathways into relevant fish health policy is thus unlikely to be limited by the generation of descriptive molecular markers. More likely, full realisation of the potential to better explain viral transmission pathways will be dependent on the ability to assimilate and analyse knowledge from a range of more traditional information sources. The development of methods to systematically record and share such epidemiologically important information thus represents a major challenge for fish health professionals in making the best future use of molecular data in supporting fish health policy and disease control.

Making best use of this generated data to better understand the molecular basis of virulence also remains an important area for future research. Significant progress has been made in developing reverse genetic approaches for fish viruses (for review see [75]) which offer great potential for better understanding viral virulence and thus the risk factors which may contribute to the emergence of new pathogens.

Despite the limitations in available knowledge, molecular epidemiology has led to an improved understanding of the origins and spread of viruses in aquaculture and wild fish, which in turn has lead to significant practical improvements in disease management strategies and policy. A common theme in the examples explored seems to be the prevalence of probably benign viruses in the environment, which undergo an adaptation in association with aquaculture, where they adopt a pathogenic lifestyle. Major challenges for sustainable aquaculture are to manage the risk associated with such occurrence by reducing the opportunity for pathogens to reside and thus potentially adapt within aquaculture systems, coupled to their rapid identification and containment following disease emergence. Both these factors can be facilitated through implementation of biosecurity measures and the use of synchonous fallowing strategies to break potential long term cycles of infection and ensure rapid and appropriate containment of disease in discrete management areas. The implementation of surveillance programs based on the best available molecular epidemiological information offers great potential to support such measures in further developing a healthy and sustainable aquaculture industry that is necessary to satisfy an increasing world demand for cultured fish products.

REFERENCES

1. Kilbourne ED: The molecular epidemiology of influenza. J Infect Dis. 1973, 127: 478-487. 10.1093/infdis/127.4.478.
2. Kumar S, Filipski A: Molecular phylogeny reconstruction. Encyclopedia of Life Sciences (ELS). 2008, Chichester: John Wiley & Sons, Ltd
3. Slatterty ML: The science and art of molecular epidemiology. J Epidemiol Community Health. 2002, 56: 728-729. 10.1136/jech.56.10.728.
4. Porta M, Malats N, Vioque J, Carrato A, Soler M, Ruiz L, Barbera V, Ayude D, Real FX: Incomplete overlapping of biological, clinical and environmental information in molecular epidemiological studies: a variety of causes and a cascade of consequences. J Epidemiol Community Health. 2002, 56: 734-738. 10.1136/jech.56.10.734.
5. Suttle CA: Viruses in the sea. Nature. 2005, 437: 356-361. 10.1038/nature04160.
6. Wilhelm SW, Weinbauer MG, Suttle CA, Pledger RJ, Mitchell DL: Measurements of DNA damage and photoreactivation imply that most viruses in marine surface waters are infective. Aquat Microb Ecol. 1998, 14: 215-222. 10.3354/ame014215.
7. Wilhelm SW, Suttle CA: Viruses and nutrient cycles in the sea. Bioscience. 1999, 49: 781-788. 10.2307/1313569.
8. Maranger R, Bird DF: Viral abundances in aquatic systems: a comparison between marine and fresh waters. Mar Ecol Prog Ser. 1995, 121: 217-226. 10.3354/meps121217.
9. Essbauer S, Ahne W: Viruses of lower vertebrates. J Vet Med B Infect Dis Vet Pub Health. 2001, 48: 403-475. 10.1046/j.1439-0450.2001.00473.x.
10. Skall HF, Olesen NJ, Mellergaard S: Prevalence of viral haemorrhagic septicaemia virus in Danish marine fishes and its occurrence in new host species. Dis Aquat Org. 2005, 66: 145-151. 10.3354/dao066145.
11. Skall HF, Olesen NJ, Mellergaard S: Viral haemorrhagic septicaemia virus in marine fish and its implications for fish farming- a review. J Fish Dis. 2005, 28: 509-529. 10.1111/j.1365-2761.2005.00654.x.
12. Plarre H, Devold M, Snow M, Nylund A: Prevalence of infectious salmon anaemia virus (ISAV) in wild salmonids in western Norway. Dis Aquat Org. 2005, 66: 71-79. 10.3354/dao066071.

13. Bull JJ: Perspective-virulence. Evolution. 1994, 48: 1423-1437. 10.2307/2410237.

14. Patterson KR: Modelling the impact of disease-induced mortality in an exploited population of the fungal parasite *Icthyophonus hoferi* in the North Sea herring (*Clupea harengus*). Can J Fish Aquat Sci. 1996, 53: 2870-2877. 10.1139/cjfas-53-12-2870.

15. Marty GD, Quinn TJ, Carpenter G, Meyers TR, Willits NH: Role of disease in abundance of a Pacific herring (*Clupea pallasi*) population. Can J Fish Aquat Sci. 2003, 60: 1258-1265. 10.1139/f03-109.

16. Reno PW: Factors involved in the dissemination of disease in fish populations. J Aquat Anim Health. 1998, 10: 160-171. 10.1577/1548-8667(1998)010<0160:FIITDO>2.0.CO;2.

17. Einer-Jensen K, Ahrens P, Forsberg R, Lorenzen N: Evolution of the fish rhabdovirus viral haemorrhagic septicaemia virus. J Gen Virol. 2004, 85: 1167-1179. 10.1099/vir.0.79820-0.

18. Winton JR: Evolution of fish rhabdoviruses. OJI International Symposium on Salmonid Diseases. 1992, Hokkaido University Press, 88-95.

19. Steinhauer DA, Holland JJ: Rapid evolution of RNA viruses. Ann Rev Microbiol. 1987, 41: 409-433. 10.1146/annurev.mi.41.100187.002205.

20. Wolf K: Viral Hemorrhagic Septicemia. Fish viruses and fish viral diseases. 1988, Ithaca: Cornell University Press, 217-248.

21. Basurco B, Benmansour A: Distant strains of the fish rhabdovirus VHSV maintain a sixth functional cistron which codes for a nonstructural protein of unknown function. Virology. 1995, 212: 741-745. 10.1006/viro.1995.1534.

22. Ross K, McCarthy U, Huntly PJ, Wood BP, Stuart D, Rough EI, Smail DA, Bruno DW: An outbreak of viral haemorrhagic septicaemia (VHS) in turbot (*Scophthalmus maximus*) in Scotland. Bull Eur Assoc Fish Pathol. 1994, 14: 213-214.

23. Schlotfeldt HJ, Ahne W, Jørgensen PEV, Glende W: Occurrence of viral haemmorhagic septicaemia in turbot (*Scophthalmus maximus*) - a natural ioutbreak. Bull Eur Assoc Fish Pathol. 1991, 11: 105-107.

24. Nishizawa T, Iida H, Takano R, Isshiki T, Nakajima K, Muroga K: Genetic relatedness among Japanese, American and European isolates of viral hemorrhagic septicemia virus (VHSV) based on partial G and P genes. Dis Aquat Org. 2002, 48: 143-148. 10.3354/dao048143.

25. Gagne N, Mackinnon AM, Boston L, Souter B, Cook-Versloot M, Griffiths S, Olivier G: Isolation of viral haemorrhagic septicaemia virus from mummichog, stickleback, striped bass and brown trout in eastern Canada. J Fish Dis. 2007, 30: 213-223. 10.1111/j.1365-2761.2007.00802.x.
26. Lumsden JS, Morrison B, Yason C, Russell S, Young K, Yazdanpanah A, Huber P, Al-Hussinee L, Stone D, Way K: Mortality event in freshwater drum *Aplodinotus grunniens* from Lake Ontario, Canada, associated with viral haemorrhagic septicemia virus, type IV. Dis Aquat Org. 2007, 76: 99-111. 10.3354/dao076099.
27. Wunner W, Calisher C, Dietzgen R, Jackson A, Kitajima E, Lafon M, Leong J, Nichol S, Peters D, Smith J, PJ W: The Rhabdoviridae. Virus Taxonomy Sixth Report of the International Committee on Taxonomy of Viruses. Edited by: FA M, CM F, DHL B, SA G, AW J, GP M, MA M, MD S. 1995, Vienna & New York: Springer-Verlag, 275-288.
28. Benmansour A, Basurco B, Monnier AF, Vende P, Winton JR, de Kinkelin P: Sequence variation of the glycoprotein gene identifies three distinct lineages within field isolates of viral haemorrhagic septicaemia virus, a fish rhabdovirus. J Gen Virol. 1997, 78: 2837-2846.
29. Kurath G, Higman KH, Bjorklund HV: Distribution and variation of NV genes in fish rhabdoviruses. J Gen Virol. 1997, 78: 113-117.
30. Walker PJ, Benmansour A, Calisher CH, Dietzgen R, Fang RX, JAckson AO, Kurath G, Leong JC, Nadin-Davies S, Tesh RB, Tordo N: Family Rhabdoviridae. The seventh report of the international committee for taxonomy of viruses. 2000, San Diego: Academic Press, 563-583.
31. Snow M, Bain N, Black J, Taupin V, Cunningham CO, King JA, Skall HF, Raynard RS: Genetic population structure of marine viral haemorrhagic septicaemia virus (VHSV). Dis Aquat Org. 2004, 61: 11-21. 10.3354/dao061011.
32. Einer-Jensen K, Ahrens P, Lorenzen N: Parallel phylogenetic analyses using the N, G or Nv gene from a fixed group of VHSV isolates reveal the same overall genetic typing. Dis Aquat Org. 2005, 67: 39-45. 10.3354/dao067039.
33. Thiery R, de Boisseson C, Jeffroy J, Castric J, de Kinkelin P, Benmansour A: Phylogenetic analysis of viral haemorrhagic septicaemia virus (VHSV) isolates from France (1971-1999). Dis Aquat Org. 2002, 52: 29-37. 10.3354/dao052029.

34. Dixon PF, Feist S, Kehoe E, Parry L, Stone DM, Way K: Isolation of viral haemorrhagic septicaemia virus from Atlantic herring *Clupea harengus* from the English Channel. Dis Aquat Org. 1997, 30: 81-89. 10.3354/dao030081.
35. Skall HF, Slierendrecht WJ, King JA, Olesen NJ: Experimental infection of rainbow trout *Oncorhynchus mykiss* with viral haemorrhagic septicaemia virus isolates from European marine and farmed fishes. Dis Aquat Org. 2004, 58: 99-110. 10.3354/dao058099.
36. Raja-Halli M, Vehmas TK, Rimaila-Parnanen E, Sainmaa S, Skall HF, Olesen NJ, Tapiovaara H: Viral haemorrhagic septicaemia (VHS) outbreaks in Finnish rainbow trout farms. Dis Aquat Org. 2006, 72: 201-211. 10.3354/dao072201.
37. Nishizawa T, Savas H, Isidan H, Ustundag C, Iwamoto H, Yoshimizu M: Genotyping and pathogenicity of viral hemorrhagic septicemia virus from free-living turbot (*Psetta maxima*) in a Turkish coastal area of the Black Sea. Appl Environ Microbiol. 2006, 72: 2373-2378. 10.1128/AEM.72.4.2373-2378.2006.
38. Benmansour A, Basurco B, Monnier AF, Vende P, Winton JR, de Kinkelin P: Sequence variation of the glycoprotein gene identifies three distinct lineages within field isolates of viral haemorrhagic septicaemia virus, a fish rhabdovirus. J Gen Virol. 1997, 78: 2837-2846.
39. Stone DM, Way K, Dixon PF: Nucleotide sequence of the glycoprotein gene of viral haemorrhagic septicaemia (VHS) viruses from different geographical areas: a link between VHS in farmed fish species and viruses isolated from North Sea cod (*Gadus morhua* L.). J Gen Virol. 1997, 78: 1319-1326.
40. Jensen NJ, Bloch B, Larsen JL: The ulcus syndrome in cod (*Gadus morhua*) III. A preliminary virological report. Nordisk Veterinaermedicin. 1979, 31: 436-442.
41. Brunson R, True K, Yancey J: VHS virus isolated at Makah National Fish Hatchery. Am Fish Soc News. 1989, 17: 3-
42. OIE Collaborating Centre for Information on Aquatic Animal Diseases. [http://www.collabcen.net]
43. Skall HF, Olesen NJ, Mellergaard S: Viral haemorrhagic septicaemia virus in marine fish and its implications for fish farming--a review. J Fish Dis. 2005, 28: 509-529. 10.1111/j.1365-2761.2005.00654.x.

44. Duesund H, Nylund S, Watanabe K, Ottem KF, Nylund A: Characterisation of a VHS virus genotype III isolated from rainbow trout (*Oncorhynchus mykiss*) at a marine site on the West coast of Norway. Virology J. 2010, 7: 19-10.1186/1743-422X-7-19.
45. Snow M, King JA, Garden A, Shanks AM, Raynard RS: Comparative susceptibility of turbot (*Scopthalmus maximus*) to different genotypes of viral haemorrhagic septicaemia virus. Dis Aquat Organ. 2005, 67: 31-38. 10.3354/dao067031.
46. Elsayed E, Faisal M, Thomas M, Whelan G, Batts W, Winton J: Isolation of viral haemorrhagic septicaemia virus from muskellunge, *Esox masquinongy* (Mitchill), in Lake St Clair, Michigan, USA reveals a new sublineage of the North American genotype. J Fish Dis. 2006, 29: 611-619. 10.1111/j.1365-2761.2006.00755.x.
47. Coon TJ: Icthyofauna of the Great Lakes Basin. Great Lakes Policy and Management. Edited by: Taylor WW, Ferreri CP. 1999, East Lansing, MI.: Michigan State University Press, 55-71.
48. Anon: Epidemiological Investigations into an Outbreak of Viral Haemorrhagic Septicaemia (VHS) in Yorkshire, United Kingdom. Second Report (Molecular Epidemiology Findings). 2007, Weymouth, UK: CEFAS
49. Matejusova I, McKay P, McBeath AJ, Collet B, Snow M: Development of a sensitive and controlled real-time RT-PCR assay for viral haemorrhagic septicaemia virus (VHSV) in marine salmonid aquaculture. Dis Aquat Org. 2008, 80: 137-144. 10.3354/dao01911.
50. Snow M, King JA, Garden A, Raynard RS: Experimental susceptibility of Atlantic cod (*Gadus morhua*) and Atlantic halibut (*Hippoglossus hippoglossus*) to different genotypes of viral haemorrhagic septicaemia virus (VHSV). J Fish Dis. 2005, 28: 737-742. 10.1111/j.1365-2761.2005.00682.x.
51. Campbell S, Collet B, Einer-Jensen K, Secombes CJ, Snow M: Identifying potential virulence determinants in viral haemorrhagic septicaemia virus (VHSV) for rainbow trout. Dis Aquat Org. 2009, 86: 205-212. 10.3354/dao02127.
52. McLoughlin MF, Nelson RN, McCormick JI, Rowley HM, Bryson DB: Clinical and histopathological features of naturally occurring pancreas disease in farmed Atlantic salmon, *Salmo salar* L. J Fish Dis. 2002, 25: 33-43. 10.1046/j.1365-2761.2002.00334.x.

53. Boucher P, Laurencin FB: Sleeping disease and pancreas disease: comparative histopathology and acquired cross-protection. J Fish Dis. 1996, 19: 303-310. 10.1111/j.1365-2761.1996.tb00708.x.
54. Weston J, Villoing S, Bremont M, Castric J, Pfeffer M, Jewhurst V, McLoughlin M, Rodseth O, Christie KE, Koumans J, Todd D: Comparison of two aquatic alphaviruses, salmon pancreas disease virus and sleeping disease virus, by using genome sequence analysis, monoclonal reactivity, and cross-infection. J Virol. 2002, 76: 6155-6163. 10.1128/JVI.76.12.6155-6163.2002.
55. Mayo MA, Maniloff J, Desselberger U, Ball LA: Virus Taxonomy: VIIIth Report of the International Committee on Taxonomy of Viruses. 2005, Elsevier Academic Press
56. McLoughlin MF, Nelson RT, Rowley HM, Cox DI, Grant AN: Experimental pancreas disease in Atlantic salmon *Salmo salar* post-smolts induced by salmon pancreas disease virus (SPDV). Dis Aquat Org. 1996, 26: 117-124. 10.3354/dao026117.
57. Fringuelli E, Rowley HM, Wilson JC, Hunter R, Rodger H, Graham DA: Phylogenetic analyses and molecular epidemiology of European salmonid alphaviruses (SAV) based on partial E2 and nsP3 gene nucleotide sequences. J Fish Dis. 2008, 31: 811-823. 10.1111/j.1365-2761.2008.00944.x.
58. Snow M, Black JA, Matejusova I, McIntosh R, Baretto E, Wallace IS, Bruno DW: Evidence for the detection of salmonid alphavirus (SAV) RNA in wild marine fish in ares remote from aquaculture activity: Implications for the origins of salmon pancreas disease (SPD) in aquaculture. Dis Aquat Org. 2010, 91: 189-200. 10.3354/dao02265.
59. Graham DA, Staples C, Wilson CJ, Jewhurst H, Cherry K, Gordon A, Rowley HM: Biophysical properties of salmonid alphaviruses: influence of temperature and pH on virus survival. J Fish Dis. 2007, 30: 533-543. 10.1111/j.1365-2761.2007.00811.x.
60. Branson E: Sleeping disease in trout. Vet Rec. 2002, 150: 759-760
61. Weston JH, Graham DA, Branson E, Rowley HM, Walker IW, Jewhurst VA, Jewhurst HL, Todd D: Nucleotide sequence variation in salmonid alphaviruses from outbreaks of salmon pancreas disease and sleeping disease. Dis Aquat Org. 2005, 66: 105-111. 10.3354/dao066105.
62. Hodneland K, Endresen C: Sensitive and specific detection of Salmonid alphavirus using real-time PCR (TaqMan). J Virol Methods. 2006, 131: 184-192. 10.1016/j.jviromet.2005.08.012.

63. Graham DA, Frost P, McLaughlin K, Rowley HMI, Gabestad I, Gordon A, McLoughlin MF: A comparative study of marine alphavirus subtypes 1-6 using an experimental cohabitation model. J Fish Dis. 2011,
64. Karlsen M, Villoing S, Rimstad E, Nylund A: Characterization of untranslated regions of the salmonid alphavirus 3 (SAV3) genome and construction of a SAV3 based replicon. Virol J. 2009, 6: 173-10.1186/1743-422X-6-173.
65. Thorud K, Djupvik HO: Infectious anaemia in Atlantic salmon (*Salmo salar* L.). Bull Eur Assoc Fish Pathol. 1988, 8: 109-111.
66. Krossoy B, Hordvik I, Nilsen F, Nylund A, Endresen C: The putative polymerase sequence of infectious salmon anemia virus suggests a new genus within the *Orthomyxoviridae*. J Virol. 1999, 73: 2136-2142.
67. Devold M, Karlsen M, Nylund A: Sequence analysis of the fusion protein gene from infectious salmon anaemia virus isolates: evidence of recombination and reassortment. J Gen Virol. 2006, 87: 2031-2040. 10.1099/vir.0.81687-0.
68. Kibenge FS, Godoy MG, Wang Y, Kibenge MJ, Gherardelli V, Mansilla S, Lisperger A, Jarpa M, Larroquete G, Avendano F, Lara M, Gallardo A: Infectious salmon anaemia virus (ISAV) isolated from the ISA disease outbreaks in Chile diverged from ISAV isolates from Norway around 1996 and was disseminated around 2005, based on surface glycoprotein gene sequences. Virol J. 2009, 6: 88-10.1186/1743-422X-6-88.
69. Nylund A, Plarre H, Karlsen M, Fridell F, Ottem KF, Bratland A, Saether PA: Transmission of infectious salmon anaemia virus (ISAV) in farmed populations of Atlantic salmon (*Salmo salar*). Arch Virol. 2007, 152: 151-179. 10.1007/s00705-006-0825-9.
70. Blake S, Bouchard D, Keleher W, Opitz M, Nicholson BL: Genomic relationships of the North American isolate of infectious salmon anemia virus (ISAV) to the Norwegian strain of ISAV. Dis Aquat Org. 1999, 35: 139-144. 10.3354/dao035139.
71. Cunningham CO, Gregory A, Black J, Simpson I, Raynard RS: A novel variant of the infectious salmon anaemia virus (ISAV) haemagglutinin gene suggests mechanisms for virus diversity. Bull Eur Ass Fish Pathol. 2002, 22: 366-374.

72. McBeath AJ, Bain N, Snow M: Surveillance for infectious salmon anaemia virus HPR0 in marine Atlantic salmon farms across Scotland. Dis Aquat Org. 2009, 87: 161-169. 10.3354/dao02128.
73. Cook-Versloot M, Griffith S, Cusack R, McGeachy S, Ritchie R: Identification and characterisation of infectious salmon anaemia virus (ISAV) haemagglutinin gene highly polymorphic region (HPR) type 0 in North America. Bull Eur Ass Fish Pathol. 2004, 24: 203-208.
74. Snow M, Mckay P, McBeath AJA, Black J, Doig F, Kerr R, Cunningham CO, Nylund A, Devold M, (Eds): Development, application and validation of a Taqman real-time RT-PCR assay for the detection of infectious salmon anaemia virus (ISAV) in Atlantic salmon (*Salmo salar*). New Diagnostic Technology: Applications in Animal Health and Biologics Controls. Dev Biol. Edited by: Vannier P, Espeseth D. 2006, Basel, Karger, 126: 133-145.
75. 75.
76. Biacchesi S: The reverse genetics applied to fish RNA viruses. Vet Res. 2011, 42: 12-10.1186/1297-9716-42-12.
77. Hall TA: BioEdit: a user-friendly biological sequence alignment editor and analysis program for Windows 95/98/NT. Nucl. Acids. Symp. Ser. Nucl Acids Symp Ser. 1999, 41: 95-98.
78. Larkin MA, Blackshields G, Brown NP, Chenna R, McGettigan PA, McWilliam H, Valentin F, Wallace IM, Wilm A, Lopez R, Thompson JD, Gibson TJ, Higgins DG: Clustal W and Clustal × version 2.0. Bioinformatics. 2007, 23: 2947-2948. 10.1093/bioinformatics/btm404.
79. Swofford DL: Phylogenetic Analysis Using Parsimony (*and Other Methods). 2000, Sinauer Associates, Sunderland, Massachusetts, 4
80. Calendini F, Martin J-F: PaupUP v1.0.3.1 A free graphical frontend for Paup* Dos software. 2005
81. Posada D: jModelTest: phylogenetic model averaging. Mol Biol Evol. 2008, 25: 1253-1256. 10.1093/molbev/msn083.
82. Akaike H: A new look at the statistical model indentification' IEEE Transactions on Automatic Control. IEEE Trans Aut Control. 1974, 19: 716-722. 10.1109/TAC.1974.1100705.
83. Felsenstein J: Confidence limits on phylogenies: an approach using the bootstrap. Evolution. 1985, 39: 783-791. 10.2307/2408678.

CHAPTER 7

Role of Wild Ruminants in the Epidemiology of Bluetongue Virus Serotypes 1, 4 and 8 in Spain

Ignacio García-Bocanegra[1], Antonio Arenas-Montes[1], Cristina Lorca-Oró[2], Joan Pujols[2], Miguel Ángel González[3], Sebastián Napp[2], Félix Gómez-Guillamón[4], Irene Zorrilla[5], Elena San Miguel[6] and Antonio Arenas[1]

[1] Departamento de Sanidad Animal. Facultad de Veterinaria, UCO, Campus Universitarios de Rabanales 14071, Córdoba, Spain.

[2] Centre de Recerca en Sanitat Animal (CReSA), UAB-IRTA, Campus de la Universitat Autònoma de Barcelona, 08193 Bellaterra, Barcelona, Spain.

[3] Laboratorio de Producción Animal. Consejería de Agricultura y Pesca, Sevilla, Spain.

[4] Consejería de Medio Ambiente, Junta de Andalucía, 29010, Málaga, Spain.

Citation: García-Bocanegra, I., Arenas-Montes, A., Lorca-Oró, C. et al. Role of wild ruminants in the epidemiology of bluetongue virus serotypes 1, 4 and 8 in Spain. Vet Res 42, 88 (2011). https://doi.org/10.1186/1297-9716-42-88

[5] Centro de Análisis y Diagnóstico de la Fauna Silvestre (CAD), Conserjería de Medio Ambiente (EGMASA), Junta de Andalucía, Málaga, Spain.

[6] Laboratorio Central de Veterinaria, Algete (Madrid), Spain.

ABSTRACT

Although the importance of wild ruminants as potential reservoirs of bluetongue virus (BTV) has been suggested, the role played by these species in the epidemiology of BT in Europe is still unclear. We carried out a serologic and virologic survey to assess the role of wild ruminants in the transmission and maintenance of BTV in Andalusia (southern Spain) between 2006 and 2010.

A total of 473 out of 1339 (35.3%) wild ruminants analyzed showed antibodies against BTV by both ELISA and serum neutralization test (SNT). The presence of neutralizing antibodies to BTV-1 and BTV-4 were detected in the four species analyzed (red deer, roe deer, fallow deer and mouflon), while seropositivity against BTV-8 was found in red deer, fallow deer and mouflon but not in roe deer. Statistically significant differences were found among species, ages and sampling regions. BTV RNA was detected in twenty-one out of 1013 wild ruminants (2.1%) tested. BTV-1 and BTV-4 RNA were confirmed in red deer and mouflon by specific rRT-PCR.

BTV-1 and BTV-4 seropositive and RNA positive wild ruminants, including juveniles and sub-adults, were detected years after the last outbreak was reported in livestock. In addition, between the 2008/2009 and the 2010/2011 hunting seasons, the seroprevalence against BTV-1, BTV-4 and BTV-8 increased in the majority of provinces, and these serotypes were detected in many areas where BTV outbreaks were not reported in domestic ruminants. The results indicate that wild ruminants seem to be implicated in the dissemination and persistence of BTV in Spain.

INTRODUCTION

Bluetongue (BT) is a vector-borne disease caused by a group of viruses belonging to the genus *Orbivirus*, which is mainly transmitted between vertebrate hosts by biting midges of the genus *Culicoides*. To date, 24 distinct BT virus (BTV) serotypes have been identified; three of which were found in Andalusia (southern Spain) during the last decade. In October 2004, BTV-

4 was detected in Cádiz Province (south-west Andalusia), and a total of 316 outbreaks were reported in livestock. The last BTV-4 outbreak was detected in October 2005, and the country was declared free of BTV-4 in March 2009 [1]. However, in October 2010 a new BTV-4 outbreak was detected in Cádiz, and until December 2010 eight further outbreaks were reported.

In July 2007, a new BT epidemic, caused by BTV-1, affected southern Spain. BTV-1 causes clinical disease to both sheep and goats and 4446 outbreaks were detected from July 2007 through December 2008 in Andalusia. While no BTV-1 outbreaks were detected during 2009, two new cases were confirmed at the end of 2010 in the Jaén Province (northern Andalusia) [2]. Furthermore, BTV-8 was detected in northern Spain in January 2008, and arrived in Andalusia in October 2008. Until March 2009, 23 BTV-8 outbreaks were reported, followed by a period without apparent cases. In November 2010, a new BTV-8 outbreak was confirmed.

A mandatory vaccination program in domestic ruminants against serotype 4 started in 2004 and was maintained until 2008, two years after the last outbreak was reported. In 2010 vaccination against BTV-4 was resumed in the southern provinces of Andalusia to prevent introduction from northern Africa [3]. Since November 2006, animals have also been vaccinated against serotype 1, and since the end of 2008 also against BTV-8. Currently, BTV-1, BTV-4 and BTV-8 are endemic in Andalusia, and therefore the southern regions of Spain are considered restriction zones for these serotypes [2]. A live attenuated vaccine against BTV-4 was used between 2004 and 2005. However, since 2006 the commercial vaccines against the three serotypes are inactivated vaccines [3].

A wide range of wild ruminant species in Europe have been affected by BTV [4–6]. In Spain, seropositivity to BTV was found in central and southern regions [7, 8]. In addition, clinically affected mouflons (*Ovis aries*) were reported in Andalusia [9, 10].

The distribution and density of wild ungulates in Spain have increased in recent decades and their populations usually share pastures with domestic ruminants [11]. Although the role played by wild ruminant species in the epidemiology of BT in Europe is still unclear, the importance of these species as both potential reservoirs and sentinels of BTV has been suggested, especially in regions where livestock and wild ruminants share the same habitat [6, 10, 12–14]. In order to investigate the possible role of wild ruminants in the epidemiology of the bluetongue virus, a large-scale serologic and virologic survey was carried out in southern Spain from 2006 through 2010.

MATERIALS AND METHODS

Sampling

The study was conducted in Andalusia, located in southern Spain (36° N - 38° 60' N, 1° 75' W - 7° 25' W) (Figure 1) from 2006 through 2010, a region and time period with a wide circulation of BTV in livestock (Figure 2).

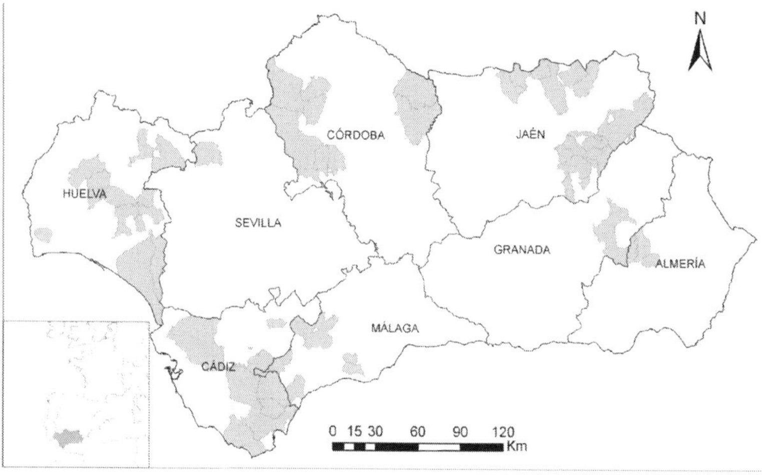

Figure 1. Map showing the location of Andalusia (southern Spain) and the provinces in which it is divided. Grey areas indicate the municipalities in which samples from wild ruminants were collected during the study period, 2006-2010. Sampling regions were grouped in western (including the Provinces of Huelva, Cádiz and Seville), central (including the Provinces of Córdoba and Málaga) and eastern regions (including the Provinces of Granada, Jaén and Almería).

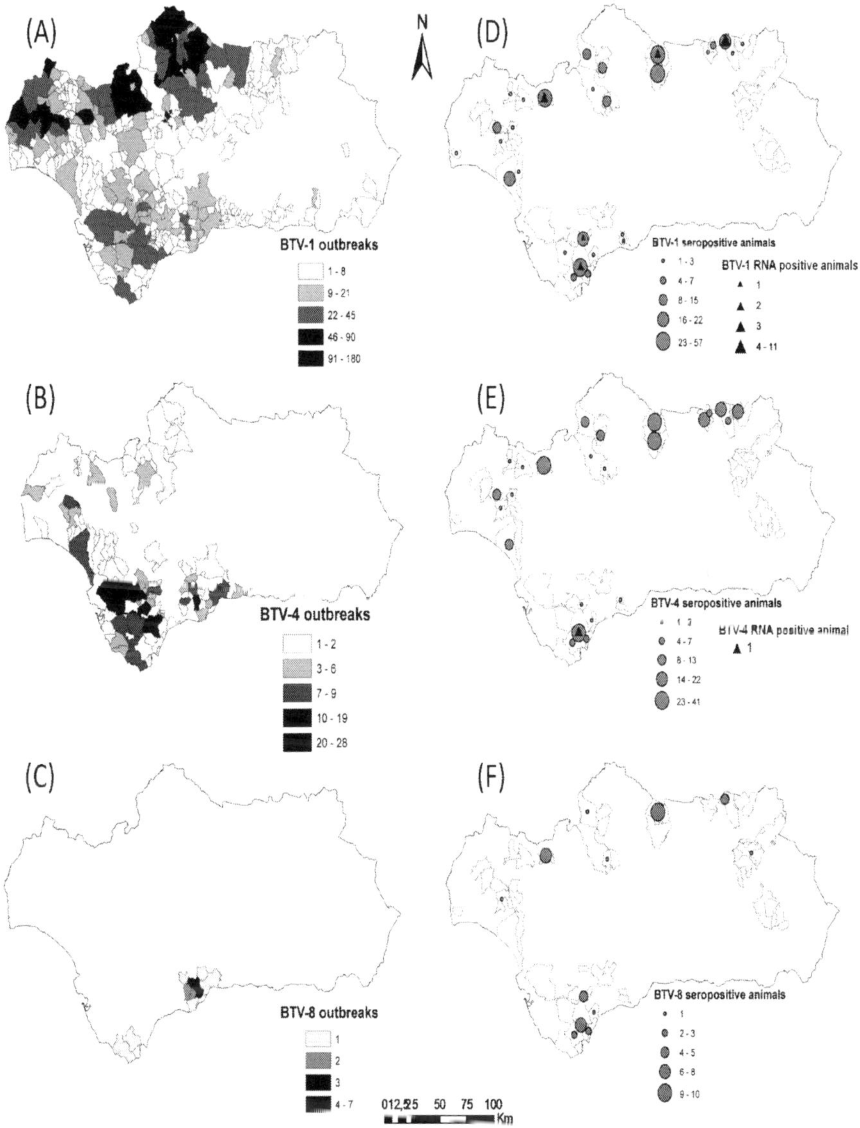

Figure 2. Map of Andalusia (southern Spain) showing the spatial distribution of BTV serotypes in livestock and wild ruminants, period 2006-2010. The gradient of grey indicates the number of BTV-1 (A), BTV-4 (B) and BTV-8 (C) infected farms by domestic ruminants. Grey dots size is proportional to the number of BTV-1 (D), BTV-4 (E) and BTV-8 (F) seropositive wild ruminants detected and the black triangles indicate the presence of BTV-1 RNA (D) and BTV-4 RNA (E) positive animals. White areas indicate the regions in which wild ruminants were sampled.

Post-mortem examinations were carried out on 1446 randomly selected free-ranging wild ruminants shot during the hunting seasons (October-March) of the five study years. A total of 1396 blood samples were collected into sterile tubes without anticoagulant and centrifuged at 400 *g* for 15 min. Blood samples from 336 animals were also placed in sterile tubes containing EDTA for RT-PCR analysis. In addition, 759 spleen samples were collected during the necropsies. All samples were stored at -20°C until analysis. Species, location, age, sex and hunting season of each animal were recorded. The animals were classified into three age groups based on tooth replacement: juveniles (< 1 year old), sub-adults (between 1 and 2 years old) and adults (> 2 years old). Unfortunately, the ages of 327 animals and the sexes of 382 could not be recorded. Data on BTV outbreaks in domestic animals in Andalusia were obtained from the Spanish Ministry of Agriculture, Fisheries and Food [2].

Serological analyses

The presence of antibodies against the VP7 protein of BTV was determined using a commercial double-antigen enzyme linked assay (ELISA; INGEZIM BTV DR 12.BTV.K0, INGENASA, Madrid, Spain) according to the manufacturer's recommendations. This ELISA has been used in previous surveys in wild ruminants [8, 15].

ELISA positive sera were then tested with the serum neutralization test (SNT) for BTV-1, BTV-4 and BTV-8 as previously described [16]. Briefly, serum samples were inactivated at 56°C for 30 min prior to analysis. Sera were diluted (1:2 - 1:256) in MEM (Eagle's minimum essential medium) and mixed with 100 $TCID_{50}$ (50% tissue culture infective doses) of each reference strain, BTV-1, BTV-4 and BTV-8. Plates were incubated for 1 h 30 min at 37°C. Finally, 100 μL of a Vero E6 cell suspension (1.5×10^4 cells/well) were added in cell growth media (MEM supplemented with 15% fetal calf serum, 300 μg L-glutamine/mL, 300 U penicillin/mL and 300 μg streptomycin/mL). The mixture was further incubated for 6 - 7 days at 37 μC until a cytopathic effect (CPE) was developed in control wells containing 100 $TCID_{50}$ of virus and no serum. Only samples that showed neutralization (absence of CPE) at dilutions ≥1:4 were considered positive [15]. Controls for cytotoxicity in the absence of virus were included for each sample at a dilution of 1:2. Seroprevalence to BTV was determined from samples positive by ELISA and SNT.

Virological Analyses

Blood and spleen samples were analyzed using a semi-quantitative real time reverse transcriptase-PCR (rRT-PCR) detecting a conserved region within segment 5 of the BTV genome [17]. Serotype specific rRT-PCR was performed on BTV positive samples, according to the following methods: for BTV-1 [18]; for BTV-4 [Agüero, not published]; for BTV-8 [19]. Field-isolated strains were used as positive controls for PCR amplification: BTV-1 (BTV-1 SPA/2007/01), BTV-4 (BTV-4 isolated in Cádiz in 2004) and BTV-8 (BTV-8 BEL/2006/01).

Statistical analyses

Association between explanatory variables (species, age, region, sex and hunting season) and BTV seropositivity was tested in three steps. Firstly, a bivariate chi-square test was performed to obtain an indication of the relevance of the explanatory variables in the risk of an animal being seropositive. Secondly, factors showing a p-value < 0.25 were further scrutinized for associations using the Pearson correlation coefficient (r) to avoid collinearity problems. If r was larger than 0.4, the variable more clearly linked to BTV infection was retained. The third step involved a multiple logistic-regression model as described by Hosmer and Lemeshow [20]. This was performed using a non-automatic backward selection of variables. Biologically plausible confounding factors were tested using Mantel-Haenszel analysis and confounding was considered to be potentially significant if odds ratio (OR) shifted appreciably. Changes in the OR greater than 30% were considered indicative of confounding. Potential two-way interactions between the variables were tested for significance in the model. The model was re-run until all the remaining variables presented statistically significant values (likelihood-ratio Wald test, $P < 0.05$), and a potential causal relationship with the response variable existed. Statistical analyses were performed using SPSS 15.0 (Statistical Package for Social Sciences (SPSS) Inc., Chicago, IL, USA).

RESULTS

A total of 639 out of 1396 wild ruminants (45.8%) analyzed showed antibodies against BTV by ELISA. Fifty-seven serum samples positive by ELISA could not be analyzed by SNT due to cytotoxicity, and 109 samples were positive by ELISA but negative by SNT. Therefore, a total of 473 out of 1339 wild ruminants (35.3%; $CI_{95\%}$: 32.7-37.9) analyzed showed antibodies against BTV by both ELISA and SNT. Seroprevalences to BTV serotypes were 25.6% to BTV-1 (343/1339; $CI_{95\%}$: 23.3-27.9), 21.6% to BTV-4 (289/1339; $CI_{95\%}$: 19.4-23.8) and 3.4% to BTV-8 (45/1339; $CI_{95\%}$: 2.4-4-4).

Seroprevalence among species, age classes, regions, sexes, and hunting seasons are shown in Table 1. Although antibodies to BTV were detected in the different species analyzed, the seroprevalence was significantly higher in red deer, fallow deer and mouflon than in roe deer. Statistically significant differences between fallow deer and mouflon were not observed (Table 2).

Table 1. Frequency of antibodies against BTV in wild ruminants.

Category	Value	No. Samples*	No. Positive (%)
Species			
	Red deer	900	381 (42.3)
	Fallow deer	188	61 (32.4)
	Mouflon	101	28 (27.7)
	Roe deer	150	3 (2.0)
Age			
	Juveniles	98	21 (21.4)
	Sub-adults	261	74 (28.4)
	Adults	653	289 (44.3)
Region			
	Western	553	190 (34.4)
	Central	430	196 (45.6)
	Eastern	356	87 (24.4)
Sex			
	Female	419	167 (39.9)
	Male	538	232 (43.1)
Hunting season			
	2006/2007	51	24 (47.1)
	2007/2008	54	23 (42.6)
	2008/2009	174	41 (23.6)
	2009/2010	744	255 (34.3)
	2010/2011	316	130 (41.1)

*Samples with no data on age or sex and cytotoxic sera were omitted.

The presence of neutralizing antibodies against BTV-1 and BTV-4 were detected in the four species analyzed, while seropositivity against BTV-8 was found in red deer, fallow deer and mouflon but not in roe deer. Antibodies against both BTV-1 and BTV-4 were detected in 145 animals of the four species analyzed. Nine animals, including red deer, fallow deer and mouflon showed antibodies against both BTV-1 and BTV-8. Seropositivity to BTV-4 and BTV-8 was detected in four red deer and two fallow deer. Twenty-two individuals, including red deer, fallow deer and mouflon showed antibodies against the three virus serotypes.

Table 2. Logistic regression analysis of potential risk factors associated with BTV seropositivity.

Variable	Category	*β*	*P*-value	OR	95% CI
Species	Red deer	0.78	0.04	2.18	(1.28-3.70)
	Roe deer	-2.80	0.00	0.06	(0.02-0.21)
	Fallow deer	0.13	0.67	1.14	(0.63-2.08)
	Mouflon	*	*	*	*
Age	Adults	0.18	0.00	1.92	(1.53-2.40)
	Sub-adults	*	*	*	*
	Juveniles	-0.49	0.10	0.61	(0.34-1.10)
Region	Western	0.83	0.00	2.30	(1.59-3.34)
	Central	1.35	0.00	3.87	(2.65-5.66)
	Eastern	*	*	*	*

*Reference category.

Adult animals presented significantly higher seroprevalence as compared to subadults and juveniles (Table 2). A significantly higher seropositivity was also found in western and central regions as compared to eastern regions (Figure 2). Distribution of BTV serotypes by hunting seasons and provinces are shown in Table 3. Seropositivity to BTV-4 was detected for all the years of sampling, neutralizing antibodies against BTV-1 were found since the 2006/2007 hunting season, while the presence of BTV-8 seropositive animals was confirmed since the 2008/2009 hunting season. BTV seroprevalence did not differ significantly between hunting season and sex.

Table 3. Distribution of BTV serotypes by hunting season and province.

	Hunting season				
Province	**2006/2007**	**2007/2008**	**2008/2009**	**2009/2010**	**2010/2011**
Almería	No samples	No samples	No samples	Negative	Negative
Cádiz	No samples	Negative	BTV1	BTV1*/ BTV4/BTV8	BTV1/ BTV4*/ BTV8
Córdoba	No samples	No samples	BTV1/ BTV4	BTV1/BTV4/ BTV8	BTV1*/ BTV4/BTV8
Granada	No samples	No samples	No samples	Negative	Negative
Huelva	No samples	No samples	No samples	BTV1/BTV4/ BTV8	BTV1/BTV4
Jaén	BTV4	BTV1/ BTV4	BTV1/ BTV4	BTV1/BTV4/ BTV8	BTV1*/ BTV4/BTV8
Málaga	No samples	BTV1*/ BTV4	BTV1	BTV1/BTV4/ BTV8	Negative
Sevilla	BTV4	BTV1*/ BTV4	BTV1/ BTV4/ BTV8	BTV1/BTV4/ BTV8	No samples

Presence of positive animals as determined by BTV serotype-specific rRT-PCR.

BTV RNA was detected in twenty-one out of 1013 wild ruminants analyzed (2.1%; $CI_{95\%}$: 1.2-2.6). The threshold cycle (Ct) RT-PCR values ranged between 30.1 and 38.1 (median = 35.5). Nineteen red deer and one mouflon were positive by BTV-1 specific rRT-PCR. BTV-1 RNA-positive results were found in the 2006/2007, the 2009/2010 and the 2010/2011 hunting seasons. One spleen sample from an adult deer sampled in the 2010/2011 hunting season was positive by BTV-4 specific rRT-PCR. No pathological lesions compatible with BT were observed in any of the BTV RNA-positive animals.

DISCUSSION

Our findings confirm that wild ruminants were actively exposed to BTV in the study area. A large number of the samples tested were positive by ELISA but negative by SNT. The differences between the tests, likely reflect that ELISA and SNT measure distinct antibody populations. The ELISA used in the present study detects both IgM and IgG antibodies to BTV, and it has been proven to be highly sensitive and specific in livestock [21]. Even though the quality of sera may also influence the results of both tests, it is particularly important for SNT. In fact, a high number of samples positive by ELISA could not be analyzed by SNT due to cytotoxicity. BTV seroprevalence was determined only from samples positive by both ELISA and SNT and therefore could have been underestimated.

To our knowledge, this is the first study on BTV serotypes in different wild ruminant species. The seroprevalence levels indicate widespread circulation of BTV in red deer, fallow deer and mouflon, which is in agreement with what was previously reported [6–8]. The results support the idea that sero-surveillance on these species would be useful to detect virus circulation, especially in areas where a vaccination program has been implemented in livestock [9, 10]. Furthermore, the low seropositivity detected in roe deer is in agreement with previous studies [7, 12, 22], suggesting that it is not a relevant species in the dissemination of BTV. Differences among species might be related to the natural resistance of the hosts, population densities, geographical distribution, sampling period or management factors.

The higher seroprevalence in adult animals was probably due to a greater exposure of this age group to the virus over time [8, 10, 12]. Interestingly, seropositivity was found in young animals (juveniles and sub-adults) sampled long after any previous outbreak in domestic ruminants. This was particularly clear in the case of BTV-4: even though the last outbreak in domestic ruminants was reported in October 2005, juveniles and sub-adults of the different wild ruminant species analyzed presented antibodies against BTV-4 between the 2008/2009 and the 2010/2011 hunting seasons in all the provinces where outbreaks had been reported in livestock. Similar findings were observed for BTV-1 and BTV-8, and evidence the ability of BTV to circulate despite no cases reported in domestic ruminants. Whether that circulation indicates the maintenance of BTV within the wild ruminant population despite vaccination of domestic ruminants, or is the consequence of repeated introductions of the virus, merits further studies.

No antibodies against BTV-4, BTV-1 or BTV-8 were detected in wild ruminants sampled prior to the detection of the first outbreak of each serotype in livestock. Between the 2008/2009 and the 2010/2011 hunting seasons, seroprevalence against the different serotypes increased in the majority of areas, even when the same species and ages were compared. These findings could indicate the persistence of antibodies against BTV for very long periods or, more likely, a longer BTV maintenance and circulation in wild ruminants compared to livestock. The lower seropositivity against BTV-8 (3.4%) was not unexpected taking into account that only 24 outbreaks had been reported in livestock in the study area [2].

The risk of being a seropositive animal was 2.3 and 3.9 times higher in the western and central regions, respectively, as compared to the eastern, which is in accordance with the geographical distribution of BTV observed in domestic ruminants (Figure 2). Vector density, host density and environmental factors are possibly implicated in the spatial distribution of BTV [23, 24]. However, BTV-1, BTV-4 and BTV-8 seropositive animals were detected in regions where these serotypes were not reported in livestock. This was particularly evident for BTV-8 (Figure 2) and confirms a different spatial distribution of BTV in wild ruminants as compared to livestock [8, 15].

The presence of BTV-1 and BTV-4 RNA in red deer and mouflon confirms the susceptibility of these species to BTV infection [9, 10]. In addition, the low cycles (Ct) RT-PCR values obtained in 12 animals (Ct lower than 35) support the potential reservoir role of these species. Three of BTV-1 RNA positive wild ruminants were found in November 2009, even though the last BTV-1 outbreak in livestock was reported in December 2008. One of these BTV-1 RNA positive animals showed no antibodies by ELISA or SNT, which could suggest a fresh infection in this animal [12, 13]. Furthermore, BTV-1 RNA positive wild ruminants sampled in the 2010/2011 hunting season were detected in locations different from those where the last BTV-1 outbreak was reported in livestock [2]. The BTV-4 RNA positive animal was detected on November 2010, one month after this serotype was first detected on a sentinel farm located at 60 km distance.

Vaccination has been suggested as one of the most important control measures for BT [25]. In this sense, the limited number of cases reported in livestock despite the circulation observed in wild ruminants suggests that vaccination seems to be effective to prevent the clinical disease. However, to prevent outbreaks in livestock, the vaccination of the domestic population would have to be maintained until virus circulation within the wild ruminant population has ceased.

Our results confirm that wild ruminant populations from southern Spain were exposed to BTV-1, BTV4 and BTV-8. The high seroprevalence to BTV-1 and BTV-4 found in the present study, the detection of both BTV seropositive and RNA positive animals, including juvenile animals, years after the last outbreak was reported in livestock, and the presence of the different BTV serotypes in areas where BTV outbreaks had never been reported in domestic ruminants, indicate that wild ruminants seem to be implicated in the dissemination and persistence of BTV, and probably play a significant role as reservoirs for BTV.

ACKNOWLEDGEMENTS

We would like to thank everyone involved in the Epidemiological Monitoring Program in Wildlife of the Regional Government of Andalusia for providing the samples. We are also grateful to Julio J Pérez for his technical assistance. This work was supported by INIA Grant FAU2008-00019-C03-01 and the Regional Government of Andalusia.

REFERENCES

1. Red de Alerta Sanitaria Veterinaria (RASVE), 2009. Ministerio de Medio Ambiente Medio Rural y Marino. Accessed November 5, 2009, [http://rasve.mapa.es/Publica/Noticias/Ficheros/Informe%20libre%20 serotipo%204%20final.pdf]
2. Red de Alerta Sanitaria Veterinaria (RASVE), 2010. Ministerio de Medio Ambiente Medio Rural y Marino. Accessed December 18, 2010, [http://rasve.mapa.es/RASVE_2005/Publica/RASVE_NET_2005/ Rasve.Presentacion/Modulo.Focos/Consultar_Focos.aspx]
3. Red de Alerta Sanitaria Veterinaria (RASVE), 2011. Ministerio de Medio Ambiente Medio Rural y Marino. Accessed April 21, 2011, [http://rasve.mapa.es/Publica/Programas/NORMATIVA%20 Y%20PROGRAMAS/PROGRAMAS/2011/LENGUA%20AZUL/ PROGRAMA%20LENGUA%20AZUL%202011.PDF]
4. De Curtis M, Bartolini C, Canonico C, Duranti A, Leoni F, Mancini P, Moscatelli F, Rocchegiani E, Gavaudan S: Serological monitoring of bluetongue virus in wild ruminants of the Pesaro-Urbino district (Italy). Webzine Sanita Pubblica Veterinaria. 2006, 40: 3-
5. Linden A, Mousset B, Gregoire F, Hanrez D, Vandenbussche F, Vandemeulebroucke E, Vanbinst T, Verheyden B, de Clerck K: Bluetongue virus antibodies in wild red deer in southern Belgium. Vet Rec. 2008, 162: 459-
6. Falconi C, López-Olvera JR, Gortázar C: BTV infection in wild ruminants, with emphasis on red deer: a review. Vet Microbiol. 2011, 151: 209-219. 10.1016/j.vetmic.2011.02.011.
7. Ruíz-Fons F, Reyes-García AR, Alcaide V, Gortázar C: Spatial and temporal evolution of bluetongue virus in wild ruminants, Spain. Emerg Infect Dis. 2008, 14: 951-953. 10.3201/eid1406.071586.
8. García I, Napp S, Casal J, Perea A, Allepuz A, Alba A, Carbonero A, Arenas A: Bluetongue epidemiology in wild ruminants from southern Spain. Eur J Wildl Res. 2009, 55: 173-178. 10.1007/s10344-008-0231-6.
9. Fernández-Pacheco P, Fernández-Pinero J, Agüero M, Jiménez-Clavero MA: Bluetongue virus serotype 1 in wild mouflons in Spain. Vet Rec. 2008, 162: 659-660. 10.1136/vr.162.20.659.
10. Rodríguez-Sánchez B, Sánchez-Cordón PJ, Molina V, Risalde MA, de Diego AC, Gómez-Villamandos JC, Sánchez-Vizcaíno JM: Detection of bluetongue serotype 4 in mouflons (*Ovis aries musimon*) from Spain. Vet Microbiol. 2010, 141: 164-167. 10.1016/j.vetmic.2009.08.003.

11. Gortázar C, Acevedo P, Ruiz-Fons F, Vicente J: Disease risk and overabundance of game species. Eur J Wildl Res. 2010, 52: 81-87.
12. Linden A, Gregoire F, Nahayo A, Hanrez D, Mousset B, Massart L, De Leeuw I, Vandemeulebroucke E, Vandenbussche F, De Clercq K: Bluetongue virus in wild deer, Belgium, 2005-2008. Emerg Infect Dis. 2010, 16: 833-836.
13. López-Olvera JR, Falconi C, Fernández-Pacheco P, Fernández-Pinero J, Sánchez MA, Palma A, Herruzo I, Vicente J, Jiménez-Clavero MA, Arias M, Sánchez-Vizcaíno JM, Gortázar C: Experimental infection of European red deer (*Cervus elaphus*) with bluetongue virus serotypes 1 and 8. Vet Microbiol. 2010, 145: 148-152. 10.1016/j.vetmic.2010.03.012.
14. Durand B, Zanella G, Biteau-Coroller F, Locatelli C, Baurier F, Simon C, Le Dréan E, Delaval J, Prengère E, Beauté V, Guis H: Anatomy of bluetongue virus serotype 8 epizootic wave, France, 2007-2008. Emerg Infect Dis. 2010, 16: 1861-1868.
15. Lorca-Oró C, Pujols J, Arenas A, Gómez-Guillamón F, Zorrilla I, Domingo M, Arenas-Montés A, Ruano MJ, García-Bocanegra I: Epidemiological surveillance of bluetongue virus serotypes 1, 4 and 8 in Spanish ibex (*Capra pyrenaica hispanica*) in southern Spain. Vet Microbiol. 2010, 149 (1-2): 230-235.
16. World Organization for Animal Health (OIE). Bluetongue and Epizootic Haemorrhagic Disease. 2009, Chapter 2.1.3: [http://www.oie.int/fileadmin/Home/eng/Health_standards/tahm/2.01.03_BLUETONGUE.pdf]
17. Toussaint JF, Sailleau C, Breard E, Zientara S, De Clercq K: Bluetongue virus detection by two real-time RT-qPCRs targeting two different genomic segments. J Virol Methods. 2007, 140: 115-123. 10.1016/j.jviromet.2006.11.007.
18. Agüero M, Jiménez-Clavero MA, El Harrak M, Gómez-Tejedor C: Real time RT-PCR method for specific detection of bluetongue virus serotype 1 strain Algeria 2006. Detection of the first BTV-1 outbreak in Spain. Abstract of the Bluetongue Satellite Symposium "Bluetongue in Europe, back to the future"; Brescia, Italy. 2008, 59-
19. Hoffmann B, Eschbaumer M, Beer M: Real-time quantitative reverse transcription-PCR assays specifically detecting bluetongue virus serotypes 1, 6, and 8. J Clin Microbiol. 2009, 47: 2992-2994. 10.1128/JCM.00599-09.

20. Hosmer DW, Lemeshow S: Applied logistic regression. 1989, Johns Wiley & Sons, New York, USA
21. Niedbalski W: Comparison of four ELISAs for the detection of antibodies against bluetongue virus. Medycyna Wet. 2010, 66: 106-108.
22. Boadella M, Carta T, Oleaga A, Pajares G, Muñoz M, Gortázar C: Serosurvey for selected pathogens in Iberian roe deer. BMC Vet Res. 2010, 6: 51-10.1186/1746-6148-6-51.
23. Allepuz A, García-Bocanegra I, Napp S, Casal J, Arenas A, Saez M, González MA: Monitoring bluetongue disease (BTV-1) epidemic in southern Spain during 2007. Prev Vet Med. 2010, 96: 263-271. 10.1016/j.prevetmed.2010.06.005.
24. Acevedo P, Ruiz-Fons F, Estrada R, Márquez AL, Miranda MA, Gortázar C, Lucientes J: A broad assessment of factors determining *Culicoides imicola* abundance: modelling the present and forecasting its future in climate change scenarios. PLoS One. 2010, 5: e14236-10.1371/journal.pone.0014236.
25. Bhanuprakash V, Indrani BK, Hosamani M, Balamurugan V, Singh RK: Bluetongue vaccines: the past, present and future. Expert Rev Vaccines. 2009, 8: 191-204. 10.1586/14760584.8.2.191.

CHAPTER 8

Epidemiology of the Emergent Disease Paridae Pox in an Intensively Studied Wild Bird Population

Shelly Lachish[1], Becki Lawson[2], Andrew A. Cunningham[2] , Ben C. Sheldon[1]

[1]Edward Grey Institute, Department of Zoology, University of Oxford, Oxford, United Kingdom,

[2]Institute of Zoology, Zoological Society of London, London, United Kingdom

ABSTRACT

Paridae pox, a novel avipoxvirus infection, has recently been identified as an emerging infectious disease affecting wild tit species in Great Britain. The incursion of Paridae pox to a long-term study site where populations of wild tits have been monitored in detail for several decades provided a

Citation: Lachish S, Lawson B, Cunningham AA, Sheldon BC (2012) Epidemiology of the Emergent Disease Paridae pox in an Intensively Studied Wild Bird Population. PLoS ONE 7(11): e38316. https://doi.org/10.1371/journal.pone.0038316

unique opportunity to obtain information on the local-scale epidemiological characteristics of this novel infection during a disease outbreak. Using captures of >8000 individual birds, we show that, within two years of initial emergence, Paridae pox had become established within the population of great tits (*Parus major*) reaching relatively high peak prevalence (10%), but was far less prevalent (<1%) in sympatric populations of several other closely related, abundant Paridae species. Nonlinear smoothing models revealed that the temporal pattern of prevalence among great tits was characterised by within-year fluctuations indicative of seasonal forcing of infection rates, which was likely driven by multiple environmental and demographic factors. There was individual heterogeneity in the course of infection and, although recovery was possible, diseased individuals were far less likely to be recaptured than healthy individuals, suggesting a survival cost of infection. This study demonstrates the value of long-term monitoring for obtaining key epidemiological data necessary to understand disease dynamics, spread and persistence in natural populations.

INTRODUCTION

Avian pox is a well-known viral disease of birds caused by dsDNA viruses in the genus *Avipoxvirus*, with a worldwide distribution and a wide host range [1]. Avipoxvirus infections cause proliferative 'wart-like' skin lesions, which are most commonly restricted to the featherless regions of the body (so-called 'dry' pox), but can also develop in the upper alimentary and respiratory tracts ('wet' or 'diptheric' pox) [1]. The incubation period and duration of avian poxvirus infection is variable, but affected birds with mild lesions frequently recover and this is considered to be the most common situation in endemic areas [2]. Indeed, where avian pox is endemic, the disease exists at very low prevalence and is presumed to have little impact on affected hosts [1], [3]. In naïve populations and species, however, the disease can attain a much higher prevalence and has been associated with substantial population declines [4], [5], [6].

A recent analysis of opportunistic reports of garden bird mortality and morbidity collected from 2006 to 2010 throughout Great Britain (GB) has revealed that an unusually severe form of avian pox affecting Paridae species (tits) is an emerging infectious disease in GB [7]. Although reports of avipoxvirus infections exist from multiple wild bird families and orders [1] and the disease is considered endemic in GB in non-Paridae species (e.g. common wood-pigeon *Columba palumbus*, dunnock *Prunella modularis*), avian pox was unknown within the Paridae family in GB prior to the index case observed in Sussex, England, in 2006 [7]. In addition, pox lesions in Paridae species are frequently larger and more florid than the small wart-like lesions characteristic of infections in non-Paridae species, and were considered a significant contributory factor to the cause of death of birds examined post mortem [7]. Spatio-temporal analysis of the nationwide incidence data, coupled with phylogenetic analysis of avian poxvirus core 4b gene sequences, showed that this novel 'Paridae pox' originated in south-east England following likely viral incursion from continental Europe and has since spread extensively throughout south and central England and into Wales [7].

Assessing the implications of novel infectious disease epizootics and accurately forecasting their epidemic trends requires a thorough understanding of disease epidemiology and dynamics in natural populations [8], [9], [10]. Monitoring epizootics in natural populations from their outset, though, is inherently difficult, and is further hindered by the financial and logistical constraints involved in regularly capturing and testing large numbers of wild animals [11]. For this reason, empirical data describing the epidemiological characteristics of emerging diseases in free-living wild animals are rare. The appearance of a novel pathogen in a long-term monitored population of wild animals, however, provides a unique opportunity to collect the crucial data necessary to aid our understanding of disease epidemiology, spread and persistence [12], [13].

In May 2009, the first case of Paridae pox in Wytham Woods, south-east England, was detected in a great tit. Wytham Woods is a long-term study site where populations of both great tits (*Parus major*) and blue tits (*Cyanistes caeruleus*) have been intensively monitored, including ringing of all breeding adults and nestlings, for more than 50 years. A substantial increase in the number of great tit cases of pox throughout 2010, coupled with the first observed cases in other species captured at this site, indicated the establishment of the disease. In this study, we take advantage of this rare opportunity to characterise the epidemiological patterns of Paridae pox among tits at a local, within-population scale. Our aims were to: (a) describe the temporal and spatial patterns in the prevalence of this viral disease from its initial appearance in the great tit population; (b) evaluate the influence of demographic factors on patterns of prevalence of the disease in great tits (host age, sex, or immigrant status: locally-born residents, versus immigrants born elsewhere); and (c) assess species-specific differences in disease prevalence.

METHODS

Field Sampling Techniques

From May 2009 to November 2011, great tits were monitored in Wytham Woods near Oxford, UK (51°46′N, 1°20′W), as part of several on-going ecological studies [14], [15], [16]. This 385-ha study site is a continuous mixed semi-deciduous forest, in which approximately 1200 nestboxes are distributed at variable densities [17]. During the annual, synchronous breeding season of this species (May/June), individuals were captured between days 6 and 14 of the nestling phase, either within the nestbox by hand or using traps, or with mist nets in front of the nest entrance. Outside the breeding season, individuals were captured with mist nets set throughout the woodland in association with seed feeders. All captured individuals and all locally-born nestlings that survive to day 15 were marked with unique metal rings. Capture sessions outside the breeding season occurred on an approximately monthly basis from October 2009 to early March 2010, from October 2010 to late March 2011, and from August 2011 to November 2011.

Table 1. Details of the capture sessions of great tits in Wytham Woods (May 2009–November 2011).

Capture Session	From	To	Trapping intensity*	'trend'†	No. Individuals Captured	No. Diseased	Prevalence (%)	No. Diseased (Other Species)
May-09*	04/04/2009	19/06/2009	0.47	0.00	512	1	0.20	0
Oct-09	08/10/2009	28/10/2009	0.30	0.42	74	0	0.00	0
Nov-09	02/11/2009	24/11/2009	0.32	0.50	35	1	2.86	0
Feb-10	26/01/2010	04/03/2010	0.57	0.75	218	1	0.46	0
May-10*	05/05/2010	26/06/2010	0.69	1.00	554	20	3.62	0
Nov-10	20/10/2010	29/11/2010	0.65	1.50	421	37	8.79	4
Dec-10	01/12/2010	17/12/2010	0.75	1.58	144	14	9.72	0
Jan-11	02/01/2011	31/01/2011	0.41	1.67	79	3	3.80	1
Feb-11	02/02/2011	25/02/2011	0.39	1.75	93	6	6.45	1
Mar-11	03/03/2011	22/03/2011	0.47	1.83	67	2	2.99	1
May-11*	04/05/2011	05/06/2011	0.78	2.00	396	6	1.52	1
Aug-11	05/08/2011	17/08/2011	0.50	2.25	34	1	2.94	0
Sep-11	02/09/2011	4/10/2011	0.29	2.33	167	6	3.59	0
Oct-11	3/10/11	25/10/11	0.29	2.42	91	3	3.30	0
Nov-11	1/11/11	28/11/11	0.48	2.50	323	11	3.40	2

***Trapping intensity gives the proportion of days within each capture session when trapping occurred.**

†'trend' = time since initial disease occurrence (in years). Also shown are the number of individuals captured, the apparent prevalence of pox per capture session (* indicates breeding seasons), and the number of diseased individuals of other Paridae species captured during the study (see Table 2 for more details).

https://doi.org/10.1371/journal.pone.0038316.t001

The prevalence of Paridae pox was determined as the proportion of individuals with visible pox lesions (see below) per 'capture session' (for the breeding seasons as a whole, and for each calendar month otherwise; with prevalence in Feb 2010, Nov 2010 and Sep 2011 determined for the calendar month plus the few trapping days conducted just outside the month assigned, see Table 1). Several other species that commonly form mixed-species foraging flocks with great were captured during field sampling (see Table 2); blue tits were captured at high rates in all capture sessions, but the remainder of species were captured predominantly outside the breeding season in mist nets; we report data from these bird species as well.

Table 2. Pox prevalence in species within the tit-nuthatch guild captured within Wytham Woods (May 2009–November 2011).

Species	Total Captures	No. Individuals Captured	No. Diseased	Pox prevalence
Great tit (*Parus major*)	3661	2076	104	5.01%
Blue tit (*Cyanistes caeruleus*)	6700	4820	3	0.06%
Coal tit (*Periparus ater*)	1013	692	4	0.58%
Marsh tit (*Poecile palustris*)	579	354	3	0.85%
Long-tailed tit (*Aegithalos caudatus*)	42	37	0	0.00%
Nuthatch (*Sitta europaea*)	107	69	0	0.00%

Values shown are the number of individuals captured, the total number of captures, and number diseased for each species, along with the species-specific prevalence of avian pox throughout the study.

https://doi.org/10.1371/journal.pone.0038316.t002

Following identification of Paridae pox within our study site, strict hygiene protocols were implemented to minimise anthropogenic transmission and spread of pox. Dedicated holding bags and measuring devices for diseased birds were used. After each diseased bird was processed field workers thoroughly disinfected their hands and all equipment (including traps, and mist nets) with F10 veterinary disinfectant (www.f10biocare.co.uk). Each captured bird was carefully examined for visual signs of skin lesions. Pox lesions were identified by the presence of swellings or proliferative skin lesions, especially around the beak and eyes, legs, and on sparsely feathered parts of the wings and body. Only individuals with one or more skin lesions characteristic of pox were classified as 'diseased'. We confirmed that this disease was caused by avian poxvirus infection by histopathologic and PCR examination of skin lesions from two great tits found dead in the woodland; see [7] for details. Previous studies have shown very good concordance between the presence of pox-like lesions and poxvirus infection; reviewed in [1]. In addition, avian pox was confirmed as the cause of the skin lesions in all suspected cases of Paridae pox in an additional 18 great tits examined as part of a larger study of the emergence and spread of this novel disease across Great Britain (via a combination of histopathology, electron microscopy and PCR) [7]. That study also reported no DNA sequence variation in the 4b core protein of the virus obtained from 20 diseased great tits [7].Hence we are confident that birds with lesions were infected with avian poxvirus and that only one virus strain was present in the population. We recognize, however, that some infected birds may have been misclassified as healthy (those that had no skin lesions or had very small lesions that were overlooked, although lesions as small as 1 mm were easily observed), and thus our estimates are of disease, rather than infection, prevalence.

Statistical Analyses

The prevalence of avian pox in species other than great tits was extremely low (see Table 2), and so statistical analyses were restricted to great tits. Since exploring only linear changes of disease prevalence through time can mask more complex temporal patterns [18], we employed a statistical approach that seeks the best linear or nonlinear fit to prevalence data. Variation in pox prevalence was examined using generalized additive modelling, essentially a generalized linear model in which a smoothed function of a covariate can be considered alongside conventional linear predictors and their interactions [19]. We incorporated a smoothed date covariate ('trend', calculated as the time since first observed occurrence of the disease in years based on the capture session, see Table 1) while examining associations between pox prevalence and linear functions of host sex, host age and host status (resident or immigrant), using binomial errors and a logit link. Host sex was determined based either on the presence (female) or absence (male) of a brood patch (breeding birds), or on plumage coloration (males have broad black breast stripes and glossy crowns) at other times of year [20]. During and following the autumn moult it is only possible to age newly captured individuals, based on their plumage characteristics, as either "juveniles" (aged 0–1 yr) or "older" (aged 1+ years) [20]. As a significant proportion of individuals in this study were only captured outside the breeding season (32.1%), we restricted age effects to these two age classes. We assigned 'resident' status to all birds that were ringed on the site as nestlings in nestboxes, and 'immigrant' status to all other individuals (we acknowledge that a small proportion of great tits may nest in natural cavities rather than nestboxes, and hence that a small proportion of individuals may be misclassified by this simple division). All interactions between factors were considered in the starting model, except those involving the smoothed 'trend' term because too few diseased individuals were captured in most capture sessions to rigorously assess temporal interactions. The full starting model was optimized by backward stepwise elimination of non-significant terms, beginning with higher order interactions. Likelihood ratio tests (LRT) were used to determine if the removal of terms caused a significant change in model deviance and to confirm that a nonlinear smoothed function of 'trend' performed better than either a linear or a quadratic function of 'trend'.

The Kuldorff spatial scan statistic implemented in SaTScan™ (http://satscan.org/) was used to test for the existence and extent of clusters of pox infection within the study site, and to identify their approximate locations and sizes [21]. Spatial analyses were restricted to the breeding season because during this period individuals are territorial and forage only in the immediate vicinity of the nest [22], so the nestbox coordinates [23] give an accurate description of an individuals› location and are likely to reflect the area where they acquire infections. Throughout the autumn and winter, birds forage and move over far greater ranges, and the locations of captured individuals (i.e. the locations of mist nets) are not likely to accurately reflect where they acquired infection. The spatial scan statistic creates a series of circular windows of variable radius around every diseased individual, each of which is set to contain from zero to a maximum proportion of the total population-at-risk (set to 25%). For each location and size of scanning window, the observed number of cases within the window is compared to the expected number of cases given by a Bernoulli model and an assumption of constant risk. A LRT is used to test the hypothesis that there is an elevated rate of infection within the window. The significance of the most likely cluster, and of secondary clusters (those that do not geographically overlap with the most likely cluster), is determined by simulated p-values using Monte Carlo methods with 999 replications, and are adjusted for the multiple testing inherent in both the many cluster locations considered, as well as the many possible collections of circles used for the scanning window [21].

RESULTS

Course of Infection in Great Tit Hosts

From May 2009 to November 2011 we made 3661 captures of 2076 individual great tits, of which 104 were diagnosed as diseased (total prevalence of

5.01%; Table 2). Diseased individuals possessed pox lesions ranging in size from 1 mm to 21 mm in diameter (median=6.0 mm). Only 14 diseased individuals were captured with pox lesions on two separate occasions; none more often than this (Table 3). These individuals were mostly captured a few days apart with little change in the size of their lesions, but three birds were captured at 17, 22 and 36 days apart, the latter individual displaying substantial growth in lesion size over this period (>10 mm growth, constituting a five-fold increase in the size of one lesion; Table 3). Recovery from pox was apparent (in one case within as little as one week), with 15 diseased individuals recaptured at a later date without pox lesions (Table 4). Recovered individuals formed an even mix of males and females, of both age classes, possessed lesions varying in size and location on the body and, in all but one case, displayed no visible scarring at the site of the original lesion. Only one of the 15 diseased individuals that recovered from pox infection was ever subsequently recaptured again (one year later, apparently still pox free). Nevertheless, the majority of diseased individuals (77.8%; 81/104) were never recaptured. Indeed the return rate (i.e. the proportion of marked individuals ever recaptured) of diseased individuals was far lower than for healthy individuals. For example, healthy individuals captured in May 2010 were 3.4 times more likely to be recaptured at least once after marking than were diseased individuals (odds ratio 95% CI=1.03–12.99; Fisher's exact test P=0.045), with stronger results for the November 2010 capture 'cohort' (healthy individuals were 4.61 times more likely to be recaptured than diseased individuals; odds ratio 95% CI=1.59–18.32, P=0.001).

Table 3. Details of birds captured twice with pox lesions.

Recapture Interval (days)	Change in size (mm) of lesion(s)
2	no change
2	no change
2	+1
3	+1
3	NA
3	no change
3	NA
3	no change
4	no change
4	NA
8	+1
17	+2
22	−1
36	+12 & +7

Values shown are the change in the size of lesions between capture occasions for individuals captured twice with disease.

https://doi.org/10.1371/journal.pone.0038316.t003

Table 4. Details of birds that recovered from pox.

Recovery Interval (days)	Size (mm)/location of lesion(s)
7	3/head
15	2/eye
54	3/foot
82	5/beak
89	13/eye
90	3/foot
154	6/thigh
154	5/eye
160	2/beak
160	2/eye
180	7/eye
359	8/eye
359	10 & 7.5/head & wing
365[+]	"large"/head
556	3/foot

Values shown are the intervals between the last disease capture and the first healthy capture of each recovered individual, along with the size and location of the original pox lesions.
[+]This was the only recovered individual to be subsequently recaptured.

https://doi.org/10.1371/journal.pone.0038316.t004

Species-Specific Differences in Disease Prevalence

Six species within the tit-nuthatch guild were captured during this study (Table 2). Pox prevalence was substantially higher in great tits, than in any other species captured and examined (Table 2). Particularly striking is the almost one hundred-fold difference in pox prevalence in the great tit population (5.01%) compared with the blue tit population (0.06%), as this species was both abundant and frequently recaptured during the study (Table 2). None of the diseased blue tits, coal tits or marsh tits were recaptured at a later date.

Temporal and Demographic Trends in Apparent Prevalence

A nonlinear smoothed function of 'trend' (time since initial sighting of disease) was retained as the most suitable temporal predictor of Paridae pox prevalence (Table 5). The model including a nonlinear function of 'trend' was preferred over either a linear or a quadratic function of trend

(LRT *s*('trend') vs. 'trend': Δdeviance=62.49, Δdf=3.35, P<0.001; *s*('trend') vs. 'trend'+'trend^2': Δdeviance=21.661, Δdf=2.35, P<0.001), and provided a better fit to the data (R^2 for the nonlinear model=0.75, the linear model=0.32, and the quadratic model=0.54). The predicted pox prevalence over our sampling period (predictions generated from the final model) provided a generally good fit to our observed prevalence data and showed that prevalence remained relatively low for the year following the initial disease outbreak, but then increased to peak in the autumn and early winter of 2010 (Fig. 1). Over spring and summer of 2011 prevalence declined, but increased again in autumn 2011 (although not to the same extent), indicative of seasonality in poxvirus infection rates (Fig. 1). Also, the majority of the avian pox cases in other tit species were observed during these late autumn peak periods (Table 1), supporting the idea of a higher force of infection at this time of year relative to spring and summer. An interaction between host age-class and host status was also retained in the final model (at near significance P=0.055, Table 5). For older individuals (1+ years of age) pox prevalence was higher among residents than immigrants (Fig. 2a), but this was not the case for juveniles (0–1 years of age; Fig. 2b). Overall, prevalence in older residents was approximately twice as high (4.74%; CI=3.11–6.86) as in older immigrants (2.17%; CI=1.16–3.67).

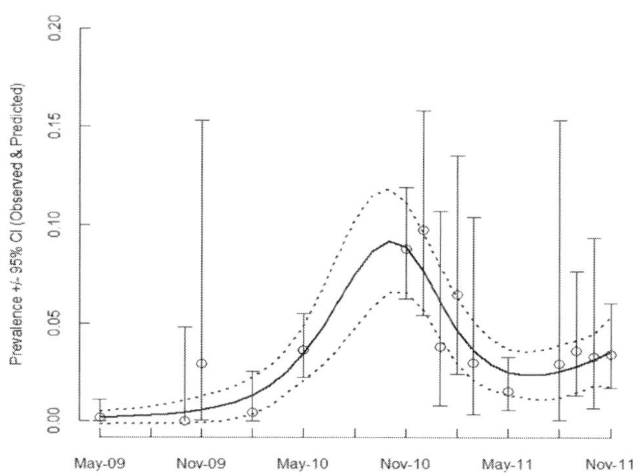

Figure 1. Observed prevalence of pox in great tits (with 95% exact binomial confidence intervals). The solid line represents the fit of the predicted values from the final generalised additive model (see Table 5). Dotted lines show 95% CI of the predicted prevalence estimates.

https://doi.org/10.1371/journal.pone.0038316.g001

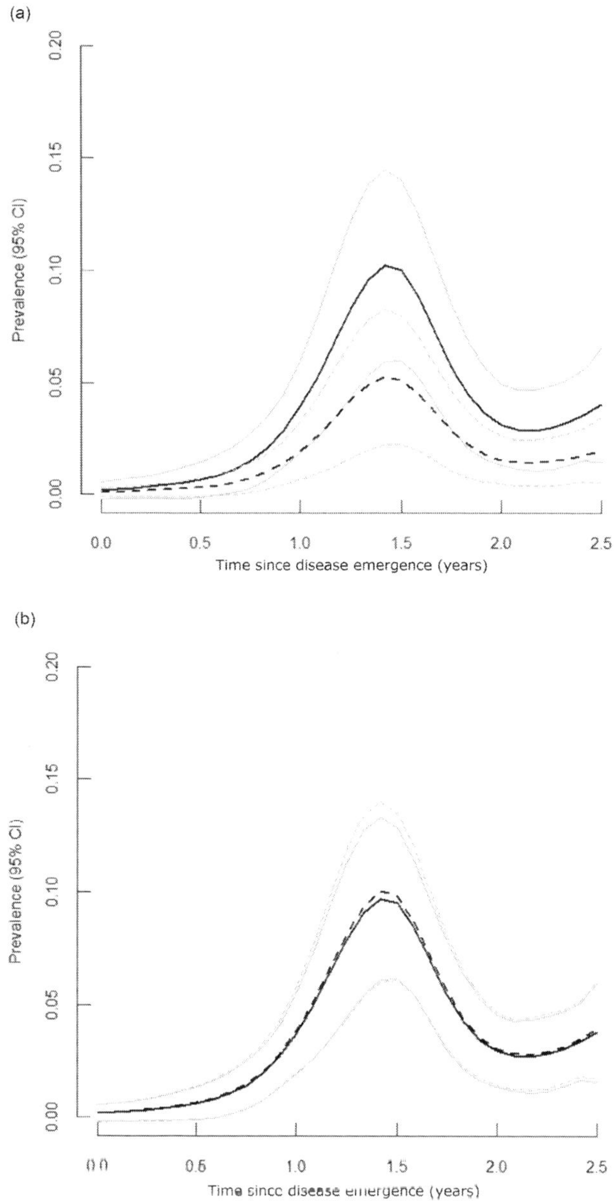

Figure 2. Predicted pox prevalence in residents (solid lines) and immigrants (dashed lines) according to host age. (a) adults (1+ years of age) and (b) juveniles (<1 year). Light grey solid and dashed lines are the 95% CI of the estimates of the predicted prevalence for residents and immigrants respectively.

https://doi.org/10.1371/journal.pone.0038316.g002

Table 5. Results of models examining the influence of temporal/demographic factors on variation in pox prevalence.

Factors	Parameter Estimate	Z/χ^2	Wald's test (P)	ΔDev/Δdf	LRT (P)
smoothed 'trend'	**estimated df = 4.461**	**26.73**	**<0.0001**	**78.862/4.56**	**<0.0001**
ageclass (older)	**−0.677±0.331**	**−2.046**	**0.041**	-	-
status (resident)	**−0.036±0.243**	**−0.148**	**0.882**	-	-
ageclass*status	**0.795±0.424**	**1.877**	**0.060**	**3.72/1**	**0.055**
sex (M)	−0.078±0.197	−0.396	0.692	0.220/1	0.642
status*sex	0.104±0.395	0.263	0.792	0.063/1	0.801
ageclass*sex	−0.278±0.416	−0.669	0.503	0.467/1	0.495
ageclass*sex*status	0.750±0.860	0.872	0.383	0.778/1	0.379

Values shown are the Z statistics (or χ^2 statistics for the smoothed term) and Wald's test P-values for coefficients, along with results of Likelihood ratio tests (LRT) (change in deviance: ΔDev, change in degrees of freedom: Δdf) comparing models with and without the specific term. Factors in bold were retained in the final model.

https://doi.org/10.1371/journal.pone.0038316.t005

Spatial Patterns of Poxvirus Infection

As expected, based on the north-westerly spread of Paridae pox from an origin in Sussex, south-east England [7], this disease first appeared in the south-east of our study site (Fig. 3a). Throughout the 2010 breeding season, cases remained clustered in the south-east region of the study site, where the relative risk of pox infection was estimated to be approximately eight times greater than in the rest of the woodland (Fig. 3b). By the following breeding season, however, no disease clustering was evident, with pox cases randomly distributed throughout the whole study site (Fig. 3c).

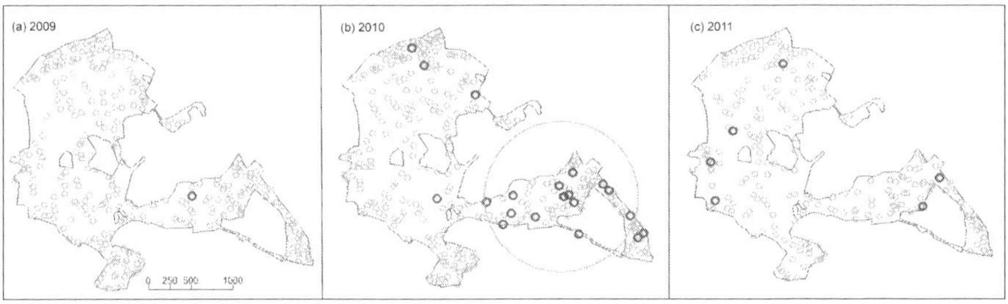

Figure 3. Locations of diseased (red circles) and healthy (grey circles) great tits captured in Wytham Woods.

https://doi.org/10.1371/journal.pone.0038316.g003

Spatial analyses were conducted for the (a) 2009, (b) 2010, and (c) 2011 breeding seasons. The location and spatial extent of the single statistically significant pox cluster identified in 2010 by the SatScan spatial cluster analysis is shown by an orange circle (log-likelihood ratio=10.55, relative risk=7.93; P=0.01).

DISCUSSION

Paridae pox is a novel, emerging infectious disease of wild tit species across Great Britain [7]. Here, we have characterised the epidemiology of the emergence and establishment of this disease at a local, within-population, scale. We have shown that there are striking differences in the prevalence of pox infections among closely-related, sympatric Paridae species with very similar ecologies, that temporal variation in pox prevalence among great tits displays regular fluctuations indicative of seasonality in infection rates and, that while recovery is possible, diseased individuals are far less likely to be recaptured than are healthy individuals.

The prevalence of Paridae pox in the great tit population (maximum 10%, mean 5%) was higher than that reported for avian pox in wild birds where the disease is endemic (0.5–1.5%) [1]. This is consistent with Paridae pox being a novel disease in a naïve host population. The maximum prevalence of pox recorded, though, was far lower than the prevalence reported for avian pox in some other naïve populations, particularly those on islands [3], [6], [24]. Island-endemic species, however, are expected to be more vulnerable to novel disease threats owing to their evolution in the face of a depauperate pathogen complement [25]. The prevalence of avian pox in great tits in Wytham Woods was much higher than in any of the other captured species within our study site (although prior reports of avian pox exist for all these species, with the exception of long-tailed tits) [7]. In particular, pox prevalence was much higher among great tits than among the closely-related blue tits, which are both abundant and frequently captured within our study site. As great and blue tits are sympatric and ecologically similar, for example often forming mixed-species flocks [17], it seems unlikely that observed differences in species-specific prevalence can be attributed to differential exposures of these species to the pathogen. Rather, it suggests that Paridae species either differ in their susceptibility to infection (a commonly reported finding in other studies) [3], [6], [26], [27], or that the virulence of this pathogen differs between host species [28]. It is also possible, however, that such species-specific patterns of disease prevalence might simply be a consequence of differences in the detectability of diseased individuals of different species [29]. Experimental infection studies will be necessary to distinguish between these alternative possibilities.

Our analyses suggest that, following the initial period of disease establishment in the great tit population, the prevalence of avian pox in great tits began to display regular temporal oscillations, indicative of seasonal forcing of infection rates. Theoretical models show that the strength and mechanisms of seasonality can alter the spread and persistence of infectious diseases, with seasonal forcing a crucial determinant of the longer-term dynamics and fluctuations of host–parasite systems [30]. This has been supported by field observations for several wildlife infections, including both directly-transmitted and vector-transmitted parasites [31], [32], [33]. In this study, the prevalence of pox in great tits was predicted to be highest throughout autumn and early winter, which is consistent with the temporal pattern of Paridae pox cases reported nationwide [7] and with the findings of many other studies of avian pox in wild birds [1], [34]. This was supported by our finding that the majority of pox cases in other species were also observed at this time of year. The higher prevalence of Paridae pox in autumn/early winter could be driven by the increased abundance of vectors in the warmer late summer and early autumn months enhancing virus transmission rates. Mosquitoes are considered a particularly important biting insect vector for avian poxvirus transmission [1]. Several studies have found strong concordance between pox prevalence in wild birds and the abundance or distribution of mosquito vectors [26], [35], or have shown strong correlations between pox prevalence and the climatic conditions preferred by mosquitoes (higher temperatures and rainfall) [36]. Preliminary investigations into mosquito ecology at our study site have found 14 species present within Wytham Woods. All of these species are apparently abundant during the summer months (June through August) but begin to decline in numbers in September (R. Alves, unpub. data), before the apparent yearly peaks in Paridae pox prevalence in this system. This temporal lag between peak mosquito abundance and peak pox prevalence might be due to a latent period between infection and disease for this poxvirus strain [37], but also could reflect a weaker role for mosquitoes in pathogen transmission in this system. A better understanding of the temporal variation in mosquito abundance and distribution within our study site, as well as the ecology and abundance of other vectors, would help to elucidate the role of vector abundance in driving seasonality, and also the magnitude of seasonal peaks in Paridae pox prevalence.

Alternatively, the seasonal increase of pox in autumn could be facilitated by the influx of immunologically naïve juveniles that follows the synchronous spring breeding period of this species; this acts to increase both host population density and the abundance of susceptible individuals in the population [30]. According to this hypothesis, the prevalence of avian pox should be higher in juveniles than in adults, a finding which is supported by many [1], [34], [36] but not all [38], [39] avian pox studies. In our study, we did not find a clear difference in Paridae pox prevalence between juveniles and older birds, but we did find a lower overall prevalence of pox in older immigrant birds. Rather than signify age-specific differences in susceptibility to infection, however, this result is more readily explained in terms of differences in the relative exposure times for the different groups. Older residents that were ringed on our study site would have been exposed to the novel pathogen for longer on average than the more-recently arrived older immigrants, which were likely born elsewhere (presumably where disease was absent). Alternatively, but perhaps less likely, older immigrants might have already been exposed/diseased and recovered before moving into Wytham Woods. The difference in exposure times (and consequently prevalence) between resident and immigrant juveniles, however, would not be as extreme, as both would only have been in the population for a matter of months. Whether this pattern persists as Paridae pox becomes established in surrounding populations remains to be seen. Clearly, though, a thorough understanding of the role of the annual cohort of naïve juveniles in the seasonal dynamics of Paridae pox will require longer-term data on age-specific prevalence.

Post-breeding flocking behaviour also may facilitate transmission by increasing conspecific and interspecific contact rates. Epidemiological models of *Mycoplasma gallisepticum* (MG) dynamics in house finches require the inclusion of both winter aggregation and pulsed breeding events to generate the observed bi-annual peaks in prevalence [40]. Similarly, the congregation of birds at feeding stations throughout the colder months, when anthropogenic provisioning of wild birds in garden habitats in Great Britain becomes more frequent [41] and natural food resources more rare, could increase inter- and intra-specific contacts and facilitate avian poxvirus transmission. The aggregation of birds at artificial seed feeders is thought to have contributed to pathogen spread in the house finch-MG system [42]. A more-thorough investigation of the strength and regularity of the seasonal peaks in Paridae pox occurrence, and the relative importance of the numerous potential factors that might contribute to driving seasonal variation in this system, is clearly warranted, but will require additional time series data. The results of this study reveal that the course of Paridae pox disease is highly variable among individual hosts. We found that some individuals could sustain the disease for relatively long periods of time (>1 month) with either no change or substantial growth in lesion size occurring during that time. Similar variability in the course of avian pox has been reported in other studies [26],[37]. In addition, the probability of recovery from disease in this study did not seem dependent on the size or location of lesions, as is commonly assumed [1], [37]. These patterns suggest that individual differences in immune system function are likely to exert significant control over the severity and duration of disease. Recovery from pox depends on cell-mediated immunity, and immunity to avipoxvirus in captive birds is known to be highly variable [37]. Very little is known, however, about the development or duration of immunity to poxvirus infections in wild birds. Given the importance of variation in individual and population immunity for predicting disease dynamics and persistence [40], this seems an urgent and fruitful area for further research.

While Paridae pox might not be invariably fatal, the disease may nonetheless increase the vulnerability of individuals to other causes of mortality, such as predation, since lesions can seriously impair an individual's vision, as well as its flight and feeding abilities [1], [3]. The majority of diseased individuals in this study were not recaptured, and diseased individuals had far lower return (recapture) rates than healthy birds. These findings suggest a survival cost of the disease, but need to be interpreted with caution because return rates confound both survival and detection, and detection rates themselves may differ depending on disease state [29]. An accurate assessment of the effect of Paridae pox on the survival rates of great tits will require the application of more rigorous mark-recapture statistical methods that overcome these limitations, and is the focus of our current research. Indeed, with this novel disease now well established within the Wytham Woods population of great tits, and continuing to spread across Great Britain [7], it is critical that we quantify the impact of the disease on both host survival and reproductive success in order to assess the consequence of this emerging infectious disease for the long-term viability of both the Wytham Woods and the national great tit populations. This information will also be vital for implementing effective disease interventions or management actions should they be required.

ACKNOWLEDGMENTS

The authors are extremely grateful to numerous Wytham Woods field workers, but particularly Amy Hinks, Ross Crates and Damien Farine for their efforts to capture the birds that formed the basis of this study.

REFERENCES

1. van Riper C, Forrester DJ (2007) Avian Pox. In: Thomas NJ, Hunter DB, Atkinson CT, editors. Infectious Diseases of Wild Birds. Oxford: Wiley Blackwell Publishing. pp. 131–176.
2. Simpson CE, Forrester DL, Nesbitt SA (1975) Avian pox in Florida Sandhill cranes. Journal of Wildlife Diseases 11: 112–115.
3. van Riper C, van Riper SG, Hansen WR (2002) Epizootiology and effect of avian pox on Hawaiian forest birds. Auk 119: 929–942.
4. Alley MR, Hale KA, Cash W, Ha HJ, Howe L (2010) Concurrent avian malaria and avipox virus infection in translocated South Island saddlebacks (Philesturnus carunculatus carunculatus). New Zealand Veterinary Journal 58: 218–223.
5. Atkinson CT, LaPointe DA (2009) Introduced Avian Diseases, Climate Change, and the Future of Hawaiian Honeycreepers. Journal of Avian Medicine and Surgery 23: 53–63.
6. Kleindorfer S, Dudaniec RY (2006) Increasing prevalence of avian poxvirus in Darwin's finches and its effect on male pairing success. Journal of Avian Biology 37: 69–76.
7. Lawson B, Lachish S, Colvile KM, Durrant C, Peck K, et al. (2012) Emergence of a novel avian pox disease in British tit species. PLoS ONE 7 (11) e40176
8. Sterner RT, Smith GC (2006) Modelling wildlife rabies: Transmission, economics, and conservation. Biological Conservation 131: 163–179.
9. Dobson A, Foufopoulos J (2001) Emerging infectious pathogens of wildlife. Philosophical Transactions of the Royal Society of London B 356: 1001–1012.
10. Smith GC (2006) Persistence of disease in territorial animals: insights from spatial models of Tb. New Zealand Journal of Ecology 30: 35–41.
11. McCallum H, Dobson A (1995) Detecting disease and parasite threats to endangered species and ecosystems. Trends in Ecology and Evolution 10: 190–194.
12. Carroll B, Russell P, Gurnell J, Nettleton P, Sainsbury AW (2009) Epidemics of squirrelpox virus disease in red squirrels (Sciurus vulgaris): temporal and serological findings. Epidemiology and Infection 137: 257–265.

13. Lachish S, Jones M, McCallum H (2007) The impact of disease on the survival and population growth rate of the Tasmanian devil. Journal of Animal Ecology 76: 926–936.
14. Charmantier A, McCleery RH, Cole LR, Perrins C, Kruuk LEB, et al. (2008) Adaptive phenotypic plasticity in response to climate change in a wild bird population. Science 320: 800–803.
15. McCleery RH, Pettifor RA, Armbruster P, Meyer K, Sheldon BC, et al. (2004) Components of variance underlying fitness in a natural population of the great tit Parus major. American Naturalist 164: E62–E72.
16. Morand-Ferron J, Quinn JL (2011) Larger groups of passerines are more efficient problem solvers in the wild. Proceedings of the National Academy of Sciences of the United States of America 108: 15898–15903.
17. Perrins CM (1979) British Tits. Glasgow: Collins.
18. Pascual M, Dobson A (2005) Seasonal patterns of infectious diseases. Plos Medicine 2: 18–20.
19. Hastie TT, Tibshirani R (1990) Generalized Additive Models. London: Chapman & Hall.
20. Svensson L (1992) Identification Guide to European Passerines. Stockholm: Natural History Museum.
21. Kulldorff M, Nagarwalla N (1995) Spatial disease clusters: Detection and inference. Statistics in Medicine 14: 799–810.
22. Stauss MJ, Burkhardt JF, Tomiuk J (2005) Foraging flight distances as a measure of parental effort in blue tits *Parus caeruleus* differ with environmental conditions. Journal of Avian Biology 36: 47–56.
23. Wilkin TA, Perrins C, Sheldon BC (2007) The use of GIS in estimating spatial variation in habitat quality: a case study of lay-date in the great tit *Parus major*. Ibis 149: 110–118.
24. Smits JE, Tella JL, Carrete M, Serrano D, Lopez G (2005) An epizootic of avian pox in endemic short-toed larks (Calandrella rufescens) and Berthelot's pipits (Anthus bertheloth) in the Canary Islands, Spain. Veterinary Pathology 42: 59–65.
25. Wikelski M, Foufopoulos J, Vargas H, Snell H (2004) Galápagos birds and diseases: invasive pathogens as threats for island species. Ecology and Society 9: 5.

26. Atkinson CT, Lease JK, Dusek RJ, Samuel MD (2005) Prevalence of pox-like lesions and malaria in forest bird communities on leeward Mauna Loa Volcano, Hawaii. Condor 107: 537–546.
27. Ritchie BW (1995) Poxviridae. In: Ritchie BW, editor. Avian viruses – Function and Control. Lake Worth, USA: Wingers Publishing Inc. pp. 285–311.
28. Jarvi SI, Triglia D, Giannoulis A, Farias M, Bianchi K, et al. (2008) Diversity, origins and virulence of Avipoxviruses in Hawaiian forest birds. Conservation Genetics 9: 339–348.
29. Jennelle CS, Cooch EG, Conroy MJ, Senar JC (2007) State-specific detection probabilities and disease prevalence. Ecological Applications 17: 154–167.
30. Altizer S, Dobson A, Hosseini P, Hudson P, Pascual M, et al. (2006) Seasonality and the dynamics of infectious diseases. Ecology Letters 9: 467–484.
31. Altizer S, Hochachka WM, Dhondt AA (2004) Seasonal dynamics of mycoplasmal conjunctivitis in eastern North American house finches. Journal of Animal Ecology 73: 309–322.
32. Randolph SE (2001) The shifting landscape of tick-borne zoonoses: tick-borne encephalitis and *Lyme borreliosis* in Europe. Philosophical Transactions of the Royal Society of London Series B-Biological Sciences 356: 1045–1056.
33. George DB, Webb CT, Farnsworth ML, O'Shea TJ, Bowen RA, et al. (2011) Host and viral ecology determine bat rabies seasonality and maintenance. Proceedings of the National Academy of Sciences 108: 10208–10213.
34. Senar JC, Conroy MJ (2004) Multi–state analysis of the impacts of avian pox on a population of Serins (*Serinus serinus*): the importance of estimating recapture rates. Animal Biodiversity and Conservation 27: 133–146.
35. Aruch S, Atkinson CT, Savage AF, LaPointe DA (2007) Prevalence and distribution of pox-like lesions, avian malaria, and mosquito vectors in Kipahulu Valley, Haleakala National Park, Hawai'i, USA. Journal of Wildlife Diseases 43: 567–575.
36. Young LC, VanderWerf EA (2008) Prevalence of avian pox virus and effect on the fledging success of Laysan Albatross. Journal of Field Ornithology 79: 93–98.

37. Gerlach H (1994) Avipoxviruses. In: Ritchie BW, Harrison GJ, Harrison LR, editors. Avian medicine: principles and application. Lake Worth, USA: Wingers Publishing Inc. pp. 865–873.
38. Buenestado F, Gortazar C, Millan J, Villafuerte R (2004) Descriptive study of an avian pox outbreak in wild red-legged partridges (*Alectoris rufa*) in Spain. Epidemiology and Infection 132: 369–374.
39. Illera JC, Emerson BC, Richardson DS (2008) Genetic characterization, distribution and prevalence of avian pox and avian malaria in the Berthelot's pipit (Anthus berthelotii) in Macaronesia. Parasitology Research 103: 1435–1443.
40. Hosseini PR, Dhondt AA, Dobson A (2004) Seasonality and wildlife disease: how seasonal birth, aggregation and variation in immunity affect the dynamics of *Mycoplasma gallisepticum* in house finches. Proceedings of the Royal Society of London Series B-Biological Sciences 271: 2569–2577.
41. Robb GN, McDonald RA, Chamberlain DE, Bearhop S (2008) Food for thought: supplementary feeding as a driver of ecological change in avian populations. Frontiers in Ecology and the Environment 6: 476–484.
42. Dhondt AA, Dhondt KV, Hawley DM, Jennelle CS (2007) Experimental evidence for transmission of Mycoplasma gallisepticum in house finches by fomites. Avian Pathology 36: 205–208.

CHAPTER 9

Molecular Epidemiology of Canine Norovirus in Dogs from Portugal, 2007–2011

João Rodrigo Mesquita[1,2] and Maria São José Nascimento[1]

[1]Departamento de Ciências Biológicas, Laboratório de Microbiologia, Universidade do Porto, Portugal, Rua Jorge Viterbo Ferreira, 228, Porto 4050-313, Portugal

[2]Secção de Ciências Veterinárias, Instituto Politécnico de Viseu, Quinta da Alagoa - Estrada de Nelas, Ranhados, Viseu 3500-606, Portugal

ABSTRACT

Background

Canine noroviruses (NoVs) have been recently described in south European countries and associated with outbreaks of diarrhea in kennels. Unlike human NoV which are known as an important cause of acute gastroenteritis, little is known about the role of canine NoV as pathogens in dogs as well as its epidemiological features.

Citation: Mesquita, J.R., Nascimento, M.S.J. Molecular epidemiology of canine norovirus in dogs from Portugal, 2007–2011. BMC Vet Res 8, 107 (2012). https://doi.org/10.1186/1746-6148-8-107

Results

From 2007–2011, 256 stool samples were collected from dogs across Portugal and tested by RT-PCR for canine NoV. Viral fecal shedding was found to be 23% (60/256). All sequences contained the GLPSG amino acid motif characteristic of the RNA-dependent RNA-polymerase gene of NoVs and had a high nucleotide identity (range 98%–100%) to the canine NoV first described in Portugal. The highest shedding rate was detected during the winter months.

Conclusions

This study shows that canine NoV infection is endemic in the dog population of Portugal. Peak shedding was detected in the winter months, a well-known epidemiologic feature of human NoV infections.

BACKGROUND

Noroviruses (NoV) are non-enveloped, single-stranded RNA viruses belonging to the genus *Norovirus* of the family *Caliciviridae* [1]. Until recently, NoV could be genetically classified into 5 genogroups (GI, GII, GIII, GIV and GV) with human NoV belonging to GI, GII, and GIV [1]. Human NoVs are the most frequent cause of outbreaks of acute gastroenteritis and the most common cause of sporadic enteric illness [2, 3]. Infections occur year round but in temperate climates outbreaks show a seasonal peak activity during the winter [4]. Recently, NoVs were reported in dogs with acute gastroenteritis [5, 6]. Canine NoV closely related to genogroup IV viruses was described in a diarrheic pup in Italy [5] and also caused an outbreak of diarrhea in a kennel in Greece [7]. A genetically different canine NoV was detected in dogs with diarrhea in Portugal [6]. Last year, we described an outbreak of gastroenteritis associated with canine NoV among kenneled dogs in Portugal, highlighting for the first time some epidemiological similarities between canine and human NoV outbreaks [8]. Despite the focus on the potential role as a pathogen in dogs, epidemiological features of canine NoV are not well described. Additionally, although a world-wide distribution of canine NoV has been suggested [9], the extent of geographical dissemination of canine NoV in different dog populations is unknown. In the present work, we tested fecal specimens from dogs collected from 2007 and 2011 in Portugal. Our data suggest seasonal variation of canine NoV infections similar to human NoV infections.

METHODS

Samples

Between 2007 and 2011, a total of 256 fecal samples were collected from dogs (regardless of their health status) housed in a variety of facilities such as municipal dog pounds, non-profit dog rescue shelters, animal shops, veterinary hospitals and veterinary clinics across Portugal.

RNA extraction and RT-PCR

Stools were diluted (10%) in phosphate-buffered saline, pH 7.2, and solids were removed by centrifugation at 8000 g for 5 min. Viral nucleic acid was extracted from 140 µl of each clarified stool suspension by the QIAamp viral RNA mini kit (Qiagen, Hilden, Germany), according to the manufacturers' instructions. Viral nucleic acid was tested for the presence of canine NoV by RT-PCR (Qiagen One-Step RT-PCR Kit) using the primers JV102 (5′-TGG GAT TCA ACA CAG CAGAG-3′) and JV103 (5′-TGC GCA ATA GAG TTG ACCTG-3′) targeting a small region of the RNA-dependent RNA-polymerase gene [6]. Briefly, 5 µl of viral nucleic acid was added to an RT-PCR mix (Qiagen One-Step RT-PCR Kit) containing RNase-free water, Qiagen RT-PCR buffer (5x), dNTP mix (10 mM), Enzyme mix (RT and Taq), 100 mM forward and reverse primer and RNase inhibitor (40 U). Cycling conditions consisted of reverse transcription for 30 min at 42°C, activation of Taq polymerase for 15 min at 95°C, 40 cycles of 1 min at 94°C, 1 min at 37°C, and 1 min at 72°C, followed by a final extension for 10 min at 72°C.

Sequencing and phylogenetic analysis

RT-PCR products were separated by electrophoresis in a 1.5% agarose gel and visualized under UV after ethidium bromide staining. Appropriately sized bands (215 bp) were excised from the gel and purified with the QIAquick gel extraction kit (QIAGEN) and sequenced in both directions using the BigDye Terminator v1.1 Cycle Sequencing kit (PE Applied Biosystems, Foster City, CA, USA). Sequence editing and multiple alignments were performed with the BioEdit software package, version 2.1. Phylogenetic analysis was drawn by using TreeCon software with bootstrap analysis (n = 1,000), and the tree topology was inferred by using neighbor-joining.

RESULTS

Sixty (23%) of the 256 samples tested positive for canine NoV. Sequences from the 60 samples contained the GLPSG amino acid motif characteristic of NoV viral RNA-dependent RNA-polymerase. Sequence analysis showed a high nucleotide identity (range 98%–100%) to the first canine NoV detected in Portugal, Viseu strain [6] (Figure 1).

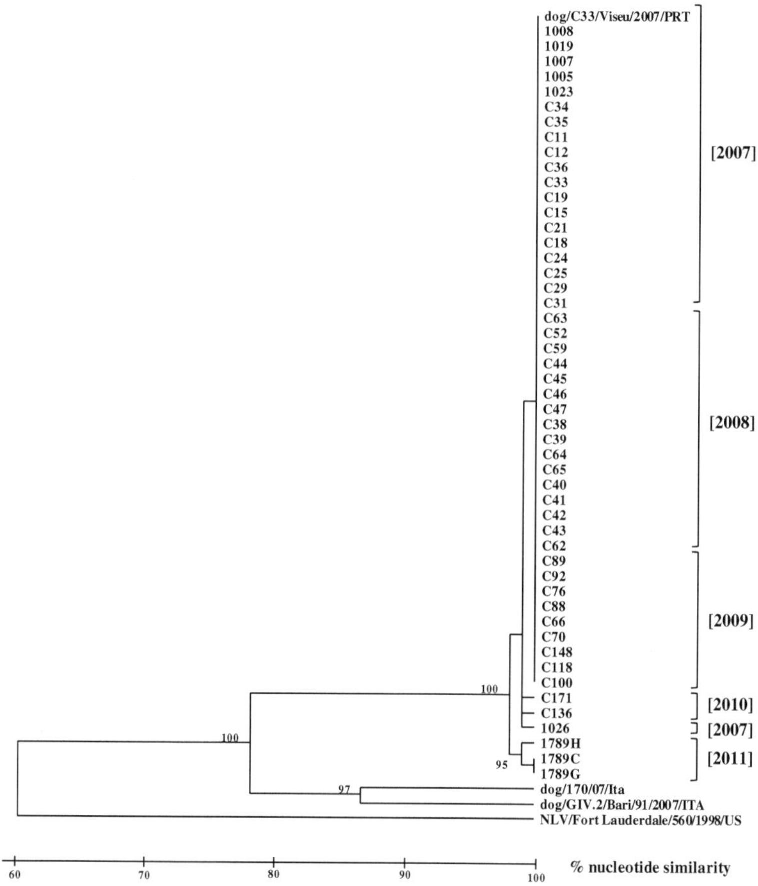

Figure 1. Phylogenetic tree of a 215-nt region of the RNA-dependent polymerase gene of one human genogroup IV strain NLV/Fort Lauderdale/560/1998/US (accession no: AF414426), the two canine noroviruses reported from Italy, dog/170/07/Ita (accession no: EU224456) and dog/GIV.2/Bari/91/2007/ITA (accession no: FJ875027), and the canine norovirus reported from Portugal dog/C33/Viseu/2007/PRT (accession no: GQ443611). Sample collection years are in brackets.

From a total of 91 dogs tested in winter months, 33 (36%) were found to be shedding canine NoV, whereas only 25% (21/84) and 7% (6/81) were shedding canine NoV in spring and autumn months, respectively (Figure 2). A chi-square test for homogeneity of proportions (SPSS 13.0, SPSS Inc., Chicago, IL, USA) showed that the percentage of positive samples in autumn, winter and spring months were significantly different ($p < 0.001$).

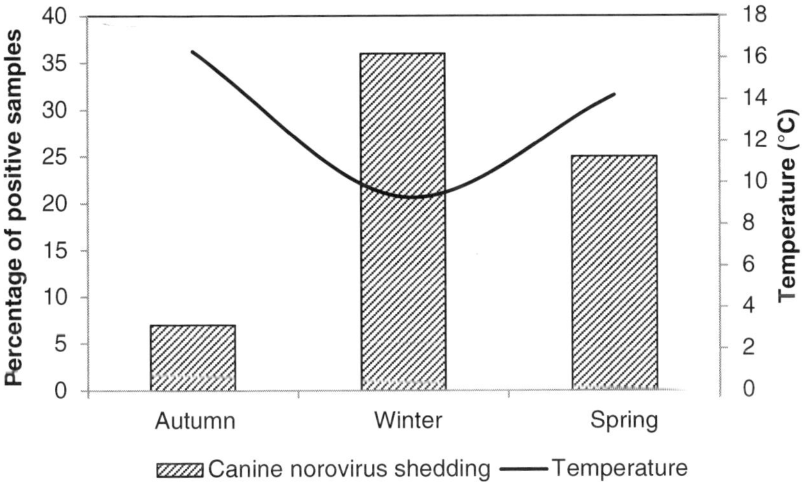

Figure 2. Seasonal variation of canine norovirus fecal shedding in dogs of Portugal from 2007 to 2011 and mean temperature in Portugal from 2000–2009 [10].

DISCUSSION

During a 4-year period 23% of stools collected from dogs across Portugal tested positive for canine NoV. Compared to the prototype canine NoV (Viseu strain), all strains had a high identity at nucleotide (range 98%–100%) and amino acid (100%) levels. At the nucleotide level several strains from 2007, 2008 and 2009 had identical sequences to the Viseu strain. Interestingly, strains detected in 2010 had an identical sequence as strain 1026 collected in 2007 tempting us to speculate that strain 1026 might have emerged as a new genetic variant and prevailed through 2010 in the dog population. However, because only a small fragment of the genome was sequenced these conclusions should be interpreted with caution and additional sequence information from the capsid regions of the genome is required to confirm this hypothesis.

The mechanism how new NoV variants emerge is currently unknown. For human NoVs, the high genetic mutation rate [11], as well as viral antigen and host receptor interactions [12] have been suggested as potential contributors. Similar events could have taken place for canine NoV in the dog population. Most canine NoV positive samples had been collected during winter months. Although no samples were collected during the summer months, our data show a peak of canine NoV activity in the colder months, an epidemiological feature typical of human NoV infections [13]. Additional studies with at least monthly sampling for several years are necessary for a more comprehensive analysis. The reason for human NoV seasonality remains unknown, but is believed to be a combination of complex multifactorial aspects that include environmental, host and virologic factors [14]. Changes in environmental conditions, such as humidity and temperature, are known to be associated with seasonality of viral diseases allowing for a winter predisposition [15]. The seasonality of rotavirus has been suggested to be due in part to the low relative humidity indoors during the cold periods, encouraging not only the persistence of infectious viruses on surfaces but also aerosolization of virus-laden dust particles [16]. Concerning NoV, it is known that cool and dry conditions are favorable for survival of infectious virus [17]. An increase in the number of canine NoV positive samples in the colder months might be associated with the overcrowding of kennels, a typical event in rainy and colder winter months where animals are more often kept indoors. Enteric virus transmission in dogs has been mentioned to be facilitated in large breeding colonies where hygiene is difficult to maintain and fecal contamination of the environment is at its maximum [18]. Moreover, seasonal variations in the host susceptibility to infections have also been suggested, possibly associated with changes in the animals' physiological status in winter [18]. Hence, more thorough kennel hygiene and disinfection procedures should be taken in the colder months when overcrowding is likely to occur, in order to decrease the chances for transmission of canine NoV.

CONCLUSIONS

This study shows that canine NoV infection is endemic in the dog population of Portugal. Peak shedding was detected in the winter months, a well-known epidemiologic feature of human NoV infections. This interesting finding is preliminary and requires to be further explored. Overall, the present work constitutes an important contribution to the knowledge of the epidemiology of canine NoV infection, still at its infancy.

REFERENCES

1. Clarke IN, Estes MK, Green KY, Hansman GS, Knowles NJ, Koopmans MK,Matson DO, Meyers G, Neill JD, Radford A, Smith AW, Studdert MJ, Thiel H-J, Vinjé J: Caliciviridae, Virus taxonomy: the classification and nomenclature of viruses. In The ninth report of the International Committee on Taxonomy of Viruses. Edited by King AMQ, Adams MJ, Carstens EB, Lefkowitz EJ. Walham, MA, USA: Springer-Verlag; 2012:977–986.
2. Glass RI, Parashar UD, Estes MK: Norovirus gastroenteritis. N Engl J Med. 2009, 361: 1776-1785. 10.1056/NEJMra0804575.
3. Patel MM, Hall AJ, Vinjé J, Parashar UD: Noroviruses: a comprehensive review. J Clin Virol. 2009, 44: 1-8. 10.1016/j.jcv.2008.10.009.
4. Siebenga JJ, Vennema H, Zheng DP, Vinjé J, Lee BE, Pang XL, Ho EC, Lim W, Choudekar A, Broor S, Halperin T, Rasool NB, Hewitt J, Greening GE, Jin M, Duan ZJ, Lucero Y, O'Ryan M, Hoehne M, Schreier E, Ratcliff RM, White PA, Iritani N, Reuter G, Koopmans M: Norovirus illness is a global problem: emergence and spread of norovirus GII.4 variants, 2001–2007. J Infect Dis. 2009, 200: 802-812. 10.1086/605127.
5. Martella V, Lorusso E, Decaro N, Elia G, Radogna A, D'Abramo M, Desario C, Cavalli A, Corrente M, Camero M, Germinario CA, Banyai K, Di Martino B, Marsilio F, Carmichael LE, Buonavoglia C: Detection and molecular characterization of a canine norovirus. Emerg Infect Dis. 2008, 14: 1306-1308.
6. Mesquita JR, Barclay L, Nascimento MSJ, Vinjé J: Novel Norovirus in Dogs with Diarrhea. Emerg Infect Dis. 2010, 16: 980-982.
7. Ntafis V, Xylouri E, Radogna A, Buonavoglia C, Martella V: An outbreak of canine norovirus infection in young dogs. J Clin Microbiol. 2010, 48: 2605-2608. 10.1128/JCM.02528-09.
8. Mesquita JR, Nascimento MSJ: Gastroenteritis outbreak associated with faecal shedding of Canine Norovirus in a Portuguese Kennel following introduction of imported dogs from Russia. Transbound Emerg Dis 2011, doi:10.1111/j.1865 1682.2011.01284.x.
9. Fenner's Veterinary Virology. Edited by: MacLachlan NJ, Dubovi EJ. 2011, Academic Press, London
10. IM: Análise Climatológica da década 2000–2009. Ministério da Ciência,Tecnologia e Ensino Superior. 2011:1–18.

11. Gallimore CI, Iturriza-Gomara M, Xerry J, Adigwe J, Gray JJ: Inter-seasonal diversity of norovirus genotypes: emergence and selection of virus variants. Arch Virol. 2007, 152: 1295-1303. 10.1007/s00705-007-0954-9.
12. Huang P, Farkas T, Zhong W, Tan M, Thornton S, Morrow AL, Jiang X: Norovirus and histo-blood group antigens: demonstration of a wide spectrum of strain specificities and classification of two major binding groups among multiple binding patterns. J Virol. 2005, 79: 6714-6722. 10.1128/JVI.79.11.6714-6722.2005.
13. Rohayem J: Norovirus seasonality and the potential impact of climate change. Clin Microbiol Infect. 2009, 15: 524-527. 10.1111/j.1469-0691.2009.02846.x.
14. Rydell GE, Kindberg E, Larson G, Svensson L: Susceptibility to winter vomiting disease: a sweet matter. Rev Med Virol. 2011, 21: 370-382. 10.1002/rmv.704.
15. Chew FT, Doraisingham S, Ling AE, Kumarasinghe G, Lee BW: Seasonal trends of viral respiratory tract infections in the tropics. Epidemiol Infect. 1998, 121: 121-128. 10.1017/S0950268898008905.
16. Rzezutka A, Cook N: Survival of human enteric viruses in the environment and food. FEMS Microbiol Rev. 2004, 28: 441-453. 10.1016/j.femsre.2004.02.001.
17. Lopman B, Armstrong B, Atchison C, Gray JJ: Host, weather and virological factors drive norovirus epidemiology: time-series analysis of laboratory surveillance data in England and Wales. PLoS One. 2009, 24: e6671.
18. Murphy F, Gibbs E, Horzinek M, Studdert MJ: Veterinary virology. London: SanDiego; 1999.

CHAPTER 10

Molecular Epidemiology of Porcine Cytomegalovirus (PCMV) in Sichuan Province, China: 2010–2012

Xiao Liu[1], Shan Liao[1], Ling Zhu[1], Zhiwen Xu[1,2], Yuancheng Zhou[1]

1 Animal Biotechnology Center, College of Veterinary Medicine, Sichuan Agricultural University, Ya' an, China, 2 Key Laboratory of Animal Disease and Human Health, College of Veterinary Medicine, Sichuan Agricultural University, Ya' an, China

ABSTRACT

Porcine cytomegalovirus (PCMV) is an immunosuppressive virus that mainly inhibits the immune function of the macrophage and T-cell lymphatic systems, and has caused huge economic losses to the porcine breeding industry. Molecular epidemiological investigation of PCMV is important

Citation: Liu X, Liao S, Zhu L, Xu Z, Zhou Y (2013) Molecular Epidemiology of Porcine Cytomegalovirus (PCMV) in Sichuan Province, China: 2010–2012. PLoS ONE 8(6): e64648. https://doi.org/10.1371/journal.pone.0064648

for prevention and treatment, and this study is the first such investigation in Sichuan Province, Southwest China. A PCMV positive infection rate of 84.4% (865/1025) confirmed that PCMV is widely distributed in Sichuan Province. A phylogenetic tree was constructed based on the PCMV glycoprotein B gene (*gB*) nucleotide and amino acid sequences from 24 novel Sichuan isolates and 18 other PCMV *gB* sequences from Genbank. PCMV does not appear to have evolved into different serotypes, and two distinct sequence groups were identified (A and B). However, whether PCMV from this region has evolved into different genotypes requires further research. Analysis of the amino acid sequences confirmed the conservation of *gB*, but amino acid substitutions in the major epitope region have caused antigenic drift, which may have altered the immunogenicity of PCMV.

INTRODUCTION

Porcine cytomegalovirus (PCMV) belongs to the genus *Cytomegalovirus*, subfamily *Betaherpesvirinae*, family *Herpesviridae* [1], and is an icosahedral virus with a double-stranded linear DNA genome. The virus particle diameter is 150–200 nm. PCMV is distributed globally, with reported cases in Britain, Japan, Germany, and the United States [2]. Rapidly progressive consumptive coagulopathy has been observed frequently in swine-to-baboon renal xenotransplantation, and there is also a risk of zoonotic transfer of PCMV in swine-to-human xenotransplantation. However, the risk of PCMV infection for human recipients needs further research [3], [4], [5].

PCMV induces fatal systemic infections in young animals, and can cause death in piglets and embryos. Infected animals show clinical symptoms of inclusion body rhinitis, pneumonia, and dysplasia, as well as disease of the nervous system. Similar to human cytomegalovirus (HCMV) and other cytomegaloviruses, lifelong latent infection often occurs in recovered pigs [6], [7], [8]. PCMV is highly host specific, and is unable to replicate in rabbits, rats, chicken embryos, or in a guinea pig cell culture system. Efforts to study PCMV have been hindered by its specificity for porcine epithelial cells. Cross-infection of PCMV and other viruses often occurs in immunosuppressed swine, and a variety of bacterial infections can occur following PCMV infection [9]. Therefore, PCMV has been recognized as a serious hazard for the pig breeding industry.

Currently, research on PCMV is lacking. Previous work has primarily concentrated on the DNA polymerase (*DPOL*) gene, glycoprotein B (*gB*) gene, and the ORF 40-like protein gene, and sequences for these regions are available in GenBank. Genetic diversity has been noted in the *DPOL* and *gB* genes of different PCMV strains, resulting in antigenic variation; however, there is no evidence to confirm that PCMV has evolved significantly different serotypes or genotypes [9], [10], [11].

The *gB* gene is highly conserved within the *Herpesviridae* family, and is widely used in species identification and phylogenetic analysis of herpesvirus species [9], [12], [13]. Several PCMV sequences from China are available in GenBank, but no systematic genome analysis of Chinese strains has been performed. In this study, we report the detection and characterization of PCMV based on sequencing and phylogenetic analysis of the *gB* nucleotide and amino acid sequences from a large quantity of samples collected from major breeding bases and rural farms in Sichuan Province, China.

MATERIALS AND METHODS

Ethical Statement

Animal welfare standard in this study was established on the basis of internationally agreed and science based principles within the World Organisation for Animal Health (OIE). All experiments were carried out in accordance with China Animal Welfare Legislation and were approved by the Sichuan Agricultural University Committee on Ethics in the Care and Use of Laboratory Animals.

Samples

A total of 670 porcine serum samples, 322 pulmonary hilar lymph node, spleen, kidney, liver, tonsil, and brain tissue samples, and 33 semen samples were randomly sampled from 1025 piglets, nursery pigs, sows, and boars from major breeding bases and rural farms in different districts of Sichuan Province between August 2010 and July 2012 (Figure 1).

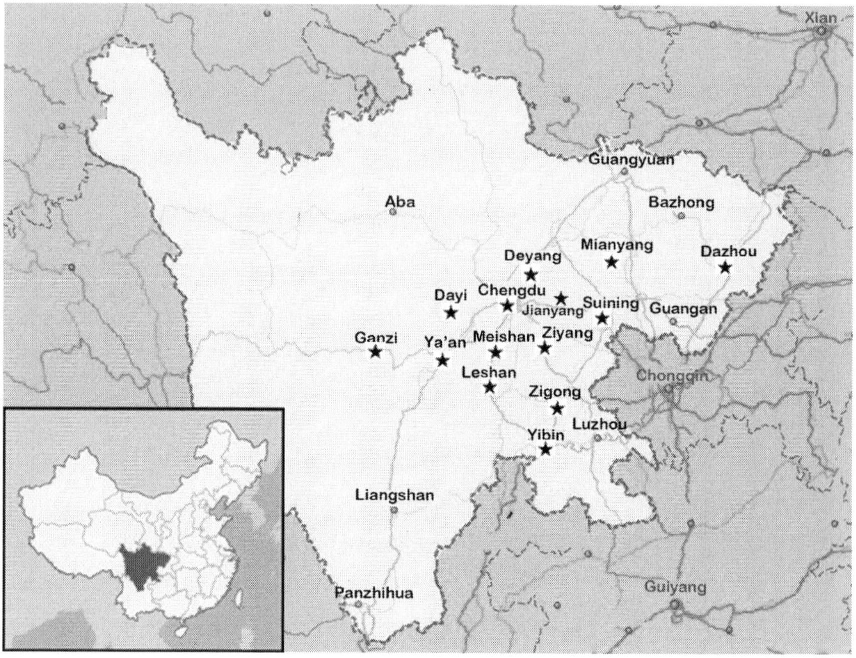

Figure 1. Geographical locations of samples collected in this study.

The stars indicate the districts of Sichuan Province where the samples were collected.

https://doi.org/10.1371/journal.pone.0064648.g001

DNA Extraction

PCMV genomic DNA was extracted from serum, semen, and tissue samples using a commercially available DNA extraction kit (Tiangen Biotech, Beijing, China) according to the manufacturer's instructions. The extracted DNA was quantitated by SmartSpec Plus Spectrophotometer (Hercules, CA, USA) and stored at −20°C until use in PCR reactions.

Virus Detection by PCR

The extracted DNA was screened by nested PCR for PCMV DNA using specific primers, P1-1:5′-CGTGGGTTACTATGCTTCTC-3′, P1-2:5′-CTTTCTAACGAGTTCTACGC-3′,P2-1:5′-TGGCTCAGGAAGzAGAAAGGAAGTG-3′,andP2-2:5′-GACGAGAGGACATTGTTGATAAAG-3′, which amplify a 236-bp region of the conserved

DPOL gene. Amplification was carried out in PCR buffer containing 200 μM of each dNTP, 10 pmol of each primer, 1.0 U *Taq* DNA polymerase (Promega, Madison, WI, USA), and 1.5 mM $MgCl_2$, in a total volume of 25 μL. PCR was performed at 94°C for 3 min, followed by 35 cycles of 94°C for 30 s, 55°C for 30 s, and 72°C for 35 s, and a final extension of 72°C for 8 min. PCR products were electrophoresed on a 1.5% agarose gel, followed by staining with ethidium bromide (Invitrogen, Carlsbad, CA, USA) and visualization under ultraviolet (UV) light on a Bio-Rad gel imaging system (Hercules, CA, USA).

Amplification of the PCMV gB Gene

PCMV genomic DNA was extracted from 24 antigen-positive samples as described above, and the primers P1:5′-**ATG**ACAGTGAGCAGTCGGAATTTATTCCGGAT-3′ and P2:5′-**TCA**CACGTCCTCGGTGGATAGCTGC-3′ were used to amplify the complete *gB* gene. The primers sets covered the entire coding region of the PCMV *gB* gene. PCR amplification was carried out as described above. The PCR products were electrophoresed on a 1.5% agarose gel, stained with ethidium bromide, and then visualized under UV light, as above.

Cloning and Sequencing of the PCMV *gB* Gene

The PCMV *gB* gene amplicons were gel-purified using a Gel Extraction Kit (Tiangen Biotech) and then cloned into pMD19-T Simple Vector (Takara, Dalian, China). The ligated products were transformed into *Escherichia coli* DH5α competent cells (Invitrogen). Colonies containing plasmids with an insert were identified by blue/white selection, and positive colonies were picked and grown in BL medium at 37°C for 8 hours. Plasmid DNA was then extracted using a Plasmid Mini Kit (Omega Bio-Tek, Norcross, GA, USA). Purified plasmid DNA was used as a template for sequencing (Invitrogen, Shanghai, China).

Phylogenetic Analyses

The amplified PCMV *gB* nucleotide and deduced amino acid sequences were assembled using LaserGene (DNAstar, Madison, WI, USA) and compared with nucleotide and amino acid sequences of 18 PCMV *gB* genes available from GenBank. The sequence homology was confirmed using the Basic Local Alignment Search Tool (BLAST) (http://blast.ncbi.nlm.nih.gov/). Multiple alignment was performed by ClustalW analysis [14], and Molecular Evolutionary Genetics Analysis (MEGA) software version 5.0 was used to conduct phylogenetic analyses and produce the neighbor-joining (NJ) and maximum-likelihood (ML) trees [15]. All analyses were based on the 2583 bp PCMV *gB* gene complete sequences and carried out using the Kimura 2-parameter model [16].

gB Nucleotide and Amino Acid Sequence Analyses

The nonsynonymous to synonymous nucleotide substitution rates were estimated by the Synonymous Nonsynonymous Analysis Program (SNAP) [17]. The transition/transversion bias (*R*) and substitution rates were estimated by MEGA 5.0. The substitution pattern and rates were estimated under the Kimura 2-parameter model, using nucleotide frequencies of A=25.00%, T/U=25.00%, C=25.00%, and G=25.00%. For estimating ML values, a user-specified topology was used. The maximum Log likelihood values for this computation were −713.948 and −737.102. Codon positions included were 1st +2nd +3rd+noncoding. All positions containing gaps and missing data were eliminated, and there were 2580 positions in the final dataset. Evolutionary analyses were conducted in MEGA 5.0 [16], [18]. The Data Analysis in Molecular Biology and Evolution program (DAMBE) was used to establish the plots of transition (s) and transversion (v) frequencies against the K80 distance, and to estimate the amino acid substitution propensity [19]. The gB protein phosphorylation sites were predicted using NetPhos 2.0 [20], analysis of the secondary structure was based on the Network Protein Sequence Analysis server (NPS@) [21], and transmembrane topology was analyzed using the Network server [22]. Three-dimensional (3-D) molecular structures were constructed by the Swiss-Model workspace (alignment mode) [23]. The T cell epitopes were predicted by NetChop 3.1 [24], and the image of the T cell epitopes was generated using Protean (DNAstar).

Accession Numbers

The novel PCMV *gB* nucleotide sequences have been submitted to NCBI GenBank and assigned accession numbers KC342266–KC342289. Further *Herpesviridae* glycoprotein B nucleotide sequences were obtained from GenBank (http://www.ncbi.nlm.nih.gov/nuccore), Specific accession numbers are given in Figure 2 and Figure 3.

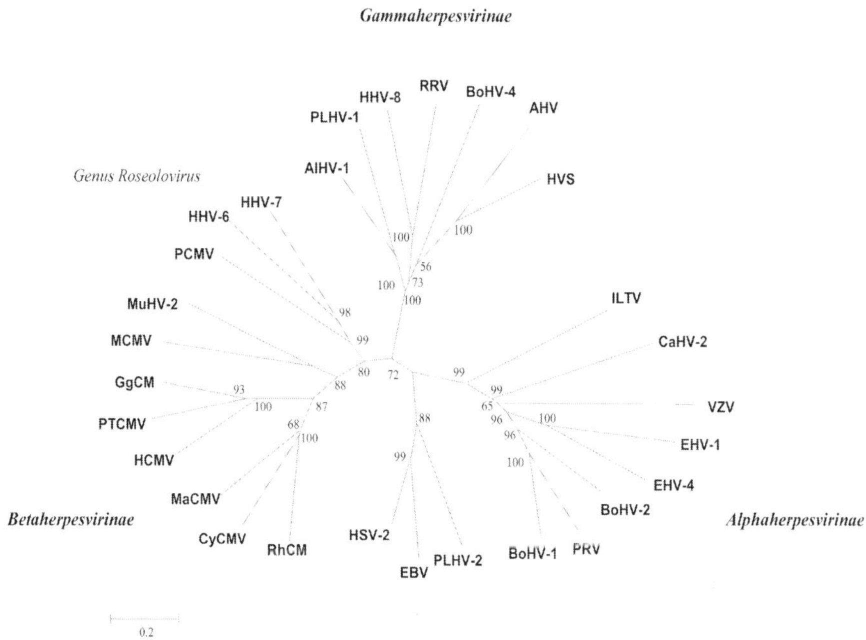

Figure 2. Phylogenetic analysis of *Herpesviridae* **viruses.**

Glycoprotein B nucleotide sequences of the alphaherpesviruses, betaherpesviruses, and gammaherpesviruses were obtained from GenBank: PCMV SC strain (accession no. JN701021.1); RhCMV (Rhesus cytomegalovirus; accession no. U76749.1); CyCMV (Cynomolgus macaque cytomegalovirus; accession no. HQ198248.1); MaCMV (Mycobacteriophage Barnyard; accession no. AY129339); PTCMV (Pan troglodytes cytomegalovirus; accession no. FJ538485); HCMV (Human cytomegalovirus; accession no. M6092.1); GgCMV (Gorilla cytomegalovirus; accession no. FJ538490); MuHV-2 (Murine herpesvirus type 2; accession no. AF232689.2); MCMV (Murine cytomegalovirus; M86302.1); HHV-6 (Human herpesvirus; accession no. M97927.1); PRV (Pseudorabies virus; accession no. AF257079.1; HHV 7 (Human herpesvirus, accession no. AF007830.1); AlHV-1 (Alcelaphineherpesvirus 1; accession no. AF005370); BoHV-1 (Bovine herpesvirus type 1; accession no. Z78205); BoHV-2 (Bovine herpesvirus type 2; accession no. Z78205); BoHV-4 (Bovine herpesvirus type 4; accession no. Z15044.1); EBV (Epstein-Barr virus; accession no. V01555.2); EHV-1 (Equine herpesvirus 1 strain Ab4; accession no. AY665713.1); EHV-4 (Equine herpesvirus 4 strain NS80567; accession no. AF030027.1); HHV-8 (Human herpesvirus 8 type M; accession no. U75698.1); VZV (Varicella-Zoster virus; accession no. X04370); PLHV-2 (porcine lymphotropicherpesvirus 2; accession no. AF191043); PLHV-1 (lymphotropicherpesvirus 1; accession no. AF478169.1); MuHV-2 (Murine herpesvirus 2; accession no. GU018179.1); HSV-2 (Herpes simplex virus 2; accession no. Z86099); AHV (Atelesherpesvirus; accession no. AF083424); ILTV (Laryngotracheitis Virus; accession no. X56093.1).Bootstrap values of >65% from 1000 pseudo-replicates are indicated at the branch nodes.

https://doi.org/10.1371/journal.pone.0064648.g002

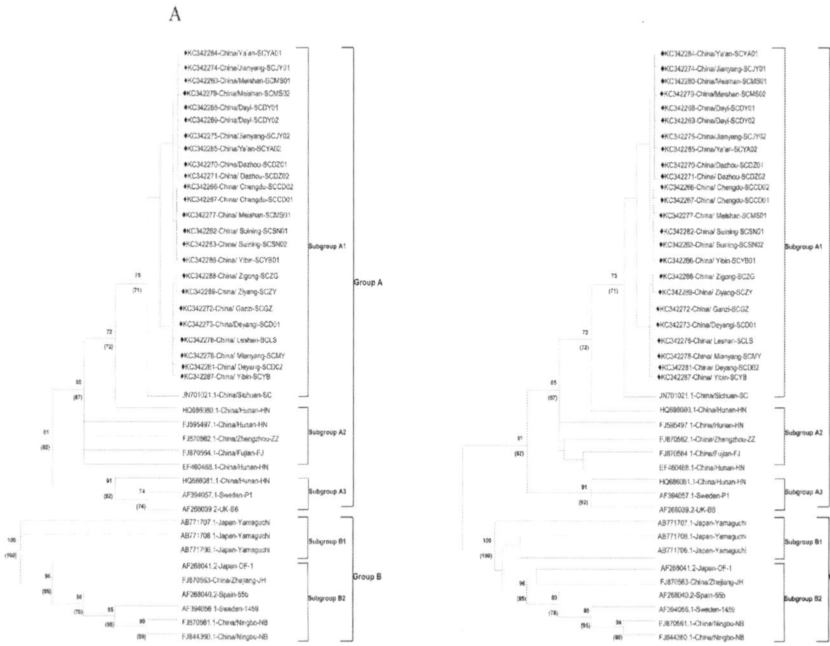

Figure 3. Phylogenetic analysis of 42 global strains of PCMV based on the 2580 bp *gB* **complete nucleotide sequence and the deduced amino acid sequence.**

https://doi.org/10.1371/journal.pone.0064648.g003

(A) Phylogenetic tree constructed from nucleotide sequences. (B) Phylogenetic tree constructed from amino acid sequences. The reference PCMV *gB* nucleotide sequences were obtained from GenBank: SC strain (accession no. JN701021.1); ZZ strain (accession no. FJ870562.1); NB strain (accession no. FJ844360.1, FJ870561.1); FJ strain (accession no. FJ870564.1); JH strain (accession no. FJ870563.1); HN isolate (accession no.HQ686081.1, HQ686080.1, FJ595497.1, EF460488.1); Spanish 55b isolate (accession no. AF268040.2); B6 strain (accession no. AF268039.2); Japanese OF-1 strain (accession no. AF268041.2); Japanese Yamaguchi strain (accession no. AB771707.1, AB771708.1, AB771706.1); Swedish isolate P1 (accession no. AF394057.1); Swedish isolate 1469 (accession no. AF394056).Multiple alignment was performed in the Clustal W program, and MEGA 5.0 software was used to construct the neighbor-joining (NJ) and maximum likelihood (ML) trees. Bootstrap values >65% from 1000 pseudo-replicates are indicated at the branch nodes. Numbers in parentheses are the maximum likelihood. Strains are designated by accession number, country/district, and strain name. The 24 strains analyzed in this study are indicated by (♦). The two major groups were identified as A and B.

RESULTS

PCMV Detection

The collection locations of the 1025 samples were distributed throughout 14 districts in Sichuan Province. Overall, the PCMV positive infection rate was 84.4% (865/1025), and the infection rate in different tissue samples was 91% (323/355). The rate of infection in serum samples was 80.9% (509/637), and was 100% (33/33) in semen samples. There were a further 25 samples that were putatively positive for PCMV but could not be confirmed, and were therefore not included in the positive rates of infection. The PCMV infection rates were greater than 80% in Yibin, Suining, Dazhou, Ya'an, Mianyang, Dayi, Meishan, Zigong, Leshan, and Ganzi districts, with rates of 66.7% (24/36), 60.5% (23/38), 77.3% (75/97), and 76.5% (52/68) detected in Chengdu, Jianyang, Deyang, and Ziyang, respectively. Except in samples collected from Ya'an, cross-infections were detected in all samples from the various districts of Sichuan province (Table 1). The PCMV antigen detection rates at different growth stages of pigs revealed that piglets have the highest PCMV infection rate (91.6%, 163/178), with PCMV infection rates of 80.4% (410/510) and 88% (220/250) in sows and nursery pigs, respectively. Boars showed the lowest rates of infection (46.7%, 42/92).

Table 1. PCMV, PRRSV, TTV1, and PRV infection and co-infection rates in different districts of Sichuan Province.

District	Infection rate (%)							
	PCMV	PRRSV	TTV1	PRV	PCMV and PRRSV Co-infection	PCMV and TTV1 Co-infection	PCMV and PRV Co-infection	PCMV, PRRSV,TTV1 and PRV Co-infection
Yibin	96.2% (50/52)	23.1% (12/52)	28.8% (15/52)	15.4% (8/52)	23.1% (12/52)	28.8% (15/52)	13.5% (7/52)	3.8% (2/52)
Suining	92.7% (102/110)	37.3% (41/110)	24.5% (27/110)	14.5% (16/110)	36.4% (40/110)	23.6% (26/110)	14.5% (16/110)	0.9% (1/110)
Chengdu	66.7% (24/36)	52.8% (19/36)	19.4% (7/36)	0	52.8% (19/36)	19.4% (7/36)	0	0
Daizhou	97.2% (105/108)	29.6% (32/108)	0	0	29.6% (32/108)	0	0	0
Yaan	100% (20/20)	0	0	0	0	0	0	0
Mianyang	93% (106/114)	46.5% (53/114)	31.6% (36/114)	36% (41/114)	43.9% (50/114)	31.6% (36/114)	34.2% (39/114)	12.3% (14/114)
Dayi	98.6% (73/74)	94.6% (70/74)	59.5% (44/74)	75.7% (56/74)	94.6% (70/74)	59.5% (44/74)	75.7% (56/74)	41.9% (31/74)
Jianyang	60.5% (23/38)	34.2% (13/38)	0	31.6% (12/38)	28.9% (11/38)	0	18.4% (7/38)	15.8% (6/38)
Meishan	94% (94/100)	92% (92/100)	42% (42/100)	84% (84/100)	92% (92/100)	40% (40/100)	84% (84/100)	38% (38/100)
Deyang	77.3% (75/97)	48.5% (47/97)	59.8% (58/97)	33% (32/97)	37.1% (36/97)	49.5% (48/97)	27.8% (27/97)	22.7% (22/97)
Zigong	92.5% (37/40)	0	15% (6/40)	0	0	15% (6/40)	0	0
Leshan	90% (54/60)	58.3% (35/60)	21.7% (13/60)	21.7% (13/60)	28.3% (17/60)	13.3% (8/60)	10% (6/60)	66.7% (4/60)
Ziyang	76.5% (52/68)	38.2% (26/68)	25% (17/68)	50% (34/68)	19.1% (13/68)	7.4% (5/68)	20.6% (14/68)	5.9% (4/68)
Ganzi	82.7% (62/75)	0	0	13.3% (10/75)	0	0	10.7% (8/75)	0

The PRRSV, TTV1, PRV infection of all samples has been detected by PCR (670 serum +322 tissue).
doi:10.1371/journal.pone.0064648.t001

https://doi.org/10.1371/journal.pone.0064648.t001

PCR Amplification, Sequence Alignment, and Phylogenic Analysis

PCRs performed with primers P1 and P2 amplified the expected 2580 bp complete PCMV *gB* gene fragment. The authenticity of the fragments cloned into the pMD19-T Simple Vector was confirmed by agarose gel electrophoresis and DNA sequencing. The 24 nucleotide and deduced amino acid sequences of the PCMV *gB* genes obtained from samples from 14 different districts of Sichuan Province were used to construct phylogenetic trees using the neighbor joining (NJ) and maximum likelihood (ML) methods. The nucleotide and amino acid sequence data of 18 PCMV and 29 *Herpesviridae* reference strains obtained from the GenBank database were used for comparison. Bootstrap values (>60) were obtained for both the NJ and ML trees.

Phylogenetic analysis of the *Herpesviridae* viral samples used in this study confirmed that PCMV strain SC (accession no. JN701021.1) is clustered with human herpes virus type 6 (HHV-6) and human herpes virus type 7 (HHV-7). PCMV strain SC, HHV-6, and HHV-7 were shown to belong to the same evolutionary branch of the subfamily *Betaherpesvirinae* (Figure 2), with nucleotide homologies of 48.5% and 49.4% for HHV-6 and HHV-7 *gB* sequences, respectively. PCMV showed less homology to Rhesus cytomegalovirus (RhCMV), *Cynomolgus macaque* cytomegalovirus (CyCMV), *Pan troglodytes* cytomegalovirus (PtCMV), murine cytomegalovirus (MCMV), gorilla cytomegalovirus (GgCMV), Mycobacteriophage Barnyard cytomegalovirus (MaCMV), and human cytomegalovirus (HCMV), with homologies 13.3–43.8%.

The phylogenetic and homology analysis results at the nucleotide (2580 bp) and amino acid (860 aa) levels showed that global PCMV strains belonged to two distinct sequence groups (A and B), which were separated with bootstrap values of 81 and 100, respectively. The PCMV *gB* gene sequences determined in this study, the 24 novel PCMV strains, the SC strain (accession no. JN701021.1), the ZZ strain (accession no. FJ870562.1), the NB strain (accession no. FJ844360.1, FJ870561.1), the FJ strain (accession no. FJ870564.1), the JH strain (accession no. FJ870563.1), the HN isolate (accession no.HQ686081.1, HQ686080.1, FJ595497.1, EF460488.1) the B6 strain from the United Kingdom (accession no. AF268039.2), and Swedish isolate P1 (accession no. AF394057.1) all belonged to A group, while isolate 55b from Spain (accession no. AF268040.2), strain OF-1 from Japan (accession no. AF268041.2), isolate 1469 from Sweden (accession no. AF394056), the JH strain (accession no. FJ870563.1), and the NB strain (accession no. FJ844360.1, FJ870561.1) belonged to B group (Figure 3). The 24 PCMV sequences identified in this study belong to the SC strain, and the nucleotide homology of the 24 *gB* gene sequences was 99.4–100%, with deduced amino acid sequence homologies of 98–100%. The nucleotide homology between the global PCMV strains and the newly obtained PCMV strains was 98–99%, with homologies of the deduced amino acid sequences of 97–99%.

Glycoprotein B Sequence Analyses

The 24 novel *gB* nucleotide sequences and their deduced amino acid sequences were also aligned. The maximum Kimura distance between the PCMV *gB* nucleotide sequences of the 24 novel Sichuan strains and the other global strains was 2.33%, in comparison with 2.13% between the *gB* nucleotide sequences of ten Chinese PCMV strains (Figure 4). The estimated transition/transversion bias (*R*) was 8.03 among the 34 PCMV strains from China, and 6.03 among global PCMV strains. The newly obtained Sichuan PCMV nucleotide sequences presented 30 common mutations (at nucleotide positions 82, 83, 85, 97, 124, 138, 308, 351, 369, 428, 457, 589, 735, 868, 1235, 1428, 1472, 1565, 1566, 1685, 2175, 2363, 2372, 2375, 2376, 2433, 2434, 2471, 2503, 2506) in the *gB* nucleotide sequence, 22 of which were found in more than one sequence. Based on a set of codon-aligned *gB* nucleotide sequences, the dN/dS (nonsynonymous to synonymous substitution) rates among the global and Sichuan strains were 2.39 (0.82/0.6) and 1.82 (0.39/0.27) respectively. The maximum likelihood estimates of the substitution matrix were also calculated (Table 2).

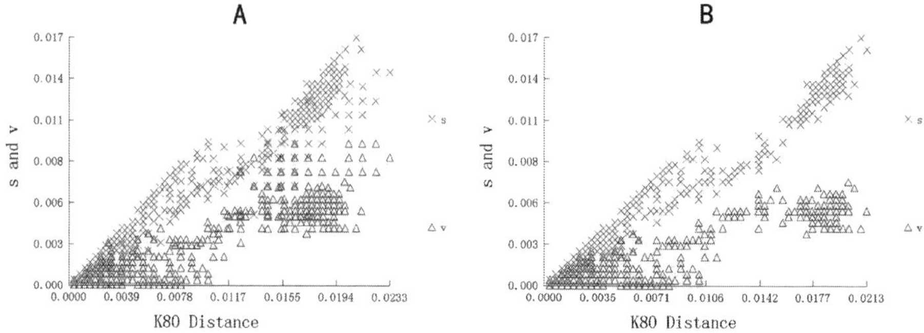

Figure 4. Plots of transition (s) and transversion (v) frequencies against the K80 distance.

S denotes transition and *V* denotes transversion. (A) Analysis of 34 PCMV strains from China. (B) Analysis of global PCMV strains.

https://doi.org/10.1371/journal.pone.0064648.g004

Table 2. Maximum likelihood estimate of substitution matrix.

34 PCMV strains from China					Global PCMV strains				
	A	T/U	C	G		A	T/U	C	G
A	-	*1.39*	*1.39*	**22.23**	A	-	*1.78*	*1.78*	**21.44**
T/U	*1.39*	-	**22.23**	*1.39*	T/U	*1.78*	-	**21.44**	*1.78*
C	*1.39*	**22.23**	-	*1.39*	C	*1.78*	**21.44**	-	*1.78*
G	**22.23**	*1.39*	*1.39*	-	G	**21.44**	*1.78*	*1.78*	-

Analysis of 34 PCMV strains from China and global PCMV strains. Substitution pattern and rates were estimated under the Kimura 2-parameter model. Rates of different transition substitutions are shown in **bold** and those of transversion substitutions are shown in *italics*.
doi:10.1371/journal.pone.0064648.t002

https://doi.org/10.1371/journal.pone.0064648.t002

The 18 deduced glycoprotein B amino acid sequences from the global strains were aligned and analyzed along with the novel 24 Sichuan strains. The maximum index of substitution propensity was 0.9% between the Spanish 55b strain (accession no. AF268040.2), the B6 strain (accession no. AF268039.2), the Japanese OF-1 strain (accession no. AF268041.2), the Japanese Yamaguchi strain (accession no. AB771707.1, AB771708.1,

AB771706.1), and the Swedish strain (accession no. AF394056, AF394057.1), in comparison with 1.3%% between the SC strain (accession no. JN701021.1), the ZZ strain (accession no. FJ870562.1), the NB strain (accession no. FJ844360.1, FJ870561.1), the JH strain (accession no. FJ870563.1), the FJ strain (accession no. FJ870564.1), and the HN isolates (accession no.HQ686081.1, HQ686080.1, FJ595497.1, EF460488.1) (Figure 5).

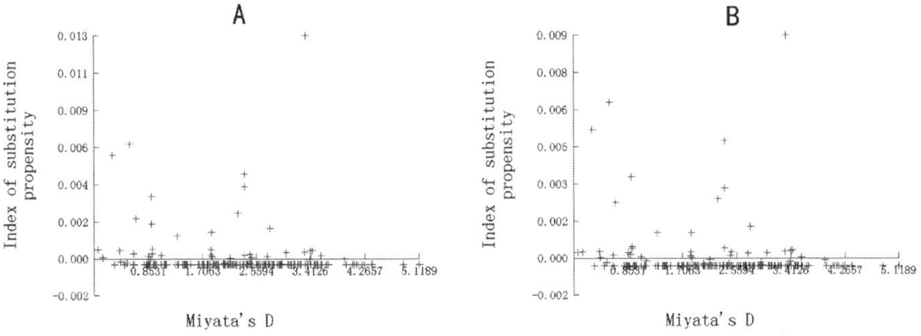

Figure 5. Index of amino acid substitution propensity.

(A) Analysis of 34 PCMV strains from China. (B) Analysis of global PCMV strains.

https://doi.org/10.1371/journal.pone.0064648.g005

Among the global PCMV strains, Most amino acid substitutions were concentrated at the C–terminus, the β–turn, and the loop region (random coil) of the gB protein (Figure 6A). Prediction of the phosphorylation sites in the gB amino acid sequence showed there are 51 potential phosphorylation sites with a threshold value of 0.5, including 29 serine phosphorylation sites, 13 threonine phosphorylation sites, and 19 tyrosine phosphorylation sites (Figure 6B). The T-cell epitopes were mainly distributed at positions 23–109 aa and 769–861 aa of the gB protein (Figure 6C). The Sichuan strains had 15 common amino acid substitutions (T^{28}→Y, P^{47}→A, D^{104}→N, S^{153}→P, N^{197}→Y, G^{245}→D, T^{290}→A, G^{412}→A, E^{491}→G, E^{791}→A, F^{792}→S, Y^{812}→D, I^{824}→T, N^{835}→D, A^{836}→T) (Table 3), including eight phosphorylation site substitutions. The 3-D conformation prediction showed that the substitutions in the loop region and β-turn altered not only the conformation of these regions, but also altered the conformation of the neighboring regions (Figure 7).

Table 3. Data from the 24 PCMV clinical isolates and prototype Sichuan strain (GenBank accession no. JN701021.1).

District	Designation	Specimen	Category	Amino acid substitutions													
				28	42	104	153	197	245	290	412	491	791	792	812	824	836
Chengdu	KC342266- SCCD01	Hepar	nursery swine	–	P	N	–	–	D	A	A	G	–	S	–	–	–
	KC342267- SCCD02	Hepar	nursery swine	Y	P	N	–	–	D	A	A	G	–	S	–	–	–
Dayi	KC342268- SCDY01	Pulmonary hilar lymph node	piglet	–	P	N	P	–	D	A	–	G	A	–	D	–	–
	KC342269- SCDY02	Pulmonary hilar lymph node	piglet	–	P	N	P	–	D	A	–	G	A	–	–	–	–
Dalzhou	KC342270- SCDZ01	Hepar	nursery swine	–	P	N	–	–	D	A	–	G	–	–	–	–	A
	KC342271- SCDZ02	Hepar	nursery swine	–	P	N	–	–	D	A	–	G	–	–	–	–	–
Ganzi	KC342272- SCGZ	Spleen	sow	–	P	N	–	–	D	A	–	G	–	–	–	–	A
Deyang	KC342273- SCD01	Hepar	piglet	–	P	N	–	–	D	A	–	G	–	–	–	–	A
	KC342281- SCD02	Hepar	piglet	–	P	N	–	–	D	A	–	G	–	–	–	–	A
Jianyang	KC342274- SCJY01	Hepar	sow	–	P	N	P	–	D	A	–	G	–	–	–	T	–
	KC342275- SCJY02	Hepar	sow	–	P	N	P	–	D	A	–	G	–	–	–	T	–
Leshan	KC342276- SCLS01	Pulmonary hilar lymph node	piglet	–	P	N	–	–	D	A	–	G	–	–	–	–	A
Mianyang	KC342277- SCMY01	Pulmonary hilar lymph node	sow	–	–	N	–	–	D	A	–	G	–	–	–	–	A
	KC342278- SCMY02	Pulmonary hilar lymph node	sow	–	–	N	–	–	D	A	–	G	–	–	–	–	–
Meishan	KC342279- SCMS01	Hepar	sow	–	P	N	P	Y	D	A	–	G	–	–	–	–	–
	KC342280- SCMS02	Hepar	sow	–	P	N	P	–	D	A	–	G	–	–	–	–	–
Suining	KC342282- SCSN01	kidney	piglet	–	A	N	P	Y	D	A	–	G	–	–	–	–	–
	KC342283- SCSN02	kidney	piglet	–	A	N	P	–	D	A	–	G	–	–	–	–	–
Yaan	KC342284- SCYA01	Pulmonary hilar lymph node	piglet	Y	–	N	P	Y	D	A	–	G	–	–	–	–	–
	KC342285- SCYA02	Pulmonary hilar lymph node	piglet	–	–	N	P	–	D	A	–	G	–	–	–	–	–
Yibin	KC342286- SCYB01	kidney	piglet	Y	–	N	–	–	D	A	–	G	–	–	–	–	A
	KC342287- SCYB02	kidney	piglet	–	.P	N	–	–	D	A	–	G	–	–	–	–	–
Zigong	KC342288- SCZG	Pulmonary hilar lymph node	piglet	–	P	N	–	–	D	A	–	G	–	–	–	–	A
Ziyang	KC342289- SCZY	Spleen	sow	–	P	N	–	–	D	A	–	G	–	–	–	–	A
	SC JN701021.1			–	P	N	–	–	D	A	–	G	–	–	–	–	–

Strains are designated according to accession number and strain name.
doi:10.1371/journal.pone.0064648.t003

https://doi.org/10.1371/journal.pone.0064648.t003

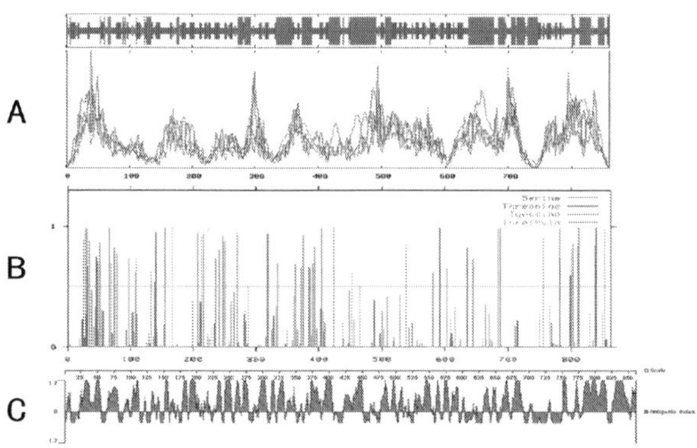

Figure 6. Illustration and sequence alignment of the gB protein.

https://doi.org/10.1371/journal.pone.0064648.g006

(A) Analysis of the secondary structure. The α-helix, extended strand, loop region, and β-turn are indicated by blue, red, yellow, and green curves, respectively. (B) Phosphorylation site prediction. (C) Prediction of T-cell epitopes.

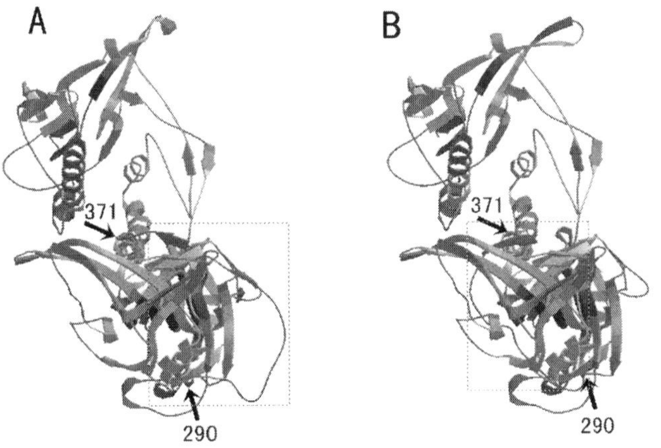

Figure 7. Three-dimensional structure prediction for alignment of the gB sequence of two viral strains, (A) JN701021.1-China-SC, (B) AF268040.2-Spain-55b.

The loop region (random coil) of the gB protein is shown in the closed box, and amino acid residues 290–371 is indicated.

https://doi.org/10.1371/journal.pone.0064648.g007

DISCUSSION

PCMV is an economically important pathogen due to its immunosuppressive effect of inhibiting the immune function of T lymphocytes and macrophages, and PCMV-infected pigs are easily infected by other pathogens [2], [6], [25], [26]. To survey the prevalence of PCMV in Sichuan Province, China, and provide a foundation for its prevention, a large number of serum, semen, and tissues samples were collected from 14 different districts of the province. The sampling area covered the entire territory of Sichuan Province, including the Chengdu Plain and mountain region of Ganzi Prefecture. The PCMV infection in different tissues of the same pig has not been detected in this study, therefore, the distribution regulation of PCMV in animal body need further research.

Nested PCR showed that PCMV infection was widespread in large-scale breeding bases and rural farms of Sichuan, and the antigen positive rate of 84.4% was close to the reported domestic and foreign rates of infection. According to the PCMV infection rates at different growth stages, piglets, nursery pigs, and expectant sows were more vulnerable to PCMV than boars. In this study, we could infer from clinical testing data that vertical transmission was the major transmission mode of PCMV. PCMV antigen detection results from different samples showed the virus is widely distributed in various organs of the body, and mainly affects the immune and respiratory systems. In this study, PCMV was detected for the first time in boar semen. The rate of PCMV infection in sows from the same farm was very high; therefore, we speculate that PCMV may spread through the reproductive system, and it may be necessary to adopt quarantine measures for infected boars. Based on the results of clinical testing, PCMV-infected pigs were always co-infected with porcine reproductive and respiratory syndrome virus (PRRSV), transfusion transmitted virus (TTV1), and pseudorabies virus (PRV). In addition, *Streptococcus*, *Haemophilusparasuis*, *Actinobacillus*, and *Pasteurellasuis* infections often occur following PCMV infection, making prevention and treatment more difficult. Therefore, the collaborative immune suppression mechanisms between PCMV and other immunosuppressive viruses are worthy of further study, and we believe that this avenue has great significance for prevention and control of immunosuppressive viruses.

PCMV glycoprotein B is an important transmembrane glycoprotein that plays a major role in the fusion and adhesion of the virus onto the membrane of host cells when the virus enters cells or transfers between cells [27]. Glycoprotein B also plays an important role in the induction of humoral and cellular immune responses. It is the target of cytotoxic T lymphocytes, and is recognized by natural killer cells that have been activated by lymphokine. Glycoprotein B induces protective immunity and has significant immunogenicity, therefore, further research could provide a new approach for the treatment and prevention of PCMV [28], [29]. Molecular epidemiological analysis of PCMV based on full-length *gB* gene sequences from different strains and further gene molecular characterization analyses would help to devise a suitable strategy to prevent losses in production.

This study is the first molecular analysis of the full-length PCMV *gB* nucleotide sequence from PCMV strains isolated worldwide. The nucleotide and amino acid sequences of 24 new Sichuan strains and 13 global strains were aligned and analysed. Phylogenetic tree and homology analysis showed there was no significant variation between *gB* nucleotide and amino acid sequences from different PCMV strains, confirming that the *gB* gene sequences from Sichuan and global PCMV strains are highly conserved, and that the viral genome is stable. These factors are conducive to the development of measures for the prevention and treatment of the virus.

We determined that the homologies of *gB* nucleotide and amino acid sequences investigated in this study were higher than 98% and 97%, respectively. According to the "75%/85% rule", if more than 75% of the nucleotide sequence or more than 85% of the amino acid sequence is homologous, the sequences are considered to be from the same serotype [30], [31]. This finding confirmed that globally, PCMV strains have not evolved into significantly different serotypes since PCMV was first isolated in 1955 in Britain.

We determined that tree strains from China, the Ningbo strain (accession no. FJ844360.1, FJ870561.1), the Zhejiang strain JH (accession no.FJ870563), isolate 55b from Spain (accession no. AF268040.2), Japanese strain OF-1 (accession no. AF268041.2), the Japanese Yamaguchi strain (accession no. AB771707.1, AB771708.1, AB771706.1), and a Swedish strain (accession no. AF394056.1) belonged to group B. Therefore, we hypothesized that the tree Chinese PCMV strains may have come from Europe or Japan. However, the prevalence of foreign strains was limited to the eastern part of China.

Phosphorylation may affect the stability of the PCMV virus complex. The level of phosphorylation may adjust the binding energy between the viral proteins in the replication complex, and affect the interaction between the viral and host proteins. The results of 3-D conformation prediction showed that the phosphorylation site substitutions of serine, threonine, and tyrosine, at $T^{28}\rightarrow Y$, $S^{153}\rightarrow P$, $N^{197}\rightarrow Y$, $T^{290}\rightarrow A$, $F^{792}\rightarrow S$, $Y^{812}\rightarrow D$, $I^{824}\rightarrow T$, and $A^{836}\rightarrow T$ (Figure 7), have likely altered the conformation of the PCMV gB protein. These amino acid substitutions were concentrated in the gB major epitope region in PCMV SC strains, and antigenic drift was affected by accumulation of point mutations among the surface proteins. Mutations in the epitope regions may change the tissue tropism of the virus, therefore, the conformational change in the β-turn and loop regions may have already lead to antigenic drift, changed the immunogenicity of the virus to the host, and affected the specific immune response in the organism. However, whether these changes have altered the capacity of gB for adhesion and penetration of the host cell membrane needs further investigation.

CONCLUSION

This is the first molecular epidemiological investigation of PCMV in Sichuan Province, China. The PCMV infection rate of 84.4% (865/1025) confirmed that PCMV has been widely distributed in southwest China. PCMV was detected for the first time in porcine semen, allowing us to conclude that it may spread through the reproductive system. The results of nucleotide and amino acid sequence analyses confirmed that the PCMV gB protein is conserved, but that amino acid substitutions in the gB epitope region may cause antigenic drift, and alter the immunogenicity of PCMV.

ETHICAL APPROVAL

Animal welfare standard in this study was established on the basis of internationally agreed and science based principles within the World Organisation for Animal Health (OIE). All experiments were carried out in accordance with China Animal Welfare Legislation and were approved by the Sichuan Agricultural University Committee on Ethics in the Care and Use of Laboratory Animals.

ACKNOWLEDGMENTS

We would like to acknowledge all members of Animal Biotechnological Center for their contribution to this study.

REFERENCES

1. Roizmann B, Desrosiers R, Fleckenstein B, Lopez C, Minson A, et al. (1992) The family Herpesviridae: an update. Arch Virol 123: 425–449.
2. Corner A, Mitchell D, Julian R, Meads E (1964) A generalized disease in piglets associated with the presence of cytomegalic inclusions. J Comp Pathol 74: 192.
3. Clark DA, Fryer JFL, Tucker AW, McArdle PD, Hughes AE, et al. (2003) Porcine cytomegalovirus in pigs being bred for xenograft organs: Progress towards control. Xenotransplantation 10: 142–148.
4. Gollackner B, Mueller NJ, Houser S, Qawi I, Soizic D, et al. (2003) Porcine cytomegalovirus and coagulopathy in pig-to-primate xenotransplantation1. Transplantation 75: 1841–1847.
5. Tucker A, Galbraith D, McEwan P, Onions D (1999) Evaluation of porcine cytomegalovirus as a potential zoonotic agent in xenotransplantation. Transplantation 31: 1460.
6. Edington N (1986) Porcine cytomegalovirus. Dis Swine 138: 330–336.
7. Edington N, Plowright W, Watt R (1976) Generalized porcine cytomegalic inclusion disease: distribution of cytomegalic cells and virus. J Comp Pathol 86: 191–202.
8. Staczek J (1990) Animal cytomegaloviruses. Microbiol Rev 54: 247–265.
9. Widen F, Goltz M, Wittenbrink N, Ehlers B, Banks M, et al. (2001) Identification and sequence analysis of the glycoprotein B gene of porcine cytomegalovirus. Virus genes 23: 339–346.
10. Goltz M, Widen F, Banks M, Belak S, Ehlers B (2000) Characterization of the DNA polymerase loci of porcine cytomegaloviruses from diverse geographic origins. Virus genes 21: 249–255.
11. Rupasinghe V, Iwatsuki-Horimoto K, Sugii S, Horimoto T (2001) Identification of the porcine cytomegalovirus major capsid protein gene. J Vet Med Sci 63: 609.
12. Cranage M, Kouzarides T, Bankier A, Satchwell S, Weston K, et al. (1986) Identification of the human cytomegalovirus glycoprotein B gene and induction of neutralizing antibodies via its expression in recombinant vaccinia virus. EMBO J 5: 3057.
13. Schleiss MR (1994) Cloning and characterization of the guinea pig cytomegalovirus glycoprotein B gene. Virology 202: 173–185.

14. Higgins DG, Sharp PM (1988) CLUSTAL: a package for performing multiple sequence alignment on a microcomputer. Gene 73: 237–244.
15. Kumar S, Stecher G, Peterson D, Tamura K (2012) MEGA-CC: Computing Core of Molecular Evolutionary Genetics Analysis program for automated and iterative data analysis. Bioinformatics 10: 1–2.
16. Kimura M (1980) A simple method for estimating evolutionary rates of base substitutions through comparative studies of nucleotide sequences. J MolEvol 16: 111–120.
17. Korber B (2000) HIV signature and sequence variation analysis. Computational analysis of HIV molecular sequences 4: 55–72.
18. Tamura K, Dudley J, Nei M, Kumar S (2007) MEGA4: molecular evolutionary genetics analysis (MEGA) software version 4. 0. MolBiolEvol 24: 1596–1599.
19. Xia X, Xie Z (2001) DAMBE: software package for data analysis in molecular biology and evolution. J Hered 92: 371–373.
20. Blom N, Gammeltoft S, Brunak S (1999) Sequence and structure-based prediction of eukaryotic protein phosphorylation sites. J Molbiol 294: 1351.
21. Combet C, Blanchet C, Geourjon C, Deleage G (2000) NPS@: network protein sequence analysis. Trends Biochem Sci. 25: 147.
22. Hofman K (1993) TMbase-A database of membrane spanning protein segments. BiolChem Hoppe-Seyler 374: 166.
23. Kiefer F, Arnold K, Künzli M, Bordoli L, Schwede T (2009) The SWISS-MODEL Repository and associated resources. Nucleic Acids Res 37: D387–D392.
24. Nielsen M, Lundegaard C, Lund O, Keşmir C (2005) The role of the proteasome in generating cytotoxic T-cell epitopes: insights obtained from improved predictions of proteasomal cleavage. Immunogenetics 57: 33–41.
25. Booth J, Goodwin R, Whittlestone P (1967) Inclusion-body rhinitis of pigs: attempts to grow the causal agent in tissue cultures. Res Vet Sci 8: 338.
26. Plowright W, Edington N, Watt R (1976) The behaviour of porcine cytomegalovirus in commercial pig herds. J Hyg 76: 125–135.
27. McGregor A, Liu F, Schleiss MR (2004) Molecular, biological, and in vivo characterization of the guinea pig cytomegalovirus (CMV) homologs of the human CMV matrix proteins pp71 (UL82) and pp65 (UL83). J Virol 78: 9872–9889.

28. Mitchell DK, Holmes SJ, Burke RL, Duliege AM, Adler SP (2002) Immunogenicity of a recombinant human cytomegalovirus gB vaccine in seronegative toddlers. Pediatr Infect Dis J 21: 133–138.

29. Mocarski E, Kemble Jr G (1996) Recombinant cytomegaloviruses for study of replication and pathogenesis. Intervirology 39: 320–330.

30. Oberste MS, Maher K, Kilpatrick DR, Flemister MR, Brown BA, et al. (1999) Typing of human enteroviruses by partial sequencing of VP1. J ClinMicrobiol 37: 1288–1293.

31. Oberste MS, Maher K, Kilpatrick DR, Pallansch MA (1999) Molecular evolution of the human enteroviruses: correlation of serotype with VP1 sequence and application to picornavirus classification. J Virol 73: 1941–1948.

SECTION 3: RABIES

CHAPTER 11

Complete Genome and Molecular Epidemiological Data Infer the Maintenance of Rabies among Kudu (Tragelaphus strepsiceros) in Namibia

Terence P. Scott[1], Melina Fischer[2], Siegfried Khaiseb[3], Conrad Freuling[2], Dirk Ho¨per[2], Bernd Hoffmann[2], Wanda Markotter[1], Thomas Mu¨ ller[2], Louis H. Nel[1]

[1]Department of Microbiology and Plant Pathology, University of Pretoria, Pretoria, South Africa,

[2]Friedrich-Loeffler-Institut, Greifswald-Insel Riems, Germany, 3 Central Veterinary Laboratories, Windhoek, Namibia

ABSTRACT

Rabies in kudu is unique to Namibia and two major peaks in the epizootic have occurred since it was first noted in 1977. Due to the large numbers of kudu that

Citation: Scott TP, Fischer M, Khaiseb S, Freuling C, Höper D, Hoffmann B, et al. (2013) Complete Genome and Molecular Epidemiological Data Infer the Maintenance of Rabies among Kudu (Tragelaphus strepsiceros) in Namibia. PLoS ONE 8(3): e58739. https://doi.org/10.1371/journal.pone.0058739

were affected, it was suspected that horizontal transmission of rabies occurs among kudu and that rabies was being maintained independently within the Namibian kudu population – separate from canid cycles, despite geographic overlap. In this study, it was our aim to show, through phylogenetic analyses, that rabies was being maintained independently within the Namibian kudu population. We also tested, through complete genome sequencing of four rabies virus isolates from jackal and kudu, whether specific mutations occurred in the virus genome due to host adaptation. We found the separate grouping of all rabies isolates from kudu to those of any other canid species in Namibia, suggesting that rabies was being maintained independently in kudu. Additionally, we noted several mutations unique to isolates from kudu, suggesting that these mutations may be due to the adaptation of rabies to a new host. In conclusion, we show clear evidence that rabies is being maintained independently in the Namibian kudu population – a unique phenomenon with ecological and economic impacts.

INTRODUCTION

Rabies virus (RABV) is typically maintained within species of the mammalian families *Carnivora* and *Chiroptera*. The classical primary host species among carnivores is the dog (*Canis familiaris*), but several other carnivores are known to be able to maintain an independent cycle of RABV, for instance the red fox (*Vulpes vulpes*) in Europe [1] and raccoons (*Procyon lotor*) [2] and skunks (*Mephitis mephitis*) in North America, among others. With these classical trends, the emergence of an extensive rabies cycle in an herbivorous host is unusual. However, in 1977 a rabies epizootic in kudu emerged in Namibia [3] and, with several peaks and troughs in the numbers of observed cases through the decades, is still ongoing today. Due to the geographical extent and the numbers of animals affected, it was suspected that this rabies cycle was maintained within the kudu population.

For rabies, spill-over infections from carnivores to domestic and wild herbivores are known to occur regularly, but such spill-overs are invariably dead-end [4] and serve as an indicator of rabid carnivore (domestic or wildlife) activity [5]. Generally, there are several factors that are necessary for the adaptation and maintenance of rabies in a new host. One factor is the need for overlapping host ranges (sympatry) of a susceptible species with that of a species already maintaining a cycle of RABV [6]. It has also been suggested that the initial infection of a new host is aided by the similarity of the cellular and immunological traits of the new host with the donor, with the number of exposures (due to geographical overlap) being a secondary factor [6]. These factors would explain the relatively high rate of transmission of rabies to new canid hosts (e.g. fox, raccoon, skunk, bat-eared fox etc.) as opposed to non-canid hosts that share the same host range. Another factor that has been suggested is that of recipient host population densities [1], [7].

A limited number of studies have aimed to determine whether specific mutations occur in the RABV genome due to the selective pressures caused by host switching and adaptation. One study noted mutations specific to certain fox species throughout Europe, specifically in the glycoprotein and nucleoprotein genes [1]. However, despite this phylogenetic clustering, different viruses predominantly clustered according to their geographical origin, suggesting that spatial and behavioral patterns of the host species play a greater role in the phylogenetic clustering of RABVs [1], [8]. A second study showed evidence that two types of adaptation could occur: 1) Post-shift adaptation, where the virus mutates in order to better suit the new host; 2) Pre-shift adaptation, in which the virus adapts through convergent evolution to better suit the shift from one host to a new host [9]. It was shown that pre-shift adaptation may have occurred in the introduction of RABVs from bats to skunks and foxes in the Flagstaff region of the USA [9].

As rabies in Namibia frequently affects canids, e.g. dogs and jackals [10], it remained unclear whether rabies cases in kudu are simply a result of a high rate of spillover from these primary hosts, or a truly independent cycle primarily driven by horizontal transmission among these herbivores. Experimental observations of RABV in kudu have suggested that transmission could occur via non-bite means and that the mucosal surfaces of kudu are particularly susceptible to infection [11]. An earlier epidemiological study did consider molecular phylogeny, but included limited sample numbers that did not allow for definitive conclusions about the independence of the kudu rabies cycle [12]. Thus, although the maintenance of an independent RABV cycle in Namibian kudu has been speculated, we intended to provide additional evidence - based on genome sequence properties - that horizontal transmission of rabies among Namibian kudu is indeed a common occurrence. Such evidence included new phylogenetic data and the identification of unique amino acid (aa) changes in viruses recovered from kudu, as opposed to those from canid host species.

METHODS

Partial N Gene Sequencing

Brain samples used in this study were received from the Central Veterinary Laboratory (CVL) in Windhoek, Namibia. The brain samples were obtained from dead, or suspected rabid animals. These samples were from kudu and jackal - from years 2008 and 2009 - that tested positive for rabies after being submitted for diagnostic testing (Table 1) via a network of farmers, interest groups and wildlife conservancies that was established for this study in Namibia. All samples obtained through this network were used in the study. These samples were from various regions throughout Namibia (Figure 1), although the majority originated from central Namibia where the game farming industry is predominant. In total, 49 fluorescent antibody test (FAT) positive samples were sequenced. RNA extraction was performed using Trizol (Invitrogen) according to the manufacturer's instructions. The RT-PCR reaction targeted a region (bases 16-646 according to the Pasteur virus rabies genome, GenBank accession number: M13215) of the nucleoprotein gene approximately 602 bp in size [13]. PCR amplicons were first confirmed by agarose gel electrophoresis, followed by PCR purification using the Wizard® SV Gel and PCR Clean-Up System (Promega), according to the manufacturer's instructions.

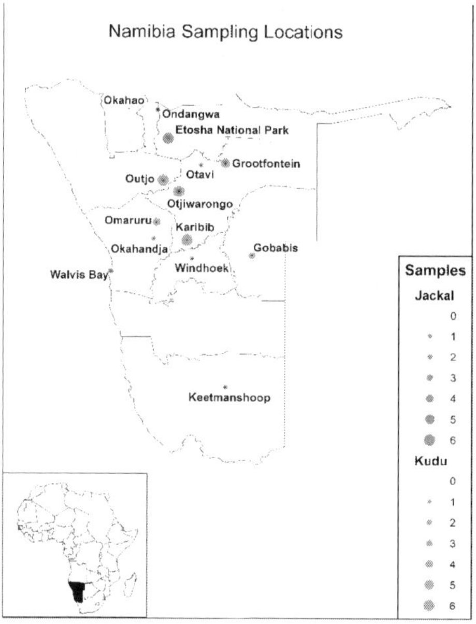

Figure 1. Kudu and jackal sample origins.

Numbers and locations of all rabies virus samples used in the partial sequencing analysis of this study from kudu (green) and jackal (red). The size of the dot increases with the number of samples from each location.

https://doi.org/10.1371/journal.pone.0058739.g001

Table 1. Brain samples from various species used in the phylogenetic analysis of partial RABV nucleoprotein gene sequences.

Sample Number	GenBank Number	Species[#]	Year of Isolation	Origin - District	Origin - Region	Country	Reference
3K08	JQ691415	Kudu	2008	Okahandja	Otjozondjupa	Namibia	This study
8K08	JQ691416	Kudu	2008	Okahandja	Otjozondjupa	Namibia	This study
18K08	JQ691417	Kudu	2008	Omaruru	Erongo	Namibia	This study
19K08	JQ691418	Kudu	2008	Gobabis	Omaheke	Namibia	This study
24K08	JQ691420	Kudu	2008	Outjo	Kunene	Namibia	This study
27K08	JQ691421	Kudu	2008	Outjo	Kunene	Namibia	This study
28K08	JQ691422	Kudu	2008	Outjo	Kunene	Namibia	This study
33K08	JQ691423	Kudu	2008	Grootfontein	Otjozondjupa	Namibia	This study
35K08	JQ691424	Kudu	2008	Grootfontein	Otjozondjupa	Namibia	This study
46K08	JQ691426	Kudu	2008	Outjo	Kunene	Namibia	This study
51K08	JQ691427	Kudu	2008	Okahandja	Otjozondjupa	Namibia	This study
52K08	JQ691428	Kudu	2008	Okahandja	Otjozondjupa	Namibia	This study
55K08	JQ691429	Kudu	2008	Gobabis	Omaheke	Namibia	This study
57K08	JQ691430	Kudu	2008	Otjiwarongo	Otjozondjupa	Namibia	This study
59K08	JQ691431	Kudu	2008	Okahandja	Otjozondjupa	Namibia	This study
60K08	JQ691432	Kudu	2008	Outjo	Kunene	Namibia	This study
95K08	JQ691437	Kudu	2008	Omaruru	Erongo	Namibia	This study
130K09	JQ691438	Kudu	2009	Outjo	Kunene	Namibia	This study
131K09	JQ691439	Kudu	2009	Otjiwarongo	Otjozondjupa	Namibia	This study
142K09	JQ691440	Kudu	2009	Otjiwarongo	Otjozondjupa	Namibia	This study
143K09	JQ691441	Kudu	2009	Windhoek	Khomas	Namibia	This study
144K09	JQ691442	Kudu	2009	Gobabis	Omaheke	Namibia	This study
146K09	JQ691443	Kudu	2009	Omaruru	Erongo	Namibia	This study
147K09	JQ691444	Kudu	2009	Otjiwarongo	Otjozondjupa	Namibia	This study
152K09	JQ691446	Kudu	2009	Okahandja	Otjozondjupa	Namibia	This study
153K09	JQ691447	Kudu	2009	Grootfontein	Otjozondjupa	Namibia	This study
158K09	JQ691448	Kudu	2009	Otjiwarongo	Otjozondjupa	Namibia	This study
172K09	JQ691449	Kudu	2009	Etosha National Park	Kunene	Namibia	This study
190K09	JQ691452	Kudu	2009	Grootfontein	Otjozondjupa	Namibia	This study
191K09	JQ691453	Kudu	2009	Otjiwarongo	Otjozondjupa	Namibia	This study
201K09	JQ691457	Kudu	2009	Omaruru	Erongo	Namibia	This study
212K09	JQ691459	Kudu	2009	Karibib	Erongo	Namibia	This study
234K09	JQ691460	Kudu	2009	Ondangwa	Ohangwena	Namibia	This study
240K09	JQ691462	Kudu	2009	Grootfontein	Otjozondjupa	Namibia	This study
244K09	JQ691463	Kudu	2009	Otavi	Otjozondjupa	Namibia	This study
20J08	JQ691419	Jackal	2008	Okahao	Omusati	Namibia	This study
38J08	JQ691425	Jackal	2008	Grootfontein	Otjozondjupa	Namibia	This study
64J08	JQ691433	Jackal	2008	Keetmanshoop	Karas	Namibia	This study
67J08	JQ691434	Jackal	2008	Etosha National Park	Kunene	Namibia	This study
89J08	JQ691435	Jackal	2008	Etosha National Park	Kunene	Namibia	This study
93J08	JQ691436	Jackal	2008	Otjiwarongo	Otjozondjupa	Namibia	This study
151J09	JQ691445	Jackal	2009	Walvis Bay	Erongo	Namibia	This study
178J09	JQ691450	Jackal	2009	Etosha National Park	Kunene	Namibia	This study
179J09	JQ691451	Jackal	2009	Etosha National Park	Kunene	Namibia	This study
192J09	JQ691454	Jackal	2009	Etosha National Park	Kunene	Namibia	This study
193J09	JQ691455	Jackal	2009	Etosha National Park	Kunene	Namibia	This study
197J09	JQ691456	Jackal	2009	Outjo	Kunene	Namibia	This study
204J09	JQ691458	Jackal	2009	Otjiwarongo	Otjozondjupa	Namibia	This study
236J09	JQ691461	Jackal	2009	Ondangwa	Ohangwena	Namibia	This study

Sample Number	GenBank Number	Species#	Year of Isolation	Origin - District	Origin - Region	Country	Reference
RV385	AY330733	Jackal	1988	Ghanzi	Ghanzi	Botswana	[16]
RV389	AY330737	Jackal	1990	Orapa	Central	Botswana	[16]
RV444	AY330749	Dog	1991	Maun	Ngamiland	Botswana	[16]
RV447	AY330752	Dog	1991	Maun	Ngamiland	Botswana	[16]
RV481	AY330761	Jackal	1991	Orapa	Central	Botswana	[16]
RV1487	DQ194855	Kudu	2003	Windhoek	Khomas	Namibia	[12]
RV1488	DQ194856	Kudu	2003	Windhoek	Khomas	Namibia	[12]
RV1489	DQ194857	Kudu	2003	Omaruru	Erongo	Namibia	[12]
RV1490	DQ194858	Kudu	2003	Okahandja	Otjozondjupa	Namibia	[12]
RV1491	DQ194859	Kudu	2003	Omaruru	Erongo	Namibia	[12]
RV1492	DQ194860	Kudu	2003	Okahandja	Otjozondjupa	Namibia	[12]
RV1493	DQ194861	Kudu	2003	Okahandja	Otjozondjupa	Namibia	[12]
RV1494	DQ194862	Kudu	2003	Omaruru	Erongo	Namibia	[12]
RV1498	DQ194865	Jackal	2000	Khorixas	Kunene	Namibia	[12]
RV1504	DQ194871	Jackal	2003	Windhoek	Khomas	Namibia	[12]
RV1508	DQ194875	Jackal	2003	Windhoek	Khomas	Namibia	[12]
RV1510	DQ194877	Dog	2000	NK	Oshikoto	Namibia	[12]
RV1511	DQ194878	Dog	2000	Oshakati	Oshana	Namibia	[12]
RV1514	DQ194880	Dog	2003	Rundu	Kavango	Namibia	[12]
RV1517	DQ194882	Dog	2003	Okahandja	Otjozondjupa	Namibia	[12]
RV1518	DQ194883	Eland	2003	Omaruru	Erongo	Namibia	[12]
RV1825	DQ489835	Bat-eared fox	1990	Gordonia	Northern Cape	South Africa	[17]
RV1826	DQ194885	Bat-eared fox	NK	Etosha	Kunene	Namibia	[17]
RV1827	DQ194886	Bat-eared fox	NK	Etosha	Kunene	Namibia	[12]
RV1828	DQ489836	Bat-eared fox	NK	Etosha	Kunene	Namibia	[17]
RV1829	DQ194887	Jackal	NK	Etosha	Kunene	Namibia	[17]
RV1830	DQ194888	Dog	NK	NK	NK	Namibia	[12]
RV1831	DQ489837	Bat-eared fox	1990	Postmasburg	Northern Cape	South Africa	[17]
RV1857	DQ489853	Bat-eared fox	1994	Namaqualand	Northern Cape	South Africa	[17]
RAVMMGN	M13215	Pasteur virus					[21]

NK - unknown information.
#Kudu – *Tragelaphus strepsiceros*; Dog – *Canis familiaris*; Bat-eared fox – *Otocyon megalotis*; Jackal – *Canis mesomelas*; Eland – *Tragelaphus oryx*.

https://doi.org/10.1371/journal.pone.0058739.t001

Purified PCR products were sequenced using the BigDye Terminator v3.1 Kit cycle sequencing protocol (Applied Biosystems) at the University of Pretoria on an ABI3300 DNA Sequencer. Sequences were edited using CLC Main Workbench v6.0 (CLCBio). For phylogenetic analysis, the newly obtained sequences - as well as various relevant sequences from previous studies (Table 1) - were aligned using the Clustal X function in BioEdit [14]. A Neighbour-joining phylogenetic tree was constructed with the use of the Kimura 2-parameter model with 1000 bootstrap replications in MEGA 5.0 [15].

Full Genome Sequencing

RNA extraction

RNA was extracted from two kudu and two jackal brain samples (Table 2) using a protocol combining Trizol (Invitrogen) and RNeasy mini kit (Qiagen) according to the manufacturer's instructions.

Table 2. Full genome RABV sequences from kudu and jackals constructed in this study.

Sample Number	GenBank accession no.	Species#	Year of Isolation	Origin - District	Origin - Region	Country
239K09	JX473840	Kudu	2009	Windhoek	Khomas	Namibia
240K09	JX473841	Kudu	2009	Grootfontein	Otjozondjupa	Namibia
178J09	JX473838	Jackal	2009	Etosha National Park	Kunene	Namibia
192J09	JX473839	Jackal	2009	Etosha National Park	Kunene	Namibia

#Kudu - *Tragelaphus strepsiceros*; Jackal - *Canis mesomelas*.
doi:10.1371/journal.pone.0058739.t002

https://doi.org/10.1371/journal.pone.0058739.t002

Reverse-Transcription Polymerase Chain Reaction (RT-PCR)

A One-Step Reverse Transcription PCR was performed using the SuperScript® III One-Step RT-PCR kit (Invitrogen) for products less than 1500 base pairs in length, according to the manufacturer›s instructions with several primer pairs (Table S1). The following cycle conditions were used on a model 2720 thermocycler (Applied Biosystems): 50°C for 30 minutes; 95°C for 2 minutes and 42 cycles of 95°C for 30 seconds, 55°C for 30 seconds, 68°C for 60 seconds; and a final extension step for 5 minutes at 68°C. A long range One-Step RT-PCR was performed using the SuperScript® One-Step RT-PCR for Long Templates (Invitrogen), according to the manufacturer›s instructions with the following conditions: 50°C for 30 minutes, 95°C for 2 minutes followed by 42 cycles of 95°C for 30 seconds, 55°C for 30 seconds, 68°C for 1 minutes/kb. A final extension step was performed for 5 minutes at 68°C.

PCR products of sizes less than 1500 bp were analysed by 1.5% agarose gel electrophoresis using 1× TAE buffer (1.6 M Tris-acetate, 40 mM EDTA). For products larger than 1500 bp, a 1% agarose gel was prepared. A 100 bp DNA molecular weight marker (Promega) was included to identify the size of the amplicons.

Purification and sequencing

PCR amplicons generated were purified using the QIAquick Gel Extraction Kit (Qiagen), according to the manufacturer's instructions. PCR products were sequenced using the BigDye Terminator v1.1 Kit cycle sequencing protocol (Applied Biosystems) with slight modifications. One microlitre 5× sequencing buffer, 1 μl 5 pmol/μl primer (Table S1), 5μl [15 ng/μl] template, 1 μl nuclease free molecular grade water (Qiagen) and 2μl BigDye Terminator mix v1.1 was added to a final volume of 10 μl. The reaction was performed using the following profile: an initial denaturation step at 96°C for 1 min; then 26 cycles of 96°C for 15 seconds; 53°C for 10 seconds; and 60°C for 4 minutes.

Following the sequencing reaction, sequence products were purified using Sigma Spin Sequencing Reaction Clean-up Post Reaction Purification Columns (Sigma) according to the manufacturer's instructions. Fifteen microlitres of the eluate were added to 15μl of Hi-Di-Formamide and subsequently analysed using an ABI 3130 DNA sequencer (Applied Biosystems).

Genetic analyses

Sequences were assembled, translated and annotated using CLC Main Workbench v6 (CLCBio) and the ClustalX function of BioEdit [14]. Multiple alignments including sequences from the public domain (Table 3) were then manually screened for nucleotide sequence differences and those that also resulted in aa changes were identified. To identify any known viruses with the same aa in the variable positions identified, we used a blast analysis, screening the protein data bases in Genbank with 7 aa sequence segments that were composed of the variable aa and the 3 aa immediately to the left and right of this position. Finally, phylogenetic analysis was performed with the inclusion of the full genome sequences presented in Table S2 and according to the methodology described for partial N gene sequences, above.

Table 3. Lyssavirus isolates used in comparative amino acid analyses.

Virus isolate	Country and area of isolation	Host species*	Year of isolation	Laboratory reference number	GenBank accession numbers
RABV (canid variant)	Sibasa, South Africa	Dog	2006	262/06	HM179504 (N), HQ266628 (P), HQ266609 (M), HQ266620 (G).
RABV (canid variant)	emKhondo, formerly Piet Retief, South Africa	Dog	2004	567/04	HM179505 (N), HQ266626 (P), HQ266607 (M), HQ266618 (G).
RABV (canid variant)	Thabazimbi, South Africa	Dog	1996	479/96	HM179506 (N), HQ266625 (P), HQ266610 (M), HQ266621 (G).
RABV (canid variant)	Soutpansberg, South Africa	Black-backed jackal	2005	819/05	HM179507 (N), HQ266629 (P), HQ266611 (M), HQ266622 (G).
RABV (canid variant)	Umtata, South Africa	Bat-eared fox	2005	31/05	HM179508 (N), HQ266627 (P), HQ266608 (M), HQ266619 (G).
RABV (canid variant)	Japan	Laboratory strain	1915	Nishigihara	AB044824
RABV (canid variant)		SAD strain derivative		ERA	EF206707
RABV (canid variant)	USA	Silver-haired bat	1983	SBBRV-18	AY705373
RABV (canid variant)	Japan	Nishigahara derivative	1918	RC-HL	AB009663
RABV (canid variant)	China	Human		Flury-LEP	FJ577895
RABV (canid variant)	USA	LEP-Fury derivative	1939	HEP-Flury	AB085828
RABV (canid variant)		Vaccine		PV	M13215
RABV (mongoose variant)	Rusape, Zimbabwe	Slender mongoose	1994	22107	FJ392391 (N), HQ266633 (P), HQ266615 (M), FJ465408 (G).
RABV (mongoose variant)	Grootgewaagd, South Africa	Yellow mongoose	1990	669/90	FJ392385 (N), HQ266616 (M), FJ465402 (G).
RABV (mongoose variant)	Kroonstad, South Africa	Yellow mongoose	1995	767/95	FJ392388 (N), HQ266630 (P), HQ266617 (M), FJ465405 (G).
RABV (mongoose variant)	Uitenhage, South Africa	Yellow mongoose	1996	364/96	FJ392379 (N), HQ266632 (P), HQ266614 (M), FJ465397 (G).
RABV (mongoose variant)	Beaufort West, South Africa	Water mongoose	1991	113/91	FJ392372 (N), HQ266631 (P), HQ266613 (M), FJ465390 (G).
LBV	Durban, South Africa	Wahlbergs Epauletted Fruit bat	2008	LBVSA2008	HM179509 (N), HQ266634 (P), HQ266612 (M), HQ266623 (G).
LBV	Amanzimtoti, South Africa	Wahlbergs Epauletted Fruit bat	2006	LBVSA2006	EF547452 (N), EF547414 (P), EF547435 (M), EF547422 (G).
LBV	Exported to France from an unknown African origin	Egyptian fruit bat	1999	LBVAFR1999	EF547447 (N), EF547418 (P), EF547445 (M), EF547432 (G).
LBV	Lagos Island, Nigeria	Straw-coloured fruit bat	1956	LBVNIG1956	EF547459 (N), EF547407 (P), EF547444 (M), EF547431 (G).
LBV	Durban, South Africa	Wahlbergs Epauletted Fruit bat	2004	LagSA2004	EF547458 (N), EF547415 (P), EF547440 (M), EF547428 (G).
LBV	Westville, South Africa	Slender mongoose	2004	Mongoose2004	EF547453 (N), EF547409 (P), EF547438 (M), EF547423 (G).
MOKV	Bulawayo, Zimbabwe	Cat	1981	12341	FJ465417 (N), GQ861350 (P), GQ472991 (M), GQ473003 (G).
MOKV	East London, South Africa	Cat	1995	543/95	FJ465415 (N), GQ500116 (P), GQ472992 (M), GQ500110 (G).
MOKV	Pinetown, South Africa	Cat	1997	252/97	Unpublished (N) AF369376 (P), GQ472997 (M), GQ500112 (G).
MOKV	East London, South Africa	Dog	2006	173/06	FJ465412 (N), GQ861351 (P), GQ472999 (M), HQ266624 (G).
MOKV	Zimbabwe	Cat	1981		S59447
DUVV	Pilanesberg, South Africa	Human	2006	DUVVSA2006	EU623111
DUVV	Louis Trichardt, South Africa	Insectivorous bat (Unidentified)	1981	DUVVSA1981	EU623438 (N), EU623439 (P), EU623441 (M), EU623443 (G).
DUVV	Bela Bela, South Africa	Human	1970	DUVVSA1970	EU623437 (N), EU623436 (P), EU623440 (M), EU623442 (G).

*Dog – *Canis familiaris*; Black-backed jackal – *Canis mesomelas*; Bat-eared fox – *Otocyon megalotis*; Human – *Homo sapiens sapiens*; Cat – *Felis domesticus*; Slender mongoose - *Galerella sanguinea*; Water mongoose - *Atilax paludinossus*; Yellow mongoose – *Cynictis penicillata*; Silver-haired bat - *Lasionycteris noctivagans*; Egyptian fruit bat – *Rousettus aegyptiacus*; Wahlbergs Epauletted Fruit bat - *Epomophorus wahlbergi*; Straw-coloured fruit bat - *Eidolon helvum*.

doi:10.1371/journal.pone.0058739.t003

https://doi.org/10.1371/journal.pone.0058739.t003

Rapid Amplification of cDNA Ends (RACE)

Rapid Amplification of cDNA Ends (RACE) determined the 5′- and 3′ -termini of viral RNA from samples 239K09, 240K09, 178J09 and 192J09. RACE was performed with freshly extracted RNA using the 5′- and the 3′-RACE system from Invitrogen (Invitrogen, Carlsbad, USA) according to the manufacturer's recommendations. Due to a lack of a poly-A tail at the 3′-terminus of the viral RNA, a poly-C tailing was done prior to 3′-RACE. Some modifications were introduced to the 3′-RACE procedure because of the poly-C tailing: for first strand cDNA synthesis, amplification of the cDNA and subsequent hemi-nested PCR of the abridged anchor primer (AAP) from the 5′ RACE kit had to be used instead of the recommended adapter primer (AP, designed for annealing to a poly-A tail). As gene specific primers (GSPs) for the 3′-RACE system RAB-JACKAL-150R (amplification of cDNA) and RAB-JACKAL-100R (hemi-nested PCR) and for the 5′-RACE procedure RAB-JACKAL-12235F (cDNA synthesis) and RAB-JACKAL-12265F (amplification of cDNA) were used for all four samples (Table S1). Amplification products were visualized on a 2% agarose gel, purified and subsequently sequenced (see Purification and sequencing).

RESULTS

Phylogenetic Analyses

A few previous studies have included partial nucleoprotein gene sequencing of Namibian and other southern African RABV isolates and a sizeable collection of sequences can be found in Genbank [12], [16], [17]. We therefore chose to continue with this sequence region in our analyses of the 49 new isolates from kudu and jackal (35 kudu and 14 jackal) (Table 1) from regions throughout Namibia (Figure 1). Subsequent phylogenetic analysis depicted that 42 from 43 kudu RABV isolates belonged to a single clade that were distinct from any other RABV isolates from Namibia, or elsewhere in southern Africa (Figure 2). The single exception was the kudu specimen 190K09 from Grootfontein, which bundled with a group consisting of 7 jackal isolates from Etosha National Park. The majority of RABV from jackals sequenced were from Etosha National Park (7/14) and these all grouped together; separate from a rabies cycle in bat-eared foxes – also from the Etosha National Park (Figure 2). Samples 151J09 and 204J09 grouped with jackal and dog RABV sequences from Botswana and central Namibia. Sample 64J08, originating from Keetmanshoop in southern Namibia, grouped closely with other RABV sequences from a South African bat-eared fox rabies cycle from areas bordering Namibia to the south. One eland (*Tragelaphus oryx*) sample from a previous study (RV1518) also grouped with the RABV sequences from kudu.

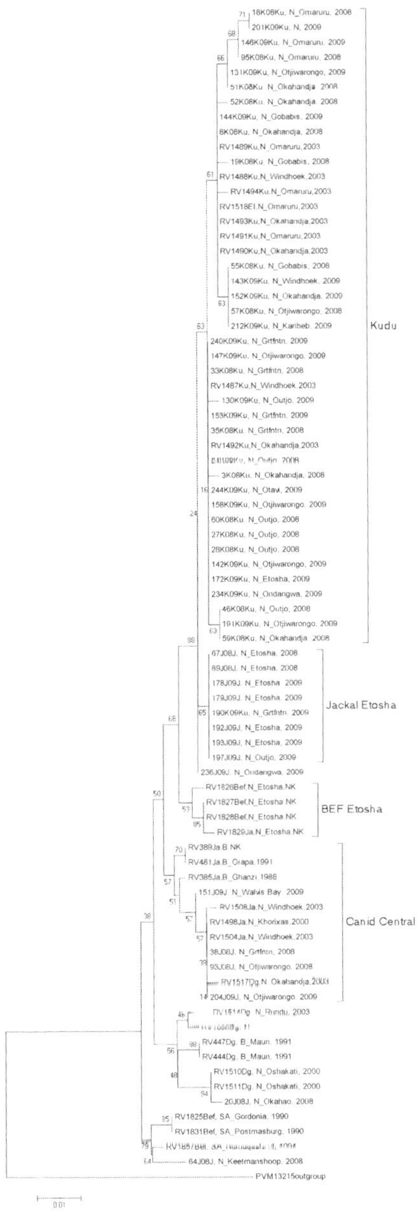

Figure 2. Partial N gene phylogenetic analysis.

Neighbour-joining phylogenetic tree with 1000 bootstrap replications of partial RABV nucleoprotein gene sequences generated in this study as well as representative sequences from South Africa and Namibia. Pasteur virus was used as an outgroup. Samples labeled as follows: Isolate number, species, country, region, and year. Ku=kudu; J=Jackal; Dg=Dog; Bef=Bat-eared fox; El=Eland; SA=South Africa; N=Namibia; B=Botswana; Grtfntn=Grootfontein; NK=Not known.

https://doi.org/10.1371/journal.pone.0058739.g002

We also wanted to compare and evaluate phylogeny based on full genome sequences, where the resultant tree (Figure 3) supported the results obtained from the partial N gene sequences shown above. The 4 isolates that were sequenced in full grouped separately from any other known sequences, and the 2 isolates from jackal (178J09 and 192J09) grouped separately from the 2 isolates from kudu (239K09 and 240K09) with 100% bootstrap support. The next closest strains were both vaccine strains from Russia and Japan respectively. The separate grouping of the viruses sequenced in this study was expected as no other full genome sequences of African RABV isolates were available, and the distinction between kudu and jackal isolates was consistent with a hypothetical genetic drift between jackal and kudu rabies cycles.

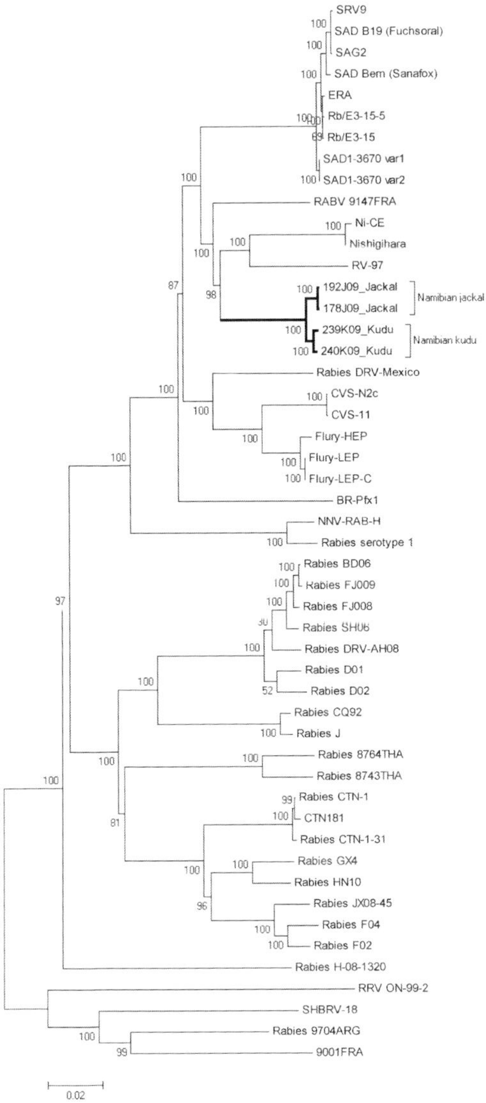

Figure 3. Complete genome phylogenetic analysis.

Neighbour-joining phylogenetic tree with 1000 bootstrap replicates, constructed using the Kimura 2-parameter model, of RABV full genomes sequenced in this study (bold branches) as well as RABV full genomes available on GenBank.

https://doi.org/10.1371/journal.pone.0058739.g003

Amino Acid Sequence Analysis of Full Genomes

The full genomes of the 2 RABV isolates from jackal were found to be identical with respect to aa sequence for all five genes, while the 2 isolates from kudu differed from one another in only one aa position, in the L gene. However, there were 13 aa differences between the kudu and the jackal RABV isolates. Three aa variations in the G gene were unique to the kudu RABV isolates and a further variation unique to both kudu and jackal RABV (L gene), even after comparing these aa to all other lyssavirus G and L sequences known (non-redundant protein and Swiss-Prot databases, Table 4).

Table 4. Amino acid variations unique to kudu and jackal RABV genomes when compared to all known lyssavirus sequences.

Gene	Host species	Position	Variable amino acid	Amino acid in other RABV (Table 3)	Charge change
Glycoprotein	Kudu	34	Asparagine	Serine	none
Glycoprotein	Kudu	112	Threonine	Alanine	non-polar to polar
Glycoprotein	Kudu	191	Serine	Asparagine	none
Polymerase	Kudu and Jackal	1140	Serine	Glycine	non-polar to polar

doi:10.1371/journal.pone.0058739.t004

https://doi.org/10.1371/journal.pone.0058739.t004

DISCUSSION

RABV is typically maintained within carnivorous hosts and maintenance within an herbivorous host population has never been proven. The apparently independent cycling of a RABV variant within Namibian kudu, known to be a non-aggressive herbivore, is therefore truly unique. In this study, we aimed to determine whether the RABVs isolated from Namibian kudu showed characteristics of host adaptation and whether they represented a distinct cycle of the disease. In general, phylogenetic analyses have shown clustering of RABVs to occur according to the geographical origin of the viruses, despite the fact that the viruses were isolated from several different species [1]. However, some exceptions have been noted, more specifically and predominantly, in bat species in the Americas [18]. Using molecular analyses, we also intended to determine whether unique mutations had occurred in the cross-species transmission from carnivores to kudu in order to: 1) further support/reject the notion of an independent cycle and; 2) determine whether the particular susceptibility of kudu to RABV infection is due to the adaptation of the virus to a new host. Historical epidemiological data is clear on the directionality of original transmission pathways for rabies in Namibia. Endemic cycles in jackal were firmly established at the turn of the 20th century, with the first recorded cases in kudu appearing several decades later [5], [7]. With the advent of molecular genetics, the close relationship between the viruses in jackal and kudu was shown – and the establishment of an independent kudu cycle is demonstrated in the present study.

Phylogenetic analyses of partial N gene sequences from 35 kudu and 14 jackal RABVs showed a significant clustering of all of the sequences from kudu RABVs separately from any other sequences from canid RABVs, with high bootstrap confidence. As the samples were all taken within the same temporal and similar spatial range, the divergence seen between RABV from canids and kudu can only be explained by the existence of a RABV cycle that is being maintained within the kudu population, separately to other RABV cycles in canid populations. An exception to the general clustering of RABV from kudu is sample 190K09, which grouped with RABV from jackals from Etosha National Park. This can be explained by a spill-over infection from a jackal to that kudu, as is commonly seen in cases of bovine infections from rabid jackals [10], [19]. It is a foregone conclusion that rabies was first introduced into the kudu population in this manner and that sporadic subsequent spillovers could be expected, but not on a scale that would account for the numbers of kudu rabies cases recorded since 1977. In addition, if a separate rabies cycle was not being maintained within the kudu population, several RABVs from canids would be interspersed within the kudu group with the assumption that one jackal could infect several kudu. It was also noted that several samples from jackals from the central regions of Namibia (where the majority of samples from kudu were taken), grouped in separate RABV cycles with other RABVs from jackals and dogs from the central region (Figure 2). This lends further support to the observation that RABV is being maintained in a separate cycle in kudu.

Sequence analyses revealed several aa variations unique to Namibian kudu RABVs. The variations specifically seen in the RABV isolates from kudu may suggest that these changes arose due to the adaptation of the virus to the host, but this notion will remain speculative until proven. Alternatively, this may suggest the divergence of the virus due to a separate RABV cycle being maintained within the kudu population (geographical and temporal drift), as kudu in Namibia may be geographically isolated due to separation of these animals by barriers such as game fences. Variations observed in viruses from both kudu and jackal may be due to the geographical isolation (on a larger scale) of the viruses from other rabies viruses and RABV cycles in Namibia and southern Africa, and support the evolutionary link between kudu and jackal rabies. In contrast, a recent study showed that RABVs did not undergo significant mutational differences after the host shift from bats to skunks, in both the coding and non-coding regions of the genomes [9]. However, the epizootic and maintenance of rabies in skunks occurred in a brief time frame in comparison with the emergence of rabies in kudu, and thus the time for divergence is greater in the kudu RABV cycle.

Complete genome sequencing is an important and useful method utilized in this study. The need for more complete genome sequences is becoming more and more important, especially with the molecular data that can only be generated through this technique. This study has shown the need for more complete genome sequencing as the sequences from this study were the first complete genome sequences for RABV from sub-Saharan Africa. These sequences will be important for a future databank of complete genomes that can be used for improved phylogenetic resolution as well as other molecular studies. Further work will need to be performed in order to determine virus evolution, the effects of certain mutations on protein folding, as well as continued insight into host-virus interactions and pathogenicity [20].

In conclusion, phylogenetic and full genome sequence analyses showed several correlations between Namibian 'street viruses' to attenuated vaccine strains, which can be explained by the introduction of the cosmopolitan RABV strain from Europe into Africa during the mid-20th century. Most importantly, several unique mutations were observed in the RABVs isolated from kudu, suggesting that these mutations may have occurred due to the adaptation of the virus to the host. Evidence from this study strongly argues for the maintenance of an independent RABV cycle in kudu, separate from the jackal cycles in Namibia. Future studies will be needed to determine the potential significance of the genomic regions identified as unique to kudu RABVs and to determine whether host specific regions occur in RABV and other lyssaviruses. The important implications of the phenomenon of the independent maintenance of RABV among kudu are that the control and eradication of a rabies cycle in this, and other species, will pose unique and unchartered challenges. From a continental perspective it is important to note that endemic canine rabies remains of primary concern and Namibia and other countries of the southern African region are no exception. Eventual effective control of rabies in kudu and jackal is likely to be subject to an effective control program for rabies in dogs.

REFERENCES

1. Bourhy H, Kissi B, Audry L, Smreczak M, Sadkowska-Todys M, et al. (1999) Ecology and evolution of rabies virus in Europe. The Journal of General Virology 80 (Pt 10: 2545–2557.
2. Beck AM, Felser SR, Glickman LT (1987) An epizootic of rabies in Maryland, 1982–84. American Journal of Public Health 77: 42–44.
3. Barnard B, Hassel R (1981) Rabies in kudus (*Tragelaphus strepsiceros*) in South West Africa/Namibia. Journal of the South African Veterinary Association 52: 309–314.
4. Petersen B, Tack D, Longenberger A (2012) Rabies in Captive deer, Pennsylvania, USA, 2007–2010. Emerging infectious 18 2007–2010 Available:http://www.newjumbo.info/go/nph-go.cgi/00010A/http/wwwnc.cdc.gov/eid/article/18/1/pdfs/11-1189.pdf. Accessed 19 April 2012.
5. Nel L, Rupprecht C (2007) Emergence of Lyssaviruses in the Old World: The Case of Africa. In: Childs JE, Mackenzie JS, Richt JA, editors. Wildlife and Emerging Zoonotic Diseases: The Biology, Circumstances and Consequences of Cross-Species Transmission. Springer Berlin Heidelberg, Vol. 315 . pp. 161–193. doi:10.1007/978-3-540-70962-6_8.
6. Streicker DG, Turmelle AS, Vonhof MJ, Kuzmin I V, McCracken GF, et al. (2010) Host phylogeny constrains cross-species emergence and establishment of rabies virus in bats. Science 329: 676–679 .
7. Hübschle O (1988) Rabies in the kudu antelope (*Tragelaphus strepsiceros*). Review of Infectious Diseases 10: s629–s633.
8. Holmes EC, Woelk CH, Kassis R, Bourhy H (2002) Genetic constraints and the adaptive evolution of rabies virus in nature. Virology 292: 247–257 .
9. Kuzmin I V, Shi M, Orciari L a, Yager P a, Velasco-Villa A, et al. (2012) Molecular Inferences Suggest Multiple Host Shifts of Rabies Viruses from Bats to Mesocarnivores in Arizona during 2001-2009. PLoS Pathogens 8: e1002786
10. Barnard B (1979) The role played by wildlife in the epizootiology of rabies in South Africa and South-West Africa. The Onderstepoort Journal of Veterinary Research 46: 155–163.
11. Barnard BJH, Hassel R, Geyer HJ, De Koker W (1982) Non-bite transmission of rabies in kudu (*Tragelaphus strepsiceros*). Onderstepoort Journal of Veterinary Research 49: 191–192.

12. Mansfield K, McElhinney L, Hübschle O, Mettler F, Sabeta C, et al. (2006) A molecular epidemiological study of rabies epizootics in kudu (*Tragelaphus strepsiceros*) in Namibia. BMC Veterinary Research 2: 2
13. Markotter W, Kuzmin I, Rupprecht CE, Randles J, Sabeta CT, et al. (2006) Isolation of Lagos bat virus from water mongoose. Emerging Infectious Diseases 12: 1913–1918.
14. Hall TA (1999) BioEdit: a user-friendly biological sequence alignment editor and analysis program for Windows 95/98/NT. Nucleic Acids Symposium Series 41: 95–98.
15. Tamura K, Peterson D, Peterson N, Stecher G, Nei M, et al. (2011) MEGA5: molecular evolutionary genetics analysis using maximum likelihood, evolutionary distance, and maximum parsimony methods. Molecular Biology and Evolution 28: 2731–2739
16. Johnson N, Letshwenyo M, Baipoledi EK, Thobokwe G, Fooks a R (2004) Molecular epidemiology of rabies in Botswana: a comparison between antibody typing and nucleotide sequence phylogeny. Veterinary Microbiology 101: 31–38
17. Sabeta CT, Mansfield KL, McElhinney LM, Fooks AR, Nel LH (2007) Molecular epidemiology of rabies in bat-eared foxes (Otocyon megalotis) in South Africa. Virus Research 129: 1–10
18. De Mattos CC, De Mattos C a, Loza-Rubio E, Aguilar-Setién A, Orciari L a, et al. (1999) Molecular characterization of rabies virus isolates from Mexico: implications for transmission dynamics and human risk. The American Journal of Tropical Medicine and Hygiene 61: 587–597.
19. Cohen C, Sartorius B, Sabeta C, Zulu G, Paweska J, et al. (2007) Epidemiology and molecular virus characterization of reemerging rabies, South Africa. Emerging Infectious Diseases 13: 1879–1886.
20. Cherian SS, Gunjikar RS, Banerjee A, Kumar S, Arankalle V a (2012) Whole genomes of Chandipura virus isolates and comparative analysis with other rhabdoviruses. PloS One 7: e30315
21. Tordo N, Poch O, Ermine A, Keith G, Rougeon F (1986) Walking along the rabies genome: is the large G-L intergenic region a remnant gene? Proceedings of the National Academy of Sciences of the United States of America 83: 3914–3918.

CHAPTER 12

Bat Rabies in France: A 24-Year Retrospective Epidemiological Study

Evelyne Picard-Meyer[1], Emmanuelle Robardet[1], Laurent Arthur[2], Ge´ rald Larcher[3], Christine Harbusch[4], Alexandre Servat[1] , Florence Cliquet[1]

[1]French Agency for Food, Environmental and Occupational Health & Safety (ANSES), Nancy Laboratory for Rabies and Wildlife, OIE Reference Laboratory for Rabies, European Union Reference Laboratory for Rabies, European Union Reference Laboratory for Rabies Serology, Malzeville, France,

[2]Museum d'Histoire Naturelle de Bourges, Bourges, France,

[3]SFEPM Chiroptera Group, Museum d'Histoire Naturelle de Bourges, Bourges, France,

[4]Nature and Biodiversity Conservation Union, Perl-Kesslingen, Germany

Citation: Picard-Meyer E, Robardet E, Arthur L, Larcher G, Harbusch C, Servat A, et al. (2014) Bat Rabies in France: A 24-Year Retrospective Epidemiological Study. PLoS ONE 9(6): e98622. https://doi.org/10.1371/journal.pone.0098622

ABSTRACT

Since bat rabies surveillance was first implemented in France in 1989, 48 autochthonous rabies cases without human contamination have been reported using routine diagnosis methods. In this retrospective study, data on bats submitted for rabies testing were analysed in order to better understand the epidemiology of EBLV-1 in bats in France and to investigate some epidemiological trends. Of the 3176 bats submitted for rabies diagnosis from 1989 to 2013, 1.96% (48/2447 analysed) were diagnosed positive. Among the twelve recognised virus species within the Lyssavirus genus, two species were isolated in France. 47 positive bats were morphologically identified as *Eptesicus serotinus* and were shown to be infected by both the EBLV-1a and the EBLV-1b lineages. Isolation of BBLV in *Myotis nattereri* was reported once in the north-east of France in 2012. The phylogenetic characterisation of all 47 French EBLV-1 isolates sampled between 1989 and 2013 and the French BBLV sample against 21 referenced partial nucleoprotein sequences confirmed the low genetic diversity of EBLV-1 despite its extensive geographical range. Statistical analysis performed on the serotine bat data collected from 1989 to 2013 showed seasonal variation of rabies occurrence with a significantly higher proportion of positive samples detected during the autumn compared to the spring and the summer period (34% of positive bats detected in autumn, 15% in summer, 13% in spring and 12% in winter). In this study, we have provided the details of the geographical distribution of EBLV-1a in the south-west of France and the north-south division of EBLV-1b with its subdivisions into three phylogenetic groups: group B1 in the north-west, group B2 in the centre and group B3 in the north-east of France.

INTRODUCTION

Bats represent about a fifth of the nearly 5000 known species of mammals with 1,232 bat species distributed worldwide and divided into two suborders: the Megachiroptera and the Microchiroptera versus Yinpterochiroptera and Yangochiroptera, respectively [1]. About 70% of bats are insectivores and most of the rest are frugivores. A few species feed on animals other than insects, one example being the hematophagous vampire bats. Bats are present throughout the world and play vital ecological roles in many ecosystems. Although much is known about the ecology and physiology of bats, they are also among the less studied mammals with regard to immunology [2]. Currently, more than 100 viruses have been isolated from bats [3]. Bats have been shown to be involved in several emergent viral diseases caused by coronaviruses, flaviviruses, astroviruses, adenoviruses, henipaviruses, filoviruses and lyssaviruses [4].

Rabies in bats is caused by negative-sense single–stranded RNA viruses of the genus lyssavirus, family *Rhabdoviridae*. Rabies virus is usually transmitted through the saliva of an infected animal during a bite. There are currently twelve recognised members of the genus lyssavirus [5]. These include classical *rabies virus (RABV), Lagos bat virus (LBV), Mokola virus (MOKV), Duvenhage virus (DUVV), European bat lyssavirus type 1 (EBLV-1), European bat lyssavirus type 2 (EBLV-2), Australian bat lyssavirus (ABLV), Irkut virus (IRKV), Aravan virus (ARAV), Khujand virus (KHUV), West Caucasian bat virus (WCBV)* and *Shimoni bat virus.* Three recent lyssaviruses have been identified but have not yet been taxonomically assessed: Ikoma bat lyssavirus (IKOV) discovered in 2009 in an African civet [6], Bokeloh bat lyssavirus (BBLV) identified in *Myotis nattereri* in Germany [7] and in France [8] and Lleida bat lyssavirus in *Miniopterus schreibersii* in Spain [9]. With the exception of IKOV and MOKV, bats are natural reservoirs for all lyssaviruses.

In Europe, bat rabies is principally caused by two rabies virus species, namely EBLV-1 and EBLV-2. EBLV-1 circulates widely throughout Europe with two variants, which are EBLV-1a and EBLV-1b. Variant a exhibits an east-west distribution from Russia to the centre of France with very little genetic variation while variant b exhibits a south-north distribution from Spain to Denmark and far more genetic diversity [10]. Most EBLV-2 cases were isolated in the United Kingdom (UK) and the Netherlands, and although EBLV-2 was detected in Finland, Germany and Switzerland, two countries neighbouring on France, up to now no cases of EBLV-2 have been reported in our country.

EBLV-1 is mainly associated with the serotine bat (*Eptesicus serotinus*) with 95% of the cases [11] and *E. isabellinus* in Spain [12] while EBLV-2 is reported mainly in Daubenton's bats (*Myotis daubentonii*) and pond bats (*Myotis dasycneme*). Single rabies cases have also been found in *Myotis myotis* and *Plecotus auritus* in Poland and occasionally in other bat species [13]. These include *P. nathusii* and *P. pipistrellus* in Germany, *R. ferrumequinum* in Turkey and *Nyctalus noctula in Yugoslavia* [*14*].

In France, bat rabies surveillance was initiated in 1989 after the discovery of the first positive case in a serotine bat in the north-eastern (NE) part of the country [15]. Since 2000, bat rabies surveillance has been improved with the consolidation of the network and involvement of local veterinary services in addition to the French national bat conservation network (SFEPM). As a result, the number of collected specimens has increased and data has enabled the estimation of rabies incidence in bat populations across the country.

In this study, we report the results of French passive surveillance of bat rabies from 1989 to 2013. We analysed the epidemiologic patterns of EBLV-1 in serotines from 1989-2012 using diagnostic laboratory data. Environmental (seasonal, annual, and geographical) and biological (sex) factors that could potentially affect the occurrence of rabies cases in serotines, were analysed in an attempt to improve knowledge of the dynamics of EBLV-1 infection. In addition, in the present study we undertook the phylogenetic analysis of the partial nucleoprotein gene from the collection of EBLV-1 samples with referenced sequences (GenBank database) isolated in Europe, to better understand the circulation of the two lineages of EBLV-1 in the country.

MATERIALS AND METHODS

Data Source and Laboratory Diagnosis Methods

The data analysed in the study included the passive bat rabies surveillance data collected from 1989 to 2013 by ANSES's Nancy Laboratory for Rabies and Wildlife, which is one of the two laboratories involved in rabies diagnosis in France.

The national passive surveillance network is based on the testing of sick bats, bats suspected of having rabies (showing clinical signs or abnormal behaviour) or bats found dead (all species), for the detection of lyssavirus infections [16]. All dead bats found in human environments as well as sick and suspicious bats with no human contact were submitted for rabies testing at ANSES›s Nancy laboratory.

The data included the diagnostic results, information on the location where the bat was isolated, date of submission, morphological identification of the bat [17] and the name of the collector (local veterinary services or bat handler from the French national bat conservation network). Bat cases linked to human contamination, which are submitted to the Pasteur Institute in Paris, were excluded from the study.

The laboratory techniques used for routine diagnosis are the ones recommended by WHO and OIE, i.e. fluorescent antibody test (FAT), rabies tissue culture infection test (RTCIT) on murine neuroblastoma cells [18] and in cases of positive rabies diagnosis an additional RT-qPCR and typing of the isolated virus by Sanger sequencing [19]. The Fluorescent Antibody Test was performed on brain tissue specimens using polyclonal fluorescent isothiocyanate-labelled rabbit anti-rabies nucleocapsid immunoglobulin G (BioRad, France) [20]. The detection of infectious particles in brain homogenate was performed using RTCIT on neuroblastoma cells with an incubation period of 48 hours at 37°C as previously described [21]. The conventional hnRT-PCR was performed with universal rabies primers JW12-JW6 as previously described [19]. Evagreen RT-qPCR was performed using rabies primers [N165-N146 and JW12], 2 µl of cDNA generated from 5 µl of total RNA and the SsoFast Evagreen Supermix kit (BioRad, France) [22].

Statistical Analysis

The proportion of rabid bats was calculated according to different environmental and biological factors including the year the animal was submitted, season, region and sex. Prevalence comparisons were carried out using the Pearson's chi-squared test ($\chi 2$) or, as indicated by the Cochran rules, the Fisher exact test when a frequency was below 5 [23]. The 95% confidence intervals (95% CI) of percentages were calculated using free open-source R software, version 2.12.2 (R Development Core Team, 2011). Animals with undetermined rabies diagnosis status were excluded from the analysis.

Phylogenetic Analysis

A total of 49 rabies viruses (48 serotines and 1 Natterer's bat), obtained from the laboratory's virus collection were included in the phylogenetic study. Of the 49 tested samples, 48 bats (i.e. 47 serotines and 1 Natterer's bat), were isolated in France and 1 in Luxembourg (a serotine bat), a neighbouring country of France. All 49 of the samples studied were initially tested by the direct fluorescent antibody test prior to genetic characterisation. The analysed samples consisted of original frozen brain material (n=42) and passaged brouse brain material (n=7).

Viral RNA was extracted from 200 µl of a 10% brain tissue homogenate using an Iprep PureLink Virus kit (Invitrogen, France) according to the manufacturer's instructions.

The cDNA synthesis and PCR amplification of the partial nucleoprotein (N) gene were performed as previously described [19]. 5 µl of PCR products were separated on a 2% agarose gel then purified on Nucleospin Extract II columns (Macherey Nagel, France) and sequenced in both directions by

Beckman Coulter Genomics (Takeley, Essex, United Kingdom). Consensus sequences were obtained by aligning the forward and reverse sequences with Vector NTI software (version 10.3.1). Duplicate sequences with a 100% match were removed from the dataset (n=15). The N gene analysis was constituted of 52 sequences (390 nucleotides, positions 71 to 460; nucleotide numbering is based on the Pasteur virus genome NC_001542) including 32 serotine isolates from France, one serotine isolate from Luxembourg, the French BBLV isolate (KC169985), 13 representative EBLV-1 sequences (Spain, Germany, Poland, Netherlands, Russia, Ukraine, Denmark), 5 representative EBLV-2 strains (United Kingdom, Switzerland, Germany, Finland, Netherlands), the German BBLV (JF311903) and one referenced sequence (WCBV (EF614258)).

Phylogenetic trees were constructed using the Neighbour-Joining (NJ) method and the Maximum-likelihood (ML) method (both with the Kimura-2 parameter calculation) with MEGA software, version 5 [24]. The bootstrap probabilities of each node were calculated using 1,000 replicates for both methods.

Bootstrap values over 70% were considered as statistically significant since the bootstrap method may be a conservative estimate for the reliability of a clade [25].

In addition, a Bayesian analysis of the nucleotide sequences including the 32 EBLV-1 isolates from France, the EBLV-1 isolate from Luxembourg, the only French BBLV isolate and 20 representative sequences was performed using the general-time reversible model (GTR, 10^7 generations) of the molecular evolutionary software package Bayesian Evolutionary Analysis by Sampling trees (BEAST), updated to version 1.7. (http://beast.bio.ed.ac.uk) [26]. BEAST outputs (MCMC log files and tree files) were examined by using LogCombiner and TreeAnnotator (http://beast.bio.ed.ac.uk/Tracer), excluding the first 25% of trees from the analysis as burn-in. The phylogenetic tree was visualised by using FigTree (version 1.3.1; http://beast.bio.ed.ac.uk/FigTree). The scale bar indicates branch length, expressed as the expected number of substitutions per site.

Net p-distances between groups (N gene sequence, 390 nucleotides) were calculated by using MEGA 5 for the estimation of evolutionary divergence.

RESULTS

Bat Rabies Surveillance

Between the inception of the bat rabies surveillance network in 1989 and 2013, 3176 bats were submitted for rabies diagnosis, 2447 of which were analysed. 1.96% (CI: 1.4–2.6) tested positive during this period. Of the 48 positive bats, 47 were morphologically identified as serotine bats (21 males, 16 females and 10 unsexed) and 1 as a Natterer's bat (female).

Figure 1 represents the total number of autochthonous dead bats submitted for rabies diagnosis and the number of positive cases per year. From 2001 to 2013, a total of 3006 autochthonous dead bats were submitted for rabies diagnosis. Of the 3006 bats, classical rabies methods were not performed on 654 bats (654/3006=21.7%), due mainly to the high level of decomposition of the carcasses or to mummification. 41 bats were shown to be infected with EBLV-1 from 2001 to 2013 (Table 1), whereas only 7 positive cases were recorded from 1989-2000 (Figure 1, Table 2). The number of submissions increased 15 fold (=3006 versus 170) following the bat rabies surveillance reinforcement (Figure 1). This increase was accompanied by an increase in mummified bats (≈20% per year).

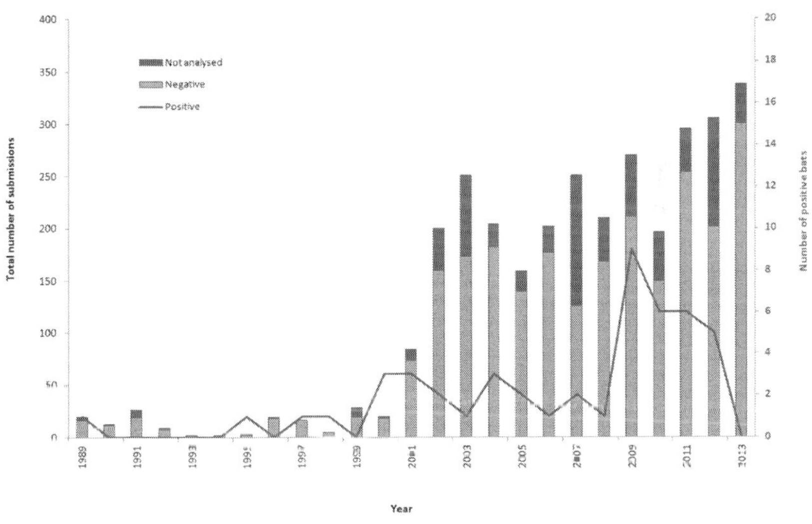

Figure 1. Changes in the number of dead autochthonous bats subjected to rabies diagnosis from 1989 to 2013.

https://doi.org/10.1371/journal.pone.0098622.g001

Table 1. Passive surveillance of bat rabies, 2001–2013.

Bats subjected to rabies diagnosis 2001–2013					
Family	Species	No. of mummified samples	No. of negative samples	No. of positive samples	Total no. of samples collected
Molossidae	Tadarida teniotis	16	6		22
Rhinolophidae	Rhinolophus hipposideros	7	41		48
	Rhinolophus ferrumequinum	4	30		34
	Rhinolophus euryale	2	7		9
Vespertilionidae	Pipistrellus pipistrellus	217	959		1176
	Eptesicus serotinus	60	217	40	317
	Pipistrellus kuhlii	8	203		211
	Myotis myotis	159	76		235
	Pipistrellus nathusii	4	115		119
	Plecotus austriacus	16	82		98
	Miniopterus schreibersii	5	36		41
	Nyctalus leisleri	7	64		71
	Myotis emarginatus	8	32		40
	Plecotus auritus	6	50		56
	Pipistrellus sp	53	108		161
	Myotis daubentonii	23	23		46
	Myotis mystacinus	3	30		33
	Pipistrellus pygmeus	8	39		47
	Myotis nattereri	2	13	1	16
	Myotis sp	6	13		19
	Pipistrellus savii		10		10
	Myotis bechsteinii	3	12		15
	Nyctalus noctula	5	76		81
	Barbastella barbastellus	4	19		23
	Eptesicus nilssonii	1	3		4
	Plecotus sp		4		4
	Myotis blythii		5		5
	Myotis brandtii		2		2

	Vespertilio murinus		2		2
	Myotis alcatoe		1		1
	Nyctalus sp	1	3		4
	Not identified	26	30		56
		654	2311	41	3006

https://doi.org/10.1371/journal.pone.0098622.t001

Table 2. Passive surveillance of bat rabies, 1989–2000.

Bats subjected to rabies diagnosis 1989–2000					
Family	Species	No. of mum mified samples	No. of negative samples	No. of positive samples	Total no. of samples collected
Molossidae	Tadarida teniotis				0
Rhinolop hidae	Rhinolophus ferrumequinum	7	6		13
Vespertili onidae	Pipistrellus pipistrellus	6	25		31
	Eptesicus serotinus	0	16	7	23
	Pipistrellus kuhlii		1		1
	Plecotus austriacus		1		1
	Myotis nattereri	0	1	0	1
	Nyctalus noctula		2		2
	Not identified	14	84		98
		27	136	7	170

Details regarding the number of negative and positive cases are given by family and

by bat species.

https://doi.org/10.1371/journal.pone.0098622.t002

Of the 3006 submitted bats, 98% (n=2950/3006) were morphologically identified by the laboratory or directly by the bat worker. 27 bat species were submitted for rabies diagnosis out of the 34 bat species reported in France, the pipistrelles being the most highly represented with 1176 samples (Table 1). The 27 collected bat species represented the three families of the order Chiroptera: *Molossidae* (n=22), *Rhinolophidae* (n=91) and *Vespertilionidae* (n=2893 collected with 317 common serotines).

Improvements in the bat surveillance network were accompanied by a strong increase in pipistrelle bat submissions (56%), while the proportion of *E. serotinus* (10.5%) and *M. daubentonii* (1.5%) bats submitted did not vary.

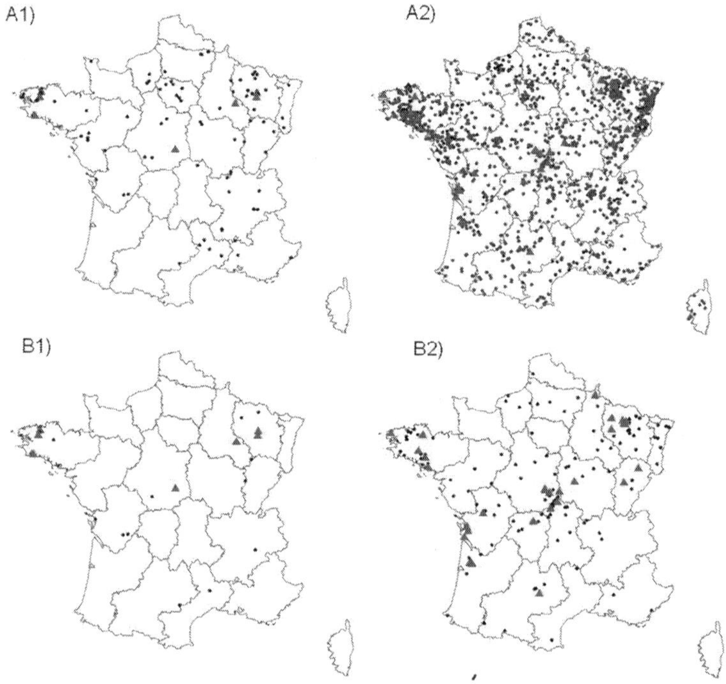

Figure 2. Bat rabies in France. Negative case: black dot, positive case: red triangle (A). Distribution of all the bat samples analysed for rabies diagnosis; A1)1989–2000; A2) 2001–2013 (B). Location of serotine (*Eptesicus serotinus*) samples analysed for rabies diagnosis; B1)1989–2000; B2)2001–2013.

https://doi.org/10.1371/journal.pone.0098622.g002

Figure 2 illustrates the geographical distribution of all bats submitted for rabies diagnosis since 1989. While the samples submitted were from throughout the country, the number of bats sampled for rabies diagnosis varied by region (Figure 2A). In this study, the north-west (NW) regions were defined as Bretagne, Normandie and Pays-de-la-Loire, the north-east (NE) regions included Alsace, Champagne-Ardenne, Ile-de-France, Franche-Comté, Lorraine, Nord-Pas-de-Calais and Picardie, The centre regions (C) were defined as Auvergne, Bourgogne and Centre, the south-east (SE) regions included Languedoc-Roussillon, Provence-Alpes-Côte d'Azur, Rhône-Alpes and Corse, and the south-west (SW) regions included Aquitaine, Midi-Pyrenées, Limousin, Pays de la Loire and Poitou-Charente. Nearly half (44.7%) of the submitted bats came from the north-east of France, followed by the north-west (NW) (17.7%), the centre (C)(14.7%), the south-east (SE)(12.5%) and the south-west (SW) (10.5%).

analysis of the environmental (spatio-temporal) and biological (sex) factors that could potentially explain the occurrence of rabies in serotine bats was carried out. Temporally, the proportion of positive samples detected over 1989–2000 (30.4%; CI: 14.1–53.0) (i.e. before network reinforcement), 2001–2003 (15.0%; CI: 6.2–30.5), 2004–2006 (10.9%; CI: 4.5–22.9), 2007–2009 (30.8%; CI: 17.5–47.7) and 2010–2013 (16.2%; CI: 9.8–25.2) did not differ significantly (Figure 3a). When comparing rabies incidence during the different seasons (seasons defined as follows: Winter: December-February; Spring: March-May, Summer: June-August, Autumn: September-November), significant variations were observed (Fisher exact test, p=0.005). Rabies cases were more frequent in autumn (34.0%; CI: 21.9–48.4) than in summer (15.1%; CI: 9.9–22.2) and in spring (12.5%; CI: 4.7–27.6) (Figure 3b). The test failed to detect any difference in rabies incidence between the winter (11.8%; CI: 2.1–37.8) and other seasons because of the low statistical power due to the low number of samples submitted between December and February.

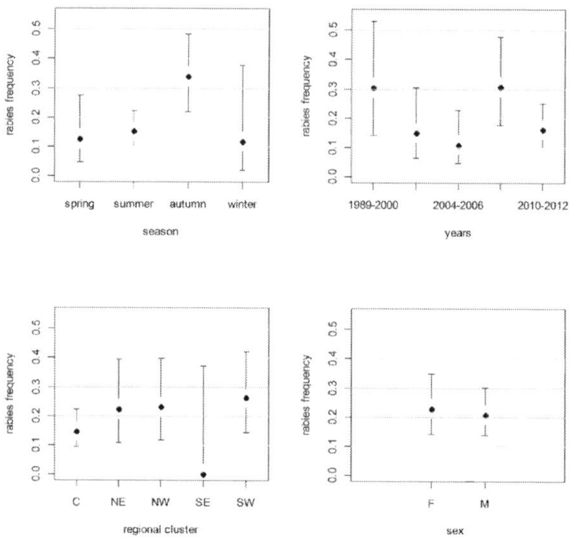

Figure 3. Rabies occurrence rate detected in serotine bats in France from 1989–2013 in the frame of the bat rabies surveillance network and according different environmental factors: season (a), year (b), regional cluster (c) and sex (d).

https://doi.org/10.1371/journal.pone.0098622.g003

The epidemiological data regions for the serotine were merged into NE, SE, C, SW and SE clustering regions to enable statistical spatial analysis. The probability of detecting rabies did not differ significantly by the geographical origin of bat sampling. Rabies prevalence was 26.2% (CI: 14.4–42.3) in C, 23.1% (CI: 11.7–39.7) in SW, 22% (CI: 10.7–39.6) in NW, 14.7% (CI: 9.3–22.3) in NE and 0% (CI: 0–37.1) in SE (Figure 3c).

When considering the sex of the animals, the probability of reporting a positive case in serotines did not differ between males and females, with respectively 22.9% (CI: 14.0–34.7) and 20.8% (CI: 13.6–30.4) of positive cases (Figure 3d).

Phylogenetic Analysis of the French EBLV-1 isolates

The phylogenetic study using the NJ method showed that all 32 infected serotines clustered with EBLV-1 viruses (bootstrap of 100). Figure 4 shows the results of phylogenetic relationships between the partial N gene sequence of the 32 French EBLV-1 sequences and 22 representative sequences. The Neighbour-Joining tree showed that the 32 French EBLV-1 strains belonged to the two lineages a and b of EBLV-1 (bootstrap of 100).

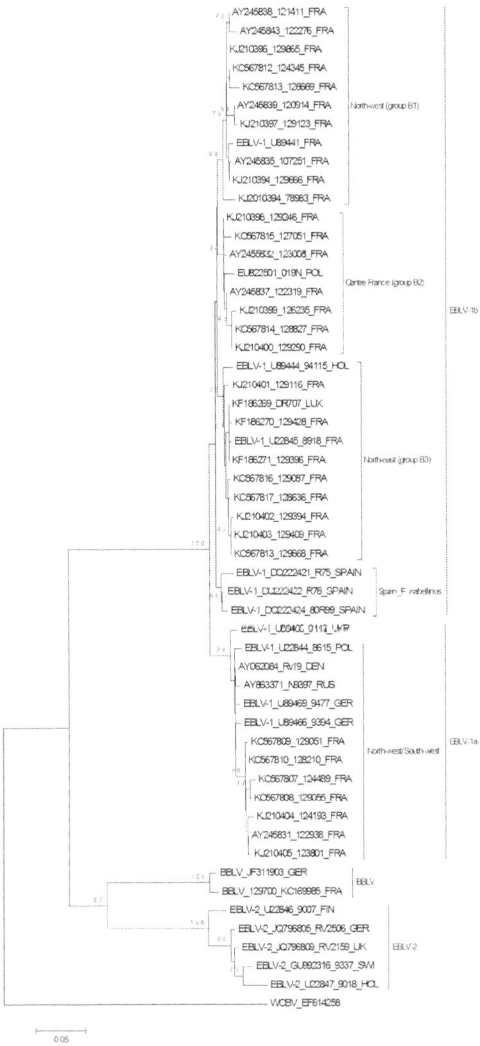

Figure 4. Neigbour-Joining and Bayesian Phylogenetics. A. NJ phylogenetic tree comparing partial N gene sequence from 32 French EBLV-1 isolates from 1989–2012, 11 referenced European EBLV-1 strains (Poland, Holland, Luxembourg, Spain, Denmark, Germany, Russia), 1 EBLV 1 isolate from Luxembourg, 5 EBLV-2 strains (Finland, Germany, United Kingdom, Switzerland, Holland) and two BBLV isolates (France, Germany). Relationships are presented as a rooted phylogram with WCBV (EF614258). Bootstraps above 70% are significant.

https://doi.org/10.1371/journal.pone.0098622.g004

Cluster A forming the EBLV-1a lineage included 7 sequences from France and 6 sequences from Poland, Ukraine, Russia, Denmark and Germany while the EBLV-1b lineage (cluster B) was made up of 25 French samples and isolates from Luxembourg (n=1), Poland (n=1) and Holland (n=1) (Figure 4). The 3 isolates from Spain belong to a distinct subgroup with a low bootstrap of 57%. The comparison between these 3 strains and the EBLV-1b cluster showed 97.1% nucleotide similarity. The nucleotide similarity was 98.6% among the three Spanish samples. The maximum likelihood method showed the same arrangements of clusters in the tree (data not shown), the EBLV-1b sequences from France being subdivided into three groups (with the bootstrap support above 70 for the three groups).

The first subgroup, B1 (bootstrap of 75), contained 7 strains isolated in the north-west of France and two strains from the centre of France (Isolates 107251 [AY245835] and 129666 [KJ210394]), respectively isolated in 1995 and in 2012 in the same city (Bourges, 47° 05′ 04″ north 2° 23′ 47″ east). One sample (78983 [KJ210395]), isolated in 1989 in the north-east of France, linked the subgroup B1 (limited bootstrap of 68). B2 (bootstrap of 84) was made up of 7 sequences from the centre of France and one referenced strain from Poland (EU822501). The last subgroup, B3 (bootstrap of 73), contained 8 strains from the north-east of France isolated between 1989 and 2012, one strain from Luxembourg isolated in 2013 (*E. serotinus*, KF186269) and one isolate from Holland (U89444, *E. serotinus* from 1992).

Cluster A (EBLV-1a) had less than 1.6% nucleotide divergence and 1% amino-acid divergence among all samples. In cluster B, 97.6% nucleotide similarity and 99.1% amino-acid similarity was shown among all sequences.

Bayesian phylogenetics with the BEAUti and the BEAST 1.7 software package showed the same division of the EBLV-1b lineage with the same geographical clusters throughout France: the centre, the north-west and the north-east (data not shown).

Figure 5 shows a map of France illustrating the location of the 47 samples according to the results of phylogenetic analysis of the N gene showing the two lineages EBLV-1a (cluster A) and EBLV-1b (cluster B), as well as the three identified sub-groups, B1, B2 and B3. The subdivision of the French sequences is geographically clustered. Taking into account the 32 unique sequences used in the phylogenetic analysis and the 15 identical sequences (with respectively 13 and 2 isolates belonging to groups B and A), the subgroups B1 and B3, composed of 11 and 18 samples respectively appeared to be restricted to the NW (n=8) and the NE of France (n=18), respectively

(Figure 5). 3 samples belonging to subgroup B1 came respectively from the centre (n=2) and from the north-east (n=1). The first positive case reported in the north-east of France (Bainville, KJ210395) in 1989 belongs to subgroup B1. Identically, the second case [AY245835] reported on 1995 in Bourges (centre of France) was also grouped in B1 with an additional case occurring in the same city (Bourges, KJ210394) on 2012. The 9 remaining samples belonging to subgroup B2 were located in the centre of France, with 6 samples in the Centre region, 2 in Auvergne and 1 in Bourgogne, while all EBLV-1a samples (group A, n=9 samples) were located in the SW of France (Figure 5).

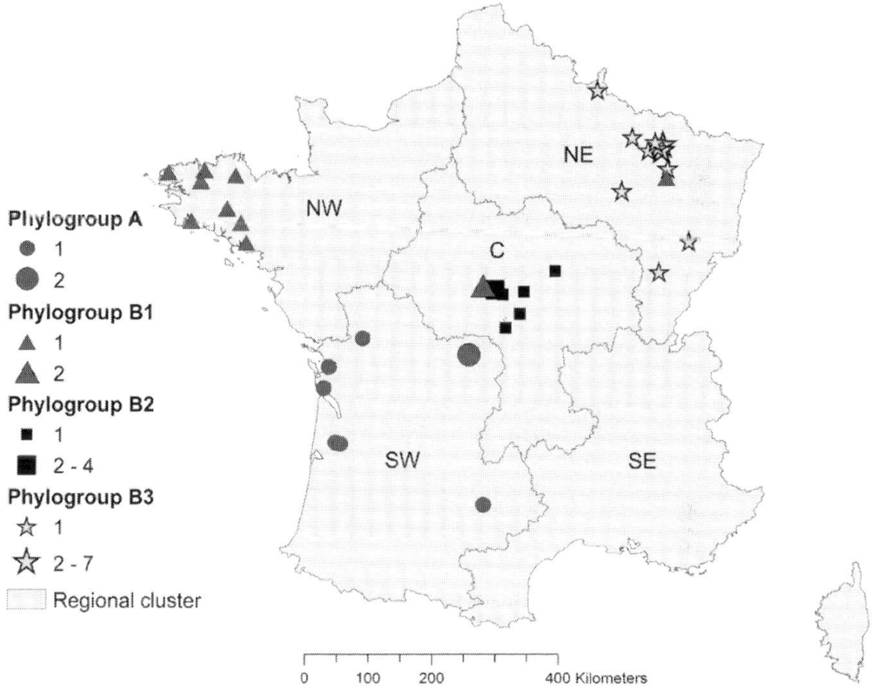

Figure 5. Regional clusters with geographical location of the 47 EBLV-1 a and b samples according to the results of the phylogenetic N gene analysis.

Samples were located geographically according to the defined lineage: [B1 (north-west), B2 (centre), B3 (north -east)] and A (south-west)]. The numbers correspond to the identification numbers of isolates

https://doi.org/10.1371/journal.pone.0098622.g005

The estimation of evolutionary divergences using MEGA5 (p-distance calculation) showed an evolutionary distance of respectively 0.024, 0.014 and 0.005 for all French EBLV-1 a and b isolates and Spanish EBLV-1b.

DISCUSSION

France has been officially free of rabies in non-flying mammals since 2001 [27]. However, rabies remains a public health concern because of the risk of importation of infected animals [27] and the natural circulation of rabies in bats. Each year, around 4000 patients receive a post-exposure treatment. In 2011, of the 4150 patients receiving post-exposure treatment, 4.8% were due to bat exposure while over 79% were due to exposure to dogs and cats [28].

Since 1954, more than 1,000 bat rabies cases have been recorded in Europe (www.who-rabies-bulletin.org). Most cases were recorded in Western European countries with a well-established rabies network, particularly in the Netherlands, Denmark, Germany and Great Britain [13], where several cases are reported each year. The passive surveillance of rabies in bats is routinely based on the collection of dead bats for diagnostic testing.

Our study showed an increase of over 17 fold in the number of total bats submitted for rabies diagnosis in France between 2001 and 2013 (n=3006 submissions) compared to the 1989–2000 period (n=170 submissions). However the increase of the total number of samples is not accompanied by an increase in submissions of serotine bats and/or infected bats. In 2013, it is interesting to note an absence of reporting of positive case in the country despite the increase of total bats and despite the fact that all regions, particularly the Centre, Bretagne and Lorraine, where the majority of the cases were found, submitted serotine bats for rabies testing.

In Europe, *E. serotinus* is mainly found infected with EBLV-1 (more than 95% of cases [14]). Of the 43 bat species reported in Europe, a few other species were also shown to be infected with EBLV-1, including *Isabellinus serotinus* in southern Spain [29], and rarely, *Pipistrellus pipistrellus* (2 cases in Germany in 1987), *Plecotus auretus* (1 case in Poland), *Pipistrellus nathusii* (1 case in Germany in 1986), *Nyctalus noctula* (3 cases in Yugoslavia in 1954), *Myotis myotis* (1 case in Germany in 1973) and *Vespertilio murinus* (1 case in Russia in 1985) [13], [30].

Serotine bats are the only bat species shown to be infected with EBLV-1 in France. Up to 2013, EBLV-1 was the only virus species that had been detected in France, with the first case having occurred in 1989 in the north-east [30]. In 2012, BBLV was reported for the first time in *Myotis nattereri* in France thanks to effective collaboration with bat workers [8]. Our study showed, in conjunction with the increase of bat submissions, a high proportion of pipistrelles (≈40%), one of the most abundant bat species in France and also in southern Europe, over the period 2001–2013. All 1176 submitted pipistrelles were diagnosed negative for the presence of rabies, as previously reported in the Netherlands [31]. *P. pipistrellus*, which shows strong synanthropic behaviour, might be less exposed to lyssavirus [32]. Our results, in association with the philopatric behavior of this bat species could indicate, as previously suggested, that the risk of *P. pistrellus* infection with a Lyssavirus is almost non-existent [32] although its cohabitation with serotines is well known (Christine Harbusch, personal communication). The other bat species (*Plecotus auritus*, *Nyctalus noctula)* occasionally found to be infected with EBLV-1 were also found negative in our study, demonstrating that these tree-dwelling bats are also rarely infected with EBLV-1.

As previously reported in European countries having a well-established bat rabies surveillance network [31], [33], the collection of bats may be potentially biased due to the sampling scheme. The studied samples are not representative of the natural population of bats due to the small numbers of bat submissions. Compared to the total bat population, which includes millions of individuals, the submissions in the passive surveillance do not reflect the true natural proportion in the field. Moreover, certain species are underrepresented: of the 34 bat species reported up to now in France, 27 species were collected from 2001 to 2013 and only 7 from 1993–2000.

The apparent proportion of 1.96% (48/2447 analysed *100) infected serotine bats is certainly biased and does not reflect the true prevalence and the correct epidemiological bat rabies situation in France. In addition, passive surveillance could lack sensitivity due to the small numbers of serotine bats submitted for rabies diagnosis compared to the estimated numbers for this species in France (400,000–600,000) (Laurent Arthur, personal communication). Furthermore, high fluctuations in the numbers of bat submissions were observed between geographical areas. For example, in the centre of France, in an area of approximately 39,151 km^2 [the total area of France is 552,000 km^2], 7 serotine bats were found to be infected out of a total of 21 tested serotines and out of 48 positive cases. Up to now, 140 colonies of serotine bats have been reported in this region out of a total of 987 colonies housing other species [34]. While serotines are shown to be distributed throughout the country [35] and while some aspects of serotine bat ecology have been studied, including diet, habitat use, foraging and roosting behaviour, others are mostly unknown, especially its hibernation ecology, mating system, dispersal and population structure. An understanding of the spatial ecology, behaviour and social dynamics of serotine bat populations is essential for a better understanding of EBLV-1 prevalence patterns in bat colonies.

Previous studies showed that the incidence of rabies in bats submitted for rabies diagnosis exhibited a strong seasonal pattern with two peaks, in the spring and autumn [36], [37]. This seasonality has recognized importance for epizootiological processes [38]. The infection status of bats has already been shown to be affected by season for some viruses, including lyssaviruses, henipaviruses, coronoviruses and filoviruses [2]. Using passive surveillance data, we report for the first time in Europe the seasonality of the rabies incidence rate in bats, with a higher rabies incidence detected in autumn (September to November) reaching 34%. In our study, 81% (=38/47) of positive cases were detected between June and October. The highest probability of detecting a rabies case among submitted bats in autumn rather in summer could be explained by the natural dispersal of serotine bats which promotes the contact rates between males/females, adults/juveniles and between infected and susceptible animals. Dispersal is a poorly understood event in the serotine bat species, as is the mating period and the ecology of males. No adult males are found within maternity roosts, indicating exclusion once they reach sexual maturity [39]. The mating period, taking place in late summer/early autumn (August-October), could also promote the diffusion of EBLV-1 in bats.

As previously suggested [37], the interaction of the incubation period with hibernation and the annual birth pulse, which starts for serotines in early June but which depends on the climate, can generate complex dynamics for the rabies virus. To date, current knowledge about rabies in serotines in Europe is limited. Little is known about the virulence, dissemination or transmission of the virus among free-living insectivorous bats, or abouts the incubation period of European bat lyssaviruses. Although the incubation period is not known for bats naturally infected with EBLV-1, experimental EBLV-1 studies of *E. serotinus* and *E. fuscus* with an inoculation dose of $10^{3.2}$ LD50/ml showed that the incubation periods varied depending on the route of inoculation, and were respectively between 7 to 26 days and 9 and 58 days [40], [41]. By comparison, the duration of rabies incubation is also highly variable in bats naturally infected with classical rabies virus, ranging from less than 14 days to over 290 days, and can rarely be greater than a year [42], [43]. These variable incubation periods could also explain the temporal dynamics of rabies incidence in our study. Indeed, several previous studies have shown a higher seroprevalence in summer. Seasonal variation in European bat infection was observed through the analysis of a large blood sample panel in Spanish bats showing a significantly higher seroprevalence of EBLV-1 in the summer, at which time maternity colonies are present in most of the bat colonies [32]. This seasonal variability was also shown in Brazilian free-tailed bats in natural cave roosts [44], indicating that roosting ecology and reproductive activity could favor bat infection. This higher seroprevalence in summer reflecting higher rabies exposure could, depending on the incubation period, lead to a higher rabies mortality rate following RV exposure in the subsequent months, and mainly in autumn. The higher seroprevalence reported in summer could also be the result of the higher metabolism of serotine bats in summer than in winter and spring.

In Europe, bat rabies is caused by EBLV-1 and EBLV-2. On a few occasions, WCBV and BBLV have been reported [11]. To date, despite the strengthening of passive surveillance, neither EBLV-2 nor WCBV have been isolated in France although EBLV-2 was isolated in three neighbouring countries, i.e. Germany, the United Kingdom and Switzerland. (Reviewed by [45]). Two bat species were shown to be infected by EBLV-2, Daubenton's bat (*M. daubentonii*) and the pond bat (*M. dasycneme*). Of the two species, *M. dasycneme* is reported in the north-west of France, while Daubenton's bat is present throughout France. The absence of EBLV-2 reporting in France has yet to be explained. The number of submissions of *M. daubentonii* should be increased to provide more elements on the true epidemiological situation of EBLV-2 in France. The real absence of this virus species needs to be proven in the country.

In almost all animals, rabies is a fatal disease once clinical signs manifest. However, most bats are able to coexist with viruses and may have evolved mechanisms to control viral replication [2]. It is clear that bats can experience either a lethal or non-lethal rabies virus infection, with in the latter case the development of specific neutralising antibodies [46]. We previously reported the presence of EBLV-1 neutralising antibodies in six indigenous bat species throughout France, mainly in *E. serotinus* and *M. myotis*, suggesting previous infection with the EBLV-1 rabies virus [47]. However, no EBLV-1 RNA was detected in the RT-PCRs performed on oropharyngeal swabs collected from these six apparently healthy bats [47]. Similarly, a high percentage of seropositive bat species (65%) was recently found in Spain, suggesting that most species of bats are exposed to EBLV-1 [32]. EBLV-1 is widely distributed throughout Europe. Serotines infected with the EBLV-1a variant exhibited an east-west distribution from Russia to France with very little genetic variation [10]. Most cases of EBLV-1a were reported in France, the Netherlands, northern Germany, Denmark and Poland. EBLV-1b, which presents far more genetic diversity and a south-north distribution, was mainly reported in Spain, France, Germany, Poland and in the Netherlands [13]. Our study demonstrated that all French EBLV-1 strains belong to the two lineages, a and b, of EBLV-1. Serotines infected with EBLV-1a clustered geographically in the south-west of France while serotines infected by EBLV-1b appeared to be grouped in the centre of France, the north-west and the north-east. Surprisingly, serotines infected by EBLV-1a were geographically separated from the regions displaying EBLV-1b cases. This question of geographic separation has not yet been explained. One hypothesis is that serotines might be a paraphyletic group composed of differentiated lineages at species level, like the Natterer's bat [48], one species being specific to lineage a and the other species specific to lineage b.

Phylogenetic analysis confirmed the high variability of EBLV-1 with a complex phylogenetic structure for EBLV-1b separated into at least 4 lineages, in contrast to EBLV-1a which has low genetic diversity and only 1 lineage despite its extensive geographic distribution from France to the Netherlands and Russia.

It was successively recommended by the French expert working group on rabies in bats (AFSSA, 2003), the French Veterinary Academy (2005) and more recently by a group of European rabies experts [16], that research work and epidemiological surveillance of EBLV-1 infection be pursued and intensified. French passive bat rabies surveillance in particular should be improved with regard to targeted bat species such as serotines, Natterer's bats and Daubenton's bats. These results, complemented by active surveillance data [47], will help clarify the dynamics of EBLV-1 transmission and improve understanding of the cross-species infection process. To date, two cases of natural transmission of EBLV-1 to domestic cats have been reported in France [49], demonstrating that the two subtypes, a and b, of EBLV-1 are able to cross the species barrier. This clearly implies that bat biologists, virologists, and public health and agricultural service officials must work together closely in order to protect the public and promote bat conservation.

ACKNOWLEDGMENTS

We would like to thank the SFEPM group (Société Française pour l'Etude et la Protection des Mammifères, groupe Chiroptères), and especially the bat network workers, for their collaboration in the passive and active surveillance of bat rabies. The authors would also like to thank Audrey Tapiero (Federation des Conservatoires d'Espaces Naturels, Besançon, France) for her coordination of the national bat surveillance plan. We gratefully acknowledge the expert technical support of the diagnostics team (Estelle Litaize, Valère Brogat), the molecular biology team (Mélanie Biarnais, Jean Luc Schereffer, Carine Peytavin de Garam), Nathalie Stroucken for her contribution to the administrative coordination of the bat surveillance network and Marie José Duchêne for her work on the bibliography.

REFERENCES

1. **1.**Fenton B (2010) Convergences in the diversification of bats. Current Zoology (formerly Acta Zoologica Sinica) 56: 454–468.
2. Baker ML, Schountz T, Wang LF (2013) Antiviral immune responses of bats: a review. Zoonoses Public Health 60: 104–116.
3. Calisher CH, Childs JE, Field HE, Holmes KV, Schountz T (2006) Bats: important reservoir hosts of emerging viruses. Clin Microbiol Rev 19: 531–545.
4. Hayman DT, Bowen RA, Cryan PM, McCracken GF, O'Shea TJ, et al. (2013) Ecology of zoonotic infectious diseases in bats: current knowledge and future directions. Zoonoses Public Health 60: 2–21.
5. ICTV (2012) International Committee on Taxonomy of Viruses. Vertebrate.
6. Marston DA, Horton DL, Ngeleja C, Hampson K, McElhinney LM, et al. (2012) Ikoma lyssavirus, highly divergent novel lyssavirus in an African civet. Emerg Infect Dis 18: 664–667.
7. Freuling CM, Beer M, Conraths FJ, Finke S, Hoffmann B, et al. (2011) Novel lyssavirus in Natterer's bat, Germany. Emerg Infect Dis 17: 1519–1522.
8. Picard-Meyer E, Servat A, Robardet E, Moinet M, Borel C, et al. (2013) Isolation of Bokeloh bat lyssavirus in Myotis nattereri in France. 10. 1007/s00705-013-1747-y. Arch Virol.
9. Ceballos NA, Moron SV, Berciano JM, Nicolas O, Lopez CA, et al. (2013) Novel lyssavirus in bat, Spain. Emerg Infect Dis 19: 793–795.
10. Davis PL, Holmes EC, Larrous F, Van der Poel WH, Tjornehoj K, et al. (2005) Phylogeography, population dynamics, and molecular evolution of European bat lyssaviruses. J Virol 79: 10487–10497.
11. McElhinney LM, Marston DA, Leech S, Freuling CM, van der Poel WH, et al. (2013) Molecular epidemiology of bat lyssaviruses in europe. Zoonoses Public Health 60: 35–45.
12. Vazquez-Moron S, Juste J, Ibanez C, Aznar C, Ruiz-Villamor E, et al. (2008) Asymptomatic rhabdovirus infection in meridonial serotine bats (Eptesicus isabellinus) from Spain. Dev Biol (Basel) 131: 311–316.
13. Schatz J, Fooks AR, McElhinney L, Horton D, Echevarria J, et al. (2013) Bat rabies surveillance in Europe. Zoonoses Public Health 60: 22–34.

14. King AA, Haagsma J, Kappeler A (2004) Lyssavirus infections in European bats. In: King A. A FAR, Aubert M. , Wandeler A. I. , editor. Historical perspective of Rabies in Europe and the Mediterranean Basin. Paris: OIE (World organisation for animal health). pp. 221–241.
15. Picard-Meyer E, Barrat J, Wasniewski M, Wandeler A, Nadin-Davis S, et al. (2004) Epidemiology of rabid bats in France, 1989 to 2002. Vet Rec 155: 774–777.
16. Cliquet F, Freuling C, Smreczak M, van der Poel WHM, Horton D, et al. (2010) Development of harmonised schemes for monitoring and reporting of rabies in animals in the European Union EFSA Scientific report 60 p.
17. Dietz C, Von Helversen O (2004) Illustrated identification key to the bats of Europe. Electronic version. Fledermaus-dietz. de website. Available: http://www. fledermaus-dietz. de/publications/publications. html. Accessed May 2014
18. Meslin F, Kaplan M, Koprowski H (1996) Laboratory techniques in rabies. Geneva: World Health Organization. 476 p p.
19. Picard-Meyer E, Bruyere V, Barrat J, Tissot E, Barrat MJ, et al. (2004) Development of a hemi-nested RT-PCR method for the specific determination of European Bat Lyssavirus 1. Comparison with other rabies diagnostic methods. Vaccine 22: 1921–1929.
20. Dean D, Abelseth M, Atanasiu P (1996) The fluorescent antibody test. In: Meslin F, Kaplan M, Koprowski H, editors. Laboratory techniques in rabies. Geneva, Switzerland. pp. 88–95.
21. Servat A, Picard-Meyer E, Robardet E, Muzniece Z, Must K, et al. (2012) Evaluation of a Rapid Immunochromatographic Diagnostic Test for the detection of rabies from brain material of European mammals. Biologicals 40: 61–66.
22. Hayman DT, Banyard AC, Wakeley PR, Harkess G, Marston D, et al. (2011) A universal real-time assay for the detection of Lyssaviruses. J Virol Methods 177: 87–93.
23. Scherrer B (1984) Biostatistique. Chicoutimi, Quebec, Canada: Gaetan Morin. 850 p.
24. Tamura K, Peterson D, Peterson N, Stecher G, Nei M, et al. (2011) MEGA5: molecular evolutionary genetics analysis using maximum likelihood, evolutionary distance, and maximum parsimony methods. Mol Biol Evol 28: 2731–2739.

25. Felsenstein J (1989) PHYLIP-Phylogeny nference Package Version 3. 2. Cladistics 5: 164–166.
26. Drummond AJ, Suchard MA, Xie D, Rambaut A (2012) Bayesian phylogenetics with BEAUti and the BEAST 1. 7. Mol Biol Evol 29: 1969–1973.
27. Picard-Meyer E, Fediaevsky A, Servat A, Cliquet F (2013) Surveillance de la rage animale en France. Bulletin épidémiologique, santé animale et alimentation 60: 12–18.
28. Bourhy H, Ribadeau-Dumas F, Dacheux Laurent, Martin E, Sevin E (2011) Epidemiologie de la rage humaine en France. 2011. Available: Pasteur website. http://www. pasteur. fr/ip/resource/filecenter/document/01s-00004j-0gi/bulletin-2011. pdf. Accessed May 2014.
29. Vazquez-Moron S, Juste J, Ibanez C, Ruiz-Villamor E, Avellon A, et al. (2008) Endemic circulation of European bat lyssavirus type 1 in serotine bats, Spain. Emerg Infect Dis 14: 1263–1266.
30. Kappeler A (1989) Bat rabies surveillance in Europe. Rabies Bulletin Europe 13: 12–13.
31. Van der Poel WH, Van der Heide R, Verstraten ER, Takumi K, Lina PH, et al. (2005) European bat lyssaviruses, The Netherlands. Emerg Infect Dis 11: 1854–1859.
32. Serra-Cobo J, Lopez-Roig M, Segui M, Sanchez LP, Nadal J, et al. (2013) Ecological factors associated with European bat lyssavirus seroprevalence in spanish bats. PLoS ONE 8: e64467.
33. Muller T, Johnson N, Freuling CM, Fooks AR, Selhorst T, et al. (2007) Epidemiology of bat rabies in Germany. Arch Virol 152: 273–288.
34. Arthur L, Picard-Meyer E, Lemaire M, Barrat J, Pavisse R, et al. (2012) Rythme des entrées/sorties dans une colonie de Sérotine commune. Symbioses 28: 23–28.
35. Godineau F, Pain D. (2007) Plan de restauration des chiroptères en France métropolitaine, 2008 – 2012. Société Française pour l'Étude et la Protection des Mammifères/Ministère de l'Écologie, du Développement et de l'Aménagement Durables. Plan National d'actions en faveur des chiroptères. Plan actions chiropteres website. Available: http://www. plan-actions-chiropteres. fr/IMG/plan_restauration_chiropteres_final. pdf. Accessed May 2014.

36. Patyk K, Turmelle A, Blanton JD, Rupprecht CE (2012) Trends in national surveillance data for bat rabies in the United States: 2001-2009. Vector Borne Zoonotic Dis 12: 666–673.
37. George DB, Webb CT, Farnsworth ML, O'Shea TJ, Bowen RA, et al. (2011) Host and viral ecology determine bat rabies seasonality and maintenance. Proc Natl Acad Sci U S A 108: 10208–10213.
38. Altizcr S, Dobson A, Hosseini P, Hudson P, Pascual M, et al. (2006) Seasonality and the dynamics of infectious diseases. Ecol Lett 9: 467–484.
39. Harbusch C (2002) Aspects of the ecology of Serotine bats [Eptesicus serotinus, Schreber 1774] in contrasting landscapes in southwest Germany and Luxembourg. : University of Aberdeen, UK. 255 p.
40. Freuling C, Vos A, Johnson N, Kaipf I, Denzinger A, et al. (2009) Experimental infection of serotine bats (Eptesicus serotinus) with European bat lyssavirus type 1a. Journal of General Virology 90: 2493–2502.
41. Franka R, Johnson N, Muller T, Vos A, Neubert L, et al. (2008) Susceptibility of North American big brown bats (Eptesicus fuscus) to infection with European bat lyssavirus type 1. J Gen Virol 89: 1998–2010.
42. Jackson FR, Turmelle AS, Farino DM, Franka R, McCracken GF, et al. (2008) Experimental rabies virus infection of big brown bats (Eptesicus fuscus). J Wildl Dis 44: 612–621.
43. Kaplan MM (1969) Epidemiology of rabies. Nature 221: 421–425.
44. Turmelle AS, Allen LC, Jackson FR, Kunz TH, Rupprecht CE, et al. (2010) Ecology of rabies virus exposure in colonies of Brazilian free-tailed bats (*Tadarida brasiliensis*) at natural and man-made roosts in Texas. Vector Borne Zoonotic Dis 10: 165–175.
45. Banyard AC, Hayman D, Johnson N, McElhinney L, Fooks AR (2011) Bats and lyssaviruses. Adv Virus Res 79: 239–289.
46. Turmelle AS, Jackson FR, Green D, McCracken GF, Rupprecht CE (2010) Host immunity to repeated rabies virus infection in big brown bats. J Gen Virol 91: 2360–2366.
47. Picard-Meyer E, Dubourg Savage MJ, Arthur L, Barataud M, Becu D, et al. (2011) Active surveillance of bat rabies in France: a 5-year study (2004-2009). Vet Microbiol 151: 390–395.

48. Salicini I, Ibanez C, Juste J (2011) Multilocus phylogeny and species delimitation within the Natterer's bat species complex in the Western Palearctic. Mol Phylogenet Evol 61: 888–898.
49. Dacheux L, Larrous F, Mailles A, Boisseleau D, Delmas O, et al. (2009) European bat Lyssavirus transmission among cats, Europe. Emerg Infect Dis 15: 280–284.

CHAPTER 13

Ecological Approaches in Veterinary Epidemiology: Mapping the Risk of Bat-Borne Rabies Using Vegetation Indices and Night-Time Light Satellite Imagery

Luis E Escobar[1,2], A Townsend Peterson[3], Monica Papeş[4], Myriam Favi[5], Veronica Yung[5], Olivier Restif[6], Huijie Qiao[7] and Gonzalo Medina-Vogel[1]

[1] Facultad de Ecología y Recursos Naturales, Universidad Andres Bello, Av. República 440, Santiago, Chile.

[2] Center for Global Health and Translational Science, SUNY Upstate Medical University, Syracuse, New York, USA.

Citation: Escobar, L.E., Peterson, A.T., Papeş, M. et al. Ecological approaches in veterinary epidemiology: mapping the risk of bat-borne rabies using vegetation indices and night-time light satellite imagery. Vet Res 46, 92 (2015). https://doi.org/10.1186/s13567-015-0235-7

[3] Biodiversity Institute, University of Kansas, Lawrence, Kansas, USA.

[4] Department of Integrative Biology, Oklahoma State University, Stillwater, Oklahoma 74078, USA.

[5] Sección Rabia, Instituto de Salud Publica de Chile, Av. Maraton 1000, Ñuñoa, Chile.

[6] Disease Dynamics Unit, Department of Veterinary Medicine, University of Cambridge, Madingley Road, Cambridge CB3 0ES, UK.

[7] Key Laboratory of Animal Ecology and Conservation Biology, Institute of Zoology, Chinese Academy of Science, Beijing, China.

ABSTRACT

Rabies remains a disease of significant public health concern. In the Americas, bats are an important source of rabies for pets, livestock, and humans. For effective rabies control and prevention, identifying potential areas for disease occurrence is critical to guide future research, inform public health policies, and design interventions. To anticipate zoonotic infectious diseases distribution at coarse scale, veterinary epidemiology needs to advance via exploring current geographic ecology tools and data using a biological approach. We analyzed bat-borne rabies reports in Chile from 2002 to 2012 to establish associations between rabies occurrence and environmental factors to generate an ecological niche model (ENM). The main rabies reservoir in Chile is the bat species *Tadarida brasiliensis*; we mapped 726 occurrences of rabies virus variant AgV4 in this bat species and integrated them with contemporary Normalized Difference Vegetation Index (NDVI) data from the Moderate Resolution Imaging Spectroradiometer (MODIS). The correct prediction of areas with rabies in bats and the reliable anticipation of human rabies in our study illustrate the usefulness of ENM for mapping rabies and other zoonotic pathogens. Additionally, we highlight critical issues with selection of environmental variables, methods for model validation, and consideration of sampling bias. Indeed, models with weak or incorrect validation approaches should be interpreted with caution. In conclusion, ecological niche modeling applications for mapping disease risk at coarse geographic scales have a promising future, especially with refinement and enrichment of models with additional information, such as night-time light data, which increased substantially the model's ability to anticipate human rabies.

Keywords: Normalize Difference Vegetation Index, West Nile Virus, Rabies, Ecological Niche Modeling, Normalize Difference Vegetation Index Data

INTRODUCTION

In light of its continued threat to public health around the world, rabies has stimulated considerable research efforts for the development of techniques and tools for vaccination and diagnosis [1]. While rabies control in dogs remains the priority in Africa and Asia [2], the management of wildlife reservoirs is the major challenge for rabies control in the Americas [3,4]. Thanks to rabies-elimination efforts in some American countries, there has been recent progress in acquiring data on rabies in wildlife over large areas [5], but more research is needed to improve spatiotemporal predictions of rabies spillover and guide government interventions.

Chile has a long history of rabies management and control [6]. Improvements in vaccination campaigns, diagnosis, surveillance, data management, and education have served to eradicate dog-related rabies in the country, with consequent reduction of human cases [7,8]. At present, Chile has only a sylvatic cycle of rabies in bats [9]. The main reservoir identified in Chile is the insectivorous bat *Tadarida brasiliensis* [8], a synanthropic bat species with a geographic distribution ranging from Canada to southern South America [10]. With increasing reports of cases of bat-borne rabies in Chile in recent years, a detailed risk map is needed urgently [8]. The Instituto de Salud Pública (ISP) Rabies Laboratory in Chile keeps records of all cases since 1929 [6], including diagnosis methods and detailed geographic location, collected mostly through a national passive surveillance system, which could be the basis for detailed mapping.

Spatial epidemiology is an emerging subdiscipline of epidemiology that aims to identify geographic areas with elevated risk of disease transmission [11], whereas environmental epidemiology works in parallel to identify environmental factors linked to disease appearance. Ecological niche modeling (ENM) can be a useful tool in reaching the goals of both fields: the resulting models allow researchers to estimate environmental factors that shape spatial distributions of organisms [12]. This integration of ecology and biogeography into public health and epidemiology allows understanding the geography of past, current, and emerging disease transmission [13,14], and explaining the role of environmental changes on climate and landscape [15].

Here, we evaluate ENM performance in predicting rabies cases across Chile using environmental variables from satellite imagery to generate high-resolution maps of rabies' potential distribution across Chile. First, we evaluated the ability of ENM to predict rabies cases across space and time, using a series of environmental variables, geographic regions, and time periods to calibrate and evaluate model predictions. Second, we mapped rabies' potential distribution across Chile based on environmental variation in a vegetation greenness index derived from Moderate Resolution Imaging Spectroradiometer (MODIS) satellite imagery. Our results demonstrate the usefulness of ENM for the management of zoonotic diseases. In particular, this method could be employed to quantify effects of land-use change on disease emergence and anticipate disease transmission in areas with lack of surveillance.

MATERIALS AND METHODS

First, we assessed the usefulness of ENM to forecast bat-borne rabies using validation metrics in the geographic and the environmental spaces; once the informative capacity of environmental variables and occurrences was corroborated, we developed a final model with a post-processing step to include risk categories (Figure 1).

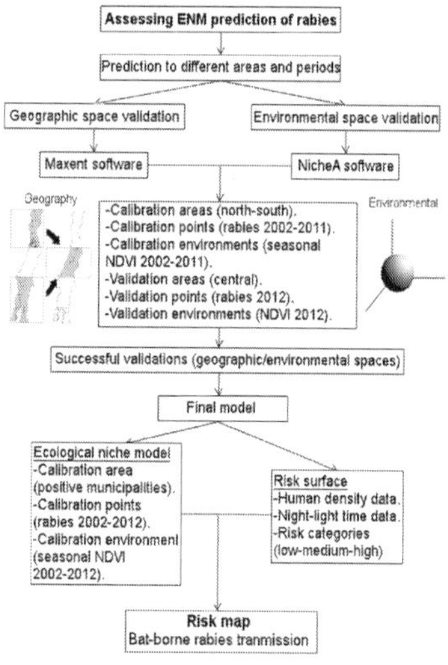

Figure 1. Flow chart of the use of ecological niche modeling for mapping bat-borne rabies risk. Before generating the final model, the robustness of the use of NDVI, ecological niche modeling algorithm, and prediction to different areas and periods were assessed in the geographic and environmental space.

Study area

Considering the critical role of the extent of the area of analysis in ENM performance [16], we limited the area for model calibration based on biogeographic barriers that included the Andes mountains (East), Pacific Ocean (West), ice fields in Patagonia (South), and the Atacama Desert (North; Figure 2). This area contained central Chile (43.5° S - 28.0° S) and was our *a priori* hypothesis regarding the extent of the accessible area, or **M** [16,17].

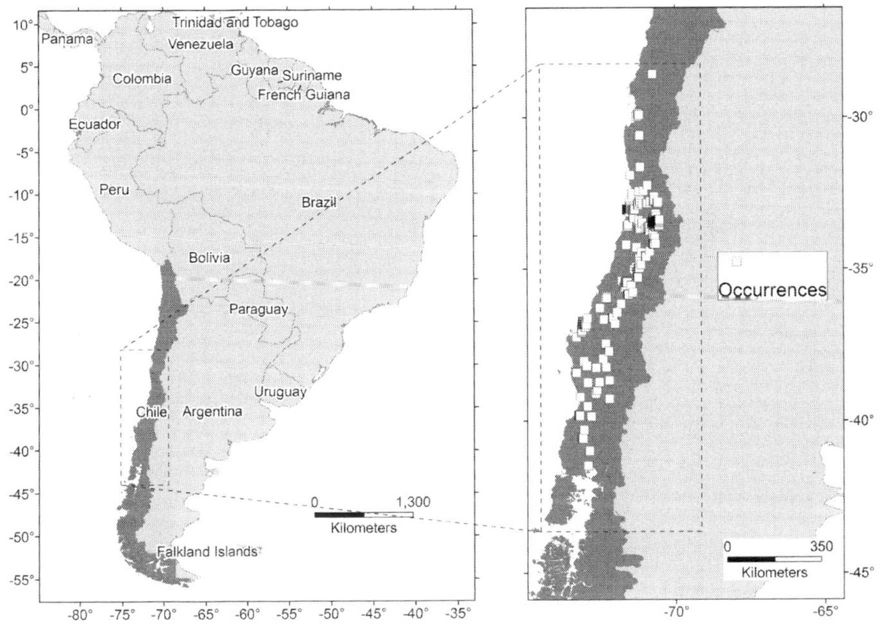

Figure 2. Study areas for model calibration and validation (M). Left: geographic position of Chile (dark gray). Right: study areas within central Chile (dark gray) and occurrence points (white squares) in the calibration area (dashed line).

Input data

Bat-rabies occurrences during 2002–2012 were obtained from samples submitted for rabies testing to ISP from passive and active surveillance. Passive surveillance is the term used for bats submitted by individuals after accidental findings of sick or dead animals, while active surveillance represents bat sampling developed by staff from the Ministry of Health in response to complaints of bats presence. Rabies diagnosis was based on direct immunofluorescence tests on bat brain tissues to confirm virus presence [18].

Identification of monoclonal antibodies and virus genotypes was carried out on positive samples [19-21]. Positive records were georeferenced using the freely available Address Validation Tool to convert textual addresses and location details into latitude and longitude coordinates with ~ 8 m error; sites with less accurate details (e.g., municipal parks) were identified using GoogleEarth with an estimated error <500 m.

In all, of 870 positive bat records, 813 samples from passive (98%) and active surveillance (2%) had address information sufficiently detailed for georeferencing. When multiple occurrences fell in the same grid cell, duplicates were removed, leaving single occurrences per cell. The final dataset included 726 occurrences (Figure 2).

As a source of environmental information, we used vegetation index data from MODIS satellite imagery. The most important features of vegetation index datasets are their relationship to primary productivity and the fine (8–16 day composites) temporal resolution [22].

We used the Normalized Difference Vegetation Index (NDVI), as this index provides a measure of herbaceous biomass and physiology through a ratio of light reflected in the red (R) and near-infrared (NIR) spectral bands [23]. This index is calculated as (NIR-R)/(NIR+R), approximating levels of photosynthetic activity, and has been associated with animal distribution and abundance in previous studies [24].

We used 16-day composites of NDVI data: data sets summarizing bi-weekly NDVI values, at ~500 m resolution, considering our accuracy in occurrences location, were downloaded for 2002–2012 from US Geological Survey ([25]; MOD13A1.005).

The original files in Hierarchical Data Format-EOS (HDF), with sinusoidal projection, were converted to GeoTIFF (Tagged Image File Format) with geographic projection using the MODIS Reprojection Tool provided by the NASA Goddard Space Flight Center [26]. NDVI values range from −2000 to 10 000, with fill values (No Data) set to −3000. NDVI layers were converted to ASCII files, and fill values changed from −3000 to −9999 to match modeling software requirements. No control quality flag layers were employed. Additionally, we included elevation information from the NASA Shuttle Radar Topographic Mission (SRTM) at ~500 m resolution [27]. The importance of parameters in explaining known cases of rabies was assessed before generating the final model (see Figure 1).

Model Calibration and Validation (geographic space)

We designed a first experiment by calibrating the model with rabies occurrences and NDVI data from 2002–2011, and validating with data from 2012, representing distinct regions in geographic space (Figure 3). Areas for calibration and validation corresponded to three regions of equal latitudinal width (Figure 3), of which we used northern and southern sectors for calibration, and the central region for validation. This framework allowed us to evaluate model performance across space and time. For this modeling experiment, we averaged 16-day NDVI values by season. Seasons were categorized as Summer for NDVI values between 15 December – 15 February, Autumn 15 March – 15 May, Winter 15 June – 15 August, and Spring 15 September – 15 November. Inter-seasonal NDVI datasets were discarded, as they represented transitional values between seasons. Minimum, mean, and maximum seasonal values were calculated for validation and calibration areas, and all 12 environmental layers (four seasons x three summary statistics) were used in model calibration. The jackknife model accuracy gain test in Maxent using all occurrences identified the contribution of each environmental variable to model performance across the calibration area.

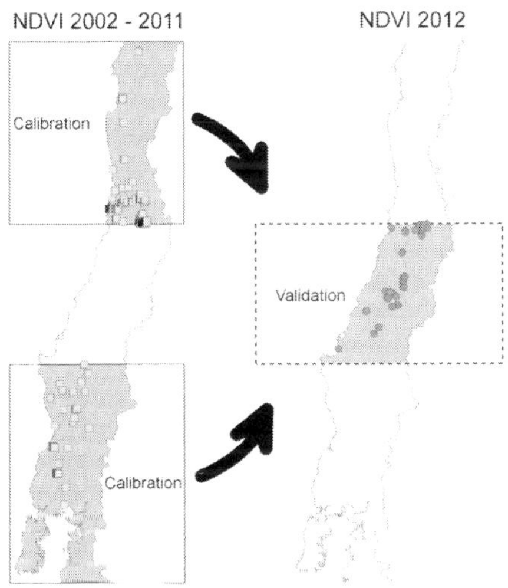

Figure 3. Model validation schema followed in this study. Calibration areas (black line boxes, left panel) contain occurrences from 2002 to 2011 (yellow squares, left panel) used to predict rabies in 2012 (green points, right panel) in the validation area (dashed box, right panel). Left panel: calibration areas based on NDVI layers and occurrences from 2002 to 2011. Right panel: validation area based NDVI layers and occurrences for 2012.

Models were generated using Maxent software version 3.3.3.k [28]. Maxent determines the probability distribution of maximum entropy (the most uniform), but constrained to the expected environmental values that correspond to the occurrence records [28]. Maxent is similar to a logistic regression algorithm for presence-background data (i.e., no true absences available), associating occurrences (presence) with environmental data across the study area (background). We selected the following settings in Maxent: random seed with 20% of occurrences set aside to evaluate models, 10 bootstrap permutations, logistic output, and the median of permutations as output. Additionally, models were calibrated with clamping and extrapolation options turned off [29]. Using ESRI ArcGIS 9.3, continuous output models from Maxent were converted to binary maps using a threshold based on omission error, finding the highest Maxent value that omitted no more than 5% of points employed during model calibration [17]; this step generated a prediction of presence and absence of environmental conditions suitable for rabies occurrence.

We used two model performance measures designed for ENM predictions, based on external sets of validation areas and occurrences [17]. First, we used a cumulative binomial test to assess whether predictions of validation occurrences across the validation region were statistically significantly better than random expectations. We used the validation occurrences as number of trials, the number of validation occurrences correctly predicted as number of successes, and the proportion of pixels predicted as suitable by the model as the probability of a success [17,30]. Second, as a complementary validation of model performance, we used a modified version of the area under the curve (AUC) of the receiving operating characteristic (ROC), the Partial ROC [31], in which ROC curves are evaluated only over ranges of values that correspond to low omission errors. Here, performance was measured as a ratio between observed prediction and a random expectation (AUC=1; e.g., $p>0.05$), where ratios above 1 represent predictions better than random expectations (AUC>1; e.g., $p<0.05$), evaluated using a bootstrap test [31]. We used Partial ROC software [32] to develop these tests, with the following settings: 50% of occurrences for bootstrap, 1000 permutations, and a threshold of 95% occurrences successfully predicted (for a detailed explanation, see [33]).

Model Calibration And Validation (environmental space)

We also validated the model in the environmental space, using environmental and occurrence datasets spatially and temporally independent of those used for model calibration (see Figure 3). A rabies' ENM calibrated in one region and period was transferred and analyzed in environments used to calibrate another rabies' ENM from a different area and period, and the shape, position, and size of both ENMs were compared to assess if our method was able to capture rabies' environmental signature across different times and geographic areas. This novel approach for model validation was developed using the software NicheA version 3.0, a powerful tool for display and analysis of ecological niches in environmental space [34]. First, we generated a model for the calibration area (see above and Figure 3) by calculating the minimum-volume ellipsoid including the occurrences for calibration from 2002–2011 against corresponding environments. We transferred this model to the validation area, occurrences, and environments (Figure 3) for 2012 to assess the predictive ability of the calibration model. We measured the proportion of overlap of the two ellipsoids for the two environmental data sets, as an estimation of niche similarity and robust prediction among areas and time frames.

ENM Projection Across Chile

After validating model predictive accuracy in space and time, we generated a country-wide model. NDVI datasets for January 2002 - December 2012 (i.e., the complete study period) were grouped by season using specific dates (see above). We used all available occurrences to generate the countrywide model.

Rabies occurrence records in Chile are influenced by biases introduced by passive surveillance [19], and this bias impacts model accuracy [35]. In this context, geographic bias correction has been suggested as an useful step in ENM [36]. Another factor affecting model output is the study area extent: the smaller the study area relative to the distribution of the species, the greater the challenge for the algorithms to produce accurate niche models [16]. In a previous study, Escobar et al. [37] found that introducing sampling effort in the form of number of rabies samples submitted reduced uncertainty, but did not improve model performance in terms of area predicted. Thus, to consider the bias sampling during our model calibration, we focused on sampled areas instead of number of bat samples from surveillance. We calibrated models in the municipalities of central Chile known to be rabies positive according to surveys. Once the final model was validated and calibrated in these specific areas, we transferred it to the whole country.

Mapping Human Risk of Rabies Infection

To assess risk of human cases, we combined the binary map (presence/absence) of environmental suitability for rabies with a surface of human population. To establish current human population at risk of rabies transmission at fine resolutions, we filtered the rabies potential distribution map by location of human settlements. The distribution of human settlements was derived from mean night-time light satellite imagery by county, as this variable has a strong association with density of human populations [38]. We calibrated our population density estimates by regressing country-averaged night-time light values against human densities from the 2012 census data for Chile [39] by county area (km^2). We used imagery at ~0.75 km resolution, specifically the band that detects light from visible to infrared acquired for 9 days in April and 13 days in October 2012 by the VIIRS sensor on the Suomi NPP satellite [40]. We used the first of the three bands that compose this image, with pixel values ranging 0–255: low values indicating darkness and high values indicating artificial light from human settlements. Although some night-time lights captured by the satellite could be wildfires or lit areas without human residents, we assumed that most values were related to artificial light.

After calibrating the satellite data against census reports, we classified light values into three categories, based on a quantile approach commonly used as a pragmatic criterion by which to define low, medium, and high risk of exposure [41]. The first quantile represents low risk, because few people are exposed in those areas; the second quantile is moderate risk; and the third quantile is high risk, as this class includes rabies-suitable cells with highest human density. To validate whether points falling in specific night-time light values could predict the human density category assigned to that light value, we generated 350 random points across the study area, and used a cumulative binomial test with the number of random points as trials, points with correct prediction of light value and human density as successes, and 1/3 (the chance of falling in the correct human density category) as the probability of a success.

RESULTS

Validation of input data

A total of 353 bat-rabies occurrences during 2002–2011 overlapped with the calibration area. When models were transferred to the validation area, only one of 46 validation occurrences from 2012 was not predicted successfully;

this was statistically better than random predictions ($P<0.001$; Table 1). This result indicated good performance of Maxent models across time, geographic areas, and environments. The most informative variables in model calibration were mean NDVI values for winter and spring, followed by maximum and minimum values in winter; least informative were minimum NDVI values for fall and summer.

Table 1. Statistical validation of rabies model performance using external occurrences and validation areas, Chile, 2002–2012

	Cumulative binomial test					Partial ROC
	Pixels predicted present	**Pixels predicted absent**	**Number of occurrences predicted present**	**Number of occurrences predicted absent**	**Binomial probability**	**AUC ratio min - max** *
Validation of input data	20 859	61 458	45	1	5.09×10^{-23}	1.52 - 1.87
ENM projection across Chile	15 532	60 630	113	27	1.10×10^{-52}	1.10 - 1.62

*min = minimum AUC ratio from 1000 permutations.

*max = maximum AUC ratio from 1000 permutations.

When occurrences from calibration and validation areas (see Figure 3) were displayed in environmental space, they showed high overlap. The NicheA algorithm uses geographic occurrences to collect environmental values, thus, analysis are based on environmental values only, excluding the geographic coordinates; this allowed us to display occurrences from different calibration and validation areas and distinct NDVI periods in a common environmental space (Figure 3). The ellipsoid of the occurrences and environments from 2002–2011 data in calibration regions was 35% larger compared to the volume of the ellipsoid of the occurrences and environments in validation areas, from 2012 (green ellipsoid). Thus occurrences from the calibration areas encompassed larger environmental variation than those available in the validation areas, which makes sense in light of the greater latitudinal diversity in the calibration subsets. These model validation exercises support the idea that selected environmental variables, study area delimitation, and available occurrences allow us to generate a robust ENM at a national extent.

ENM projection across Chile

To generate a high-confidence bat-borne rabies risk map at the country extent, we examined environmental variables across the country. Large geographic areas showed low variations in NDVI values across seasons: a large number of cells had consistently low NDVI values (500–1500), reflecting arid conditions in northern Chile, while the number of cells with very low (water) or very high (forest) NDVI values varied seasonally. The validation occurrences were predicted correctly by the model better than chance expectations, with all AUC ratio values above 1.0 ($P<0.001$; Table 1). Considering the robust model validation results, we developed a niche model using all occurrences and all positive counties, and transferred model rules to the entire country. The resulting model identified approximately 25 000 km^2 suitable for bat-borne rabies occurrence, concentrated in coastal areas of central Chile.

Mapping Human Risk

We found a strong non-linear association between human density and average night-time light intensity by county ($r^2=0.83$, $P<0.001$;). Consequently, we used this imagery to classify potential distribution of rabies obtained through ENM into areas of low (0–4 people/km^2), moderate (4–10 people/km^2), and high (10–255 people/km^2) risk. As a result, ~8600 km^2 were classified as high risk (Figure 4).

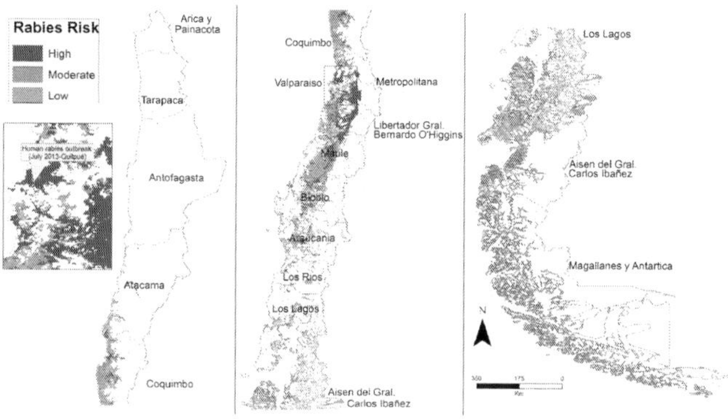

Figure 4. Classification of areas at low (gray), moderate (pink), and high (red) risk of rabies transmission to humans. Political boundaries at state level (region) are labeled for the northern (left panel), central (middle panel), and southern Chile (right panel). Insert: the last case of human rabies reported in Valparaiso in July, 2013 (green point).

Large areas of high risk were located close to the foothills of the Andes Mountains (Figure 4). During the development of this study, the first case of human rabies reported in the last 17 years was confirmed in Quilpue county, Valparaiso region, in an area classified as high risk with our methodology (Figure 4).

DISCUSSION

The most densely populated areas, in central Chile, are suitable for bat-borne rabies. We identified a subset of four environmental variables (from seasonal NDVI data) that contained most of the information by which to predict rabies distribution at the national scale. These four variables summarize approximately 2400 days of satellite data compilation. Ecological niche models generated using few climatic variables usually generate broad geographic predictions, however, our models avoided overprediction, perhaps thanks to the heterogeneous values across the study area and reduced spatial autocorrelation provided by NDVI [42].

Using 12 NDVI layers, we found that NDVI values in winter have a particular impact in model performance. Previously reported evidence of reduction in rabies reports in the winter season in Chile [19] could reflect migration or decrease in bat activity. Characterization of vegetation values that shape aspects of the species' niche in environmental space could be used to predict changes in species geographic distributions once vegetation changes occur, for example anticipating disease emergence after landscape disturbance (e.g., logging). To our knowledge, only one previous study explored the ability of ENM to anticipate disease occurrence using different time periods of vegetation indices over different geographic areas [43]: potential distribution of West Nile virus (WNV) in central United States was evaluated using vegetation indices and disease occurrence for 2002 and 2003, demonstrating that models calibrated in different geographic areas and time periods could anticipate human cases of WNV.

Ecological Niche Modeling In Epidemiology

Linking species' distributions to environmental factors has been an important goal in ecology for a century [44]. Recent developments in the field of ecological niche modeling have provided conceptual bases to select algorithms, delimit study areas, evaluate models, and identify relevant environmental variables [16,45-50]. This study applied theory and methods from the field of ecology into epidemiology to map potential bat-borne rabies risk.

Current methods to generate risk maps of infectious diseases are usually based on disease-occurrences density. This approach estimates risk based solely on spatial interpolations [51], but such estimation may fail to anticipate risk in neglected areas (i.e., lack of surveillance) [52] or may assume high risk in oversampled areas [51]. On the other hand, we argue that maps based on environmental interpolations to model the pathogen's ecological niche provide more accurate, and biologically realistic, predictions [53]. This ecological approach offers the opportunity to identify areas where the pathogen could be more abundant and genetically diverse based on suitable conditions [54,55]. Strikingly, despite its robustness, ecological niche modeling is still under-exploited in epidemiology [56].

ENM applications to infectious diseases are generally more complex than biodiversity studies [15]. Disease systems usually include several organisms: pathogens, vectors, natural reservoirs, and potential terminal hosts (e.g., humans), making application of ENM dependent on the target organisms, and on temporal and spatial scales [15]. The rabies system is clearly an example of spatial dependence and variation of environmental factors required by the virus for its persistence. At a very fine, sub-organismal scale, rabies virus shows affinity to the nervous system compared to all tissues available in the host, but with special preference for the brainstem and medulla [1]. At the other extreme of the spatial scale (i.e., continental), rabies responds to climate [14], reflecting how environments relate with all the participants in the system (the black box *sensu* Peterson [13]). Our previous exploration of ecological niche similarities between host and virus in the rabies system suggests that co-evolutionary forces may explain the close interaction between both organisms, represented in the indistinguishable occupancy of the environmental space by host and virus when explored at landscape scale [37]. Thus, species occupying the niche of another species (e.g., a parasite) may generate indistinguishable niche models between both species (i.e., the parasite and the host). Modeling rabies and bats generates indistinguishable niches, but modeling solely rabies provides more detail in the areas predicted suitable, gaining the prediction of risk [37].

Here, we focused our modeling at intermediate spatial resolution, fitting the spatial extent and available occurrences to remote sensing variables, but models at finer resolutions should be explored (e.g., free-ranging dog density, human behavior, bat abundance) to evaluate risk of rabies spillover in Chile. An area of ~8600 km^2 was classified as at high risk of rabies occurrence, representing 1.1% of the total area of Chile (756 096 km^2). This result provides a useful identification of priority areas.

Model Validation

Our validation design using different calibration and validation areas ensures statistical independence [57]. Validation is a crucial step in spatial epidemiology, especially for virulent pathogens such as rabies [42,56]. Unfortunately, in many applications of ENM to disease systems, validation procedures have been weak or even lacking [58-61]. Our assertive anticipation of rabies cases in independent validation areas is an example of how ENM of rabies may forecast rabies occurrence in unsampled areas, once the model is transferred to areas that lack data. Under the same approach, model transfer could be applied to future land-use change scenarios to predict rabies distribution under different NDVI values. The idea of an increase in rabies spread due to land-use change was proposed in previous reports of rabies in vampire bats [62-64], but has not been explored quantitatively in much detail until now. Our conservative model transfer settings avoided extrapolation into novel environmental conditions while allowing robust predictions into distinct NDVI datasets (i.e., 2012). Using extrapolation in ENM exercises generates perilous overprediction [29].

No set rules exist for selecting validation areas and occurrences. Rather, selection should be based on careful consideration of the data available and the biogeographic features of the accessible area selected *a priori* for the target species [17,65]. Selection of validation occurrences outside of areas used for model calibration (e.g., Figure 3) increases the geographic independence of validation occurrences from those used for calibration [42]. Model calibration should be developed using only areas with known disease occurrence, as this offers a means of reducing sampling bias, allowing detailed characterization of environments for rabies occurrences and avoiding uncertainty from non-sampled areas [35]. We calibrated the final model only in municipalities with known reports of rabies occurrences and found that focusing ENM calibration only in areas with surveillance reduces over-prediction and increases model accuracy, albeit potentially at the cost of precision [36]. Consideration of sampling bias is an issue of critical relevance when modeling the niche of pathogens. Our preliminary exploration of model response when negative samples were added to the analysis showed that models reduced their variability, but without significant improvement of areas predicted [37]. Thus, bias from the geographic space (e.g., clustered occurrences), should be considered different from the bias from the environmental space. To make it simple, we could have an intense and uniform sampling effort across a large area, but the environmental representation may be minimal (e.g., thousands of occurrences in a desert could represent a single environmental value). On the other hand, we could have a low geographic coverage (i.e., few occurrence points), but these points capture the complete pattern of environmental signature that shapes the species distribution. In our study, even though rabies occurrences from urbanized areas were abundant, occurrences from non-urbanized areas allowed us to generate models with high predictive performance (Figure 3). However, a well-designed sampling effort may provide more environmental information from areas not available to us, generating models with broader areas predicted suitable for bat-borne rabies occurrence.

In summary, our validation experiments confirmed that the ENM used in epidemiology can produce a robust risk map of a dangerous disease when occurrence data quality, environmental variable manipulation, and study area extent are considered carefully. As a result of these analyses, government efforts to ward prevention and control can focus in geographic areas most suitable for rabies potential distribution.

ACKNOWLEDGEMENTS

Thanks to Cristina Toledo (ISP) for collaboration in data compilation. Special thanks to Daniel J. Pons for discussions. Daniela A. Poo-Muñoz helped in preliminary data analyses. Universidad Andres Bello provided the grant DI-412-13/I for LEE. OR is supported by the Royal Society through a University Research Fellowship, and acknowledges funding from the Research and Policy on Infectious Disease Dynamics (RAPIDD) Program of the Fogarty International Center, National Institutes of Health and Science and Technology Directorate, Department of Homeland Security. This study was supported by the National Natural Science Foundation of China (A New Method to Predict the Species Distributions, 31100390). LEE postdoctoral research was supported by the Global Emerging Infectious Disease Surveillance and Response System (GEIS) grant P0435_14_UN.

REFERENCES

1. Rupprecht CE, Hanlon CA, Hemachudha T (2002) Rabies re-examined. Lancet Infect Dis 2:327–343
2. Morters MK, McKinley TJ, Horton DL, Cleaveland S, Schoeman JP, Restif O, Whay HR, Goddard A, Fooks AR, Damriyasa IM, Wood JLN (2014) Achieving population-level immunity to rabies in free-roaming dogs in Africa and Asia. PLoS Negl Trop Dis 8:e3160
3. Vigilato MAN, Cosivi O, Knöbl T, Clavijo A, Silva HMT (2013) Rabies update for Latin America and the Caribbean. Emerg Infect Dis 19:678–679
4. Escobar LE, Peterson AT, Favi M, Yung V, Medina-Vogel G (2015) Bat-borne rabies in Latin America. Rev Inst Med Trop Sao Paulo 57:63–72
5. Blanton JD, Manangan A, Manangan J, Hanlon CA, Slate D, Rupprecht CE (2006) Development of a GIS-based, real-time Internet mapping tool for rabies surveillance. Int J Health Geogr 5:1–8
6. Laval E, Lepe P (2008) Una visión histórica de la rabia en Chile. Rev Chil Infectol 25:2–7
7. Pavletic CB, Ramirez EV, De Mattos CC, De Mattos CA (1999) Rol de los murciélagos insectívoros en la transmisión de la rabia en Chile. Arch Med Vet 31:157–165
8. Escobar LE, Restif O, Yung V, Favi M, Pons DJ, Medina-Vogel G (2015) Spatial and temporal trends of bat-borne rabies in Chile. Epidemiol Infect 143:1486–1494
9. De Mattos CA, Favi M, Yung V, Pavletic C, De Mattos CC (2000) Bat rabies in urban centers in Chile. J Wildl Dis 36:231–240
10. Iriarte A (2007) Mamíferos de Chile. Lynx Edicions, Santiago
11. Ostfeld RS, Glass GE, Keesing F (2005) Spatial epidemiology: an emerging (or re-emerging) discipline. Trends Ecol Evol 20:328–336
12. Soberón J, Peterson AT (2005) Interpretation of models of fundamental ecological niches and species' distributional areas. Biodiv Inform 2:1–10
13. Peterson AT (2007) Ecological niche modelling and understanding the geography of disease transmission. Vet Ital 43:393–400
14. Peterson AT (2008) Biogeography of diseases: a framework for analysis. Naturwissenschaften 95:483–491

15. Estrada-Peña A, Ostfeld RS, Peterson AT, Poulin R, de la Fuente J (2014) Effects of environmental change on zoonotic disease risk: An ecological primer. Trends Parasitol 30:205–214
16. Barve N, Barve V, Jiménez-Valverde A, Lira-Noriega A, Maher SP, Peterson AT, Soberón J, Villalobos F (2011) The crucial role of the accessible area in ecological niche modeling and species distribution modeling. Ecol Model 222:1810–1819
17. Peterson AT, Soberón J, Pearson RG, Anderson RP, Martínez-Meyer E, Nakamura M, Bastos M (2011) Ecological Niches and Geographic Distributions. Princeton University Press, New Jersey
18. World Organisation for Animal Health (2011) Rabies. In: Fooks AR, Horton DL (eds) Man Diagnostic Test Vaccines Terr Anim, 7th edn. OIE, Paris, pp 1–20
19. Favi M, Rodríguez L, Espinosa C, Yung V (2008) Rabies in Chile: 1989–2005. Rev Chilena Infectol 25:S8–13 (in Spanish)
20. Yung V, Favi M, Fernández J (2002) Genetic and antigenic typing of rabies virus in Chile. Arch Virol 147:2197–205
21. Yung V, Favi M, Fernandez J (2012) Typing of the rabies virus in Chile, 2002–2008. Epidemiol Infect 140:2157–2162
22. Pettorelli N, Ryan S, Mueller T, Bunnefeld N, Jedrzejewska B, Lima M, Kausrud K (2011) The Normalized Difference Vegetation Index (NDVI): unforeseen successes in animal ecology. Clim Res 46:15–27
23. Horning N, Robinson J, Sterling E, Turner W, Spector S (2010) Remote Sensing for Ecology and Conservation. Oxford University Press, New York
24. Pettorelli N, Vik JO, Mysterud A, Gaillard JM, Tucker CJ, Stenseth NC (2005) Using the satellite-derived NDVI to assess ecological responses to environmental change. Trends Ecol Evol 20:503–510
25. NASA, USGS (2014) Data Pool. https://lpdaac.usgs.gov/data_access/data_pool. Accessed 11 Nov 2013
26. Land Processes DAAC (2011) MODIS Reprojection Tool 4.1. https://lpdaac.usgs.gov/tools/modis_reprojection_tool. Accessed 11 Nov 2013
27. Jarvis A, Reuter HI, Nelson A, Guevara E (2008) Hole-filled SRTM for the globe Version 4, available from the CGIAR-CSI SRTM 90m Database., http://srtm.csi.cgiar.org. Accessed 11 Nov 2013
28. Phillips SJ, Anderson RP, Schapire RE (2006) Maximum entropy modeling of species geographic distributions. Ecol Model 190:231–259

29. Owens HL, Campbell LP, Dornak LL, Saupe EE, Barve N, Soberón J, Ingenloff KK, Lira-Noriega A, Hensz CM, Myers CE, Peterson AT (2013) Constraints on interpretation of ecological niche models by limited environmental ranges on calibration areas. Ecol Model 263:10–18
30. Anderson RP, Gómez-Laverde M, Peterson AT (2002) Geographical distributions of spiny pocket mice in South America: Insights from predictive models. Glob Ecol Biogeogr 11:131–141
31. Peterson AT, Papeş M, Soberón J (2008) Rethinking receiver operating characteristic analysis applications in ecological niche modeling. Ecol Model 213:63–72
32. Barve N (2008) Tool for Partial-ROC. Version 1. Biodiversity Institute, Lawrence, KS, Available: http://kuscholarworks.ku.edu/dspace/handle/1808/10059
33. Peterson AT (2012) Niche modeling. model evaluation. Biodiv Inform 8:41
34. Qiao H, Soberón J, Escobar LE, Campbell L, Peterson AT (2015) NicheA. Version 3.0. Biodiversity Institute, Kansas
35. Phillips SJ, Dudik M, Elith J, Graham CH, Lehmann A, Leathwick J, Ferrier S (2009) Sample selection bias and presence-only distribution models: Implications for background and pseudo-absence data. Ecol Appl 19:181–197
36. Syfert MM, Smith MJ, Coomes DA (2013) The effects of sampling bias and model complexity on the predictive performance of Maxent species distribution models. PLoS One 8:e55158
37. Escobar LE, Peterson AT, Favi M, Yung V, Pons DJ, Medina-Vogel G (2013) Ecology and geography of transmission of two bat-borne rabies lineages in Chile. PLoS Negl Trop Dis 7:e2577
38. Sutton P, Roberts D, Elvidge C, Baugh K (2001) Census from heaven: An estimate of the global human population using night-time satellite imagery. Int J Remote Sens 22:3061–3076
39. Instituto Nacional de Estadistica (2012) Resultados preliminares censo de población y vivienda 2012. Santiago: INE
40. Simmon R (2013) Night Lights 2012 - Flat map., http://1.usa.gov/1FQvs5r. Accessed 11 Nov 2013
41. Winters AM, Bolling BG, Beaty BJ, Blair CD, Eisen RJ, Meyer AM, Pape WJ, Moore CG, Eisen L (2008) Combining mosquito vector and

human disease data for improved assessment of spatial West Nile virus disease risk. Am J Trop Med Hyg 78:654–665

42. Escobar LE, Peterson AT (2013) Spatial epidemiology of bat-borne rabies in Colombia. Rev Panam Salud Publica 34:135–136
43. Peterson AT, Robbins A, Restifo R, Howell J, Nasci R (2008) Predictable ecology and geography of West Nile virus transmission in the central United States. J Vector Ecol 33:63–72
44. Grinnell J (1917) The niche-relationships of the California Thrasher. Auk 34:427–433
45. Jiménez-Valverde A, Lobo JM, Hortal J (2008) Not as good as they seem: The importance of concepts in species distribution modelling. Divers Distrib 14:885–890
46. Warren DL (2012) In defense of "niche modeling". Trends Ecol Evol 27:497–500
47. Soberón J, Nakamura M (2009) Niches and distributional areas: concepts, methods, and assumptions. Proc Natl Acad Sci U S A 106:19644–19650
48. Soberón J, Peterson AT (2011) Ecological niche shifts and environmental space anisotropy: a cautionary note. Rev Mex Biodivers 82:1348–1355
49. Lobo JM, Jiménez-Valverde A, Real R (2007) AUC: a misleading measure of the performance of predictive distribution models. Glob Ecol Biogeogr 17:145–151
50. Peterson AT (2007) Why not WhyWhere: the need for more complex models of simpler environmental spaces. Ecol Model 203:527–530
51. Simarro PP, Cecchi G, Franco JR, Paone M, Diarra A, Ruiz-Postigo JA, Fèvre EM, Mattioli RC, Jannin JG (2012) Estimating and mapping the population at risk of sleeping sickness. PLoS Negl Trop Dis 6:e1859
52. Lash RR, Carroll DS, Hughes CM, Nakazawa Y, Karem K, Damon IK, Peterson AT (2012) Effects of georeferencing effort on mapping monkeypox case distributions and transmission risk. Int J Health Geogr 11:23
53. Peterson AT (2006) Ecologic niche modeling and spatial patterns of disease transmission. Emerg Infect Dis 12:1822–1826
54. Martínez-Meyer E, Díaz-Porras D, Peterson AT, Yáñez-Arenas C (2012) Ecological niche structure and rangewide abundance patterns of species. Biol Lett 9:20120637

55. Lira-Noriega A, Manthey JD (2014) Relationship of genetic diversity and niche centrality: A survey and analysis. Evolution 68:1082–1093
56. Peterson AT (2014) Mapping Disease Transmission Risk: Enriching Models Using Biogeography and Ecology. Johns Hopkins University Press, Baltimore
57. Hurlbert SH (1984) Pseudoreplication and the design of ecological field experiments. Ecol Monogr 54:187–211
58. Brito-Hoyos DM, Sierra EB, Álvarez V (2013) Distribución geográfica del riesgo de rabia de origen silvestre y evaluación de los factores asociados con su incidencia. Rev Panam Salud Publica 33:8–14
59. Suzán G, Giermakowski JT, Marcé E, Suzán-Azpiri H, Armién B, Yates TL (2006) Modeling hantavirus reservoir species dominance in high seroprevalence areas on the Azuero Peninsula of Panama. Am J Trop Med Hyg 74:1103–1110
60. Nieto P, Malone JBJ, Bavia M (2006) Ecological niche modeling for visceral leishmaniasis in the state of Bahia, Brazil, using genetic algorithm for rule-set prediction and growing degree day-water budget. Geospat Health 1:115–126
61. Hernández J, Núñez I, Bacigalupo A, Cattan PE (2013) Modeling the spatial distribution of Chagas disease vectors using environmental variables and people's knowledge. Int J Health Geogr 12:29
62. Lee DN, Papeş M, Van Den Bussche R (2012) Present and potential future distribution of common vampire bats in the Americas and the associated risk to cattle. PLoS One 7:e42466
63. McCarthy T (1989) Human depredation by vampire bats (*Desmodus rotundus*) following a hog cholera campaign. Am J Trop Med Hyg 40:320–322
64. Voigt C, Kelm D (2006) Host preference of the common vampire bat (*Desmodus rotundus*; Chiroptera) assessed by stable isotopes. J Mammal 87:1–6
65. Peterson AT, Papeş M, Kluza DA (2003) Predicting the potential invasive distributions of four alien plant species in North America. Weed Sci 51:863–868
66. Doll CNH, Muller J-P, Morley JG (2006) Mapping regional economic activity from night-time light satellite imagery. Ecol Econ 57:75–92
67. Mazor T, Levin N, Possingham HP, Levy Y, Rocchini D, Richardson AJ, Kark S (2013) Can satellite-based night lights be used for

conservation? The case of nesting sea turtles in the Mediterranean. Biol Conserv 159:63–72

68. Doll CNH, Pachauri S (2010) Estimating rural populations without access to electricity in developing countries through night-time light satellite imagery. Energy Policy 38:5661–5670
69. Bharti N, Tatem AJ, Ferrari MJ, Grais RF, Djibo A, Grenfell BT (2011) Explaining seasonal fluctuations of measles in Niger using nighttime lights imagery. Science 334:1424–1427
70. Weng Q (2008) Remote sensing of impervious surfaces. CRC Press, Boca Raton

SECTION 4: BACTERIA

CHAPTER 14

Molecular Epidemiology of Clinical Isolates of Pseudomonas Aeruginosa Isolated from Horses in Ireland

Tazumi A[1,2], Maeda Y[1,3], Buckley T[4,5], Millar BC[1], Goldsmith CE[1], Dooley JSG[3], Elborn JS[6,7], Matsuda M[2] and Moore JE[1,3]

[1] Northern Ireland Public Health Laboratory, Department of Bacteriology, Belfast City Hospital, Belfast BT9 7AD, Northern Ireland

[2] Laboratory of Molecular Biology, School of Environmental Health Sciences, Azabu University, Fuchinobe 1-17- 71, Sagamihara 229-8501, Japan

[3] School of Biomedical Sciences, Centre for Molecular Biosciences, University of Ulster, Coleraine, BT52 1SA, Northern Ireland

[4] Irish Equine Centre, Johnstown, Naas, Co. Kildare, Ireland

Citation: Tazumi, A., Maeda, Y., Buckley, T. et al. Molecular epidemiology of clinical isolates of Pseudomonas aeruginosa isolated from horses in Ireland. Ir Vet J 62, 456 (2009). https://doi.org/10.1186/2046-0481-62-7-456

[5] Department of Equine Studies, University of Limerick, Ireland

[6] Regional Adult Cystic Fibrosis Unit, Belfast City Hospital, Lisburn Road, Belfast, Northern Ireland, BT9 7AB

[7] Department of Respiratory Medicine, The Queen's University of Belfast, Respiratory Medicine, Belfast City Hospital, Lisburn Road, Belfast, Northern Ireland, BT9 7AB

ABSTRACT

Clinical isolates (n = 63) of *Pseudomonas aeruginosa* obtained from various sites in 63 horses were compared using ERIC2 RAPD PCR to determine their genetic relatedness. Resulting banding patterns (n = 24 genotypes) showed a high degree of genetic heterogeneity amongst all isolates examined, indicating a relative non-clonal relationship between isolates from these patients, employing this genotyping technique. This study characterised 63 clinical isolates into 24 distinct genotypes, with the largest cluster (genotype E) accounting for 10/63 (15.9%) of the isolates. ERIC2 RAPD PCR proved to be a highly discriminatory molecular typing tool of P. aeruginosa in isolates recovered from horses. With the adoption of several controls to aid reproducibility, this technique may be useful as an alternative to PFGE, particularly in epidemiological investigations of outbreaks where speed may be a significant parameter. This is the first report of clonal heterogeneity amongst P. aeruginosa from horses and demonstrated that ERIC RAPD PCR is a rapid method for the examination of this species in horses, which may be useful in outbreak analysis.

INTRODUCTION

Pseudomonas aeruginosa is an important Gram-negative bacterial pathogen in horses, particularly associated with the genital tract [9], although infection due to this causal agent may occur at virtually any anatomical site in the horse. Molecular genotyping of *P. aeruginosa* is of major importance in the elucidation of transmission routes. Genetic variability of clinical isolates from horses may be compared by examination of phylogenetic distances, which can give an important insight into the interrelationship of these bacterial isolates. Detailed genetic analysis at the subspecies (strain) level gives insights into the variability within a bacterial population and generates evidence on genome evolution, which in turn leads to bacterial adaptation to various environmental conditions. This information can be used in a clinical setting to help identify clinical isolates that occur frequently in animals, so that further work may be undertaken to identify common virulence characteristics in equine strains and help, for example, in directing choice of strains to use in vaccine development.

For the last two decades, PCR-based genotyping methods have played an important role in bacterial typing schemes. One of the PCR-based methods, namely random amplification of polymorphic DNA (RAPD)-PCR, also known as arbitrarily primed-polymerase chain reaction (AP-PCR), has been described to be useful on account of its simplicity and utility for analysis of large throughput samples [4]. This technique utilises a variable short length arbitrary primer, and is advantageous, as it does not require any previous knowledge of the target DNA sequence data. The primer is amplified arbitrarily at low stringency, where the oligonucleotide binds at complementary and partially mismatched sites and generates bands which differ in length and nucleotide composition. RAPD-PCR has been successfully applied for typing a wide variety of both Gram-positive and Gram-negative bacteria. Various genus specific and universal primers have been used, including the enterobacterial repetitive intergenic consensus (ERIC2) sequence and M13 primers, to successfully type both Gram-positive bacteria e.g. MRSA [5] and Gram-negative organisms e.g. *P. aeruginosa* [2], as a means of rapid strain differentiation and as an alternative to Pulsed-Field Gel Electrophoresis (PFGE). To date, there have been no reports in the literature examining the molecular epidemiology of *P. aeruginosa* from clinically significant isolates in horses. Therefore, the aim of this study was to examine the genetic relatedness of *P. aeruginosa* isolated from a variety of sites in Irish horses, over the five year period, 2003-2007, through employment of ERIC2-RAPD PCR.

MATERIALS AND METHODS

Description of Isolates Employed

P. aeruginosa isolates (n = 63) were obtained from the bacteriological culture archive of the Irish Equine Centre, Johnstown, Naas, Co. Kildare, Ireland. These strains were originally isolated from several sites in horses, including abscess, ear, eye, faecal, genito-urinary, guttural pouch, lung, nasal, pharynx, semen, skin scrapings, throat, tissue, urine and wound, during the five year period, 2003-2007, including 2007 (n = 19 isolates), 2006 (n = 15 isolates), 2005 (n = 17 isolates), 2004 (n = 8 isolates) and

2003 (n = 4 isolates). All isolates were revived from frozen storage and were identified phenotypically employing a combination of conventional identification methods (e.g. oxidase), as well as the API Identification scheme (API 20NE) (BioMérieux, Les Halles, France).

DNA extraction

Genomic DNA was extracted from confirmed isolates of *P. aeruginosa*. Isolates were cultured on Columbia Blood Agar (Oxoid CM0331) supplemented with 5% (v/v) defibrinated horse blood, for 24 hours at 37°C. All DNA isolation procedures were carried out in a Class II Biological Safety Cabinet (MicroFlow, England) in a room physically separated from that used to set up nucleic acid amplification reaction mixes and also from the 'post-PCR' room. This was performed in accordance with the Good Molecular Diagnostic Procedures (GMDP) guidelines of Millar *et al.* [6], in order to minimise contamination and hence the possibility of false positive results. Bacterial genomic DNA was extracted from a single colony of *P. aeruginosa* from each isolate, under standard conditions, by employment of the Roche High Purity PCR Template Preparation Kit (Roche, England), in accordance with the manufacturer›s instructions. Extracted DNA was stored at -80°C prior to PCR amplification. For each batch of extractions, a negative extraction control containing all reagents minus organism, was performed, as well as an extraction positive control with *P. aeruginosa*.

ERIC2 RAPD PCR Amplification

All reaction mixes were set up in a PCR hood in a room separate from that used to extract DNA and the amplification and post-PCR room in order to minimise contamination. Initially, PCR amplification conditions were optimised by separately varying magnesium chloride concentration, annealing temperature, primer concentration and DNA template concentration. Following optimisation, reaction mixes (25 µl) were set up as follows: -10 mM Tris- HCl, pH 8.3, 50 mM KCl, 2.5 mM $MgCl_2$, 200 µM (each) dATP, dCTP, dGTP and dTTP; 1.25U of *Taq* DNA polymerase (Amplitaq; Perkin Elmer), 0.1 µM of the RAPD primer, ERIC2 (enterobacterial repetitive intergenic consensus sequence) (5' - AAG TAA GTG ACT GGG GTG AGC G - 3') and 1 µl of DNA template. The reaction mixtures following a 'hot start' were subjected to the following empirically optimised thermal cycling parameters in a Perkin Elmer 2400 thermocycler: 94°C for five minutes, followed by 40 cycles of 94°C for one minute, 50°C for one minute, 72°C for one minute, followed by a final extension at 72°C for 10 minutes. Multiple negative (water) amplification controls were included in every set of PCR reactions. Following amplification, aliquots (10 μl) were removed from each reaction mixture and examined by electrophoresis (80 V, 45 min) in gels composed of 2% (w/v) agarose (Gibco, UK) in TAE buffer (40 mM Tris, 20 mM acetic acid, 1 mM EDTA, pH 8.3), stained with ethidium bromide (5 μg/100 ml). In order to improve resolution of images, resulting PCR amplicons were resolved on 2% (w/v) metaphor agarose (Flowgen Ltd., UK), as described above. Ethidium bromide stained agarose gels were visualised under UV illumination using a gel image analysis system (UVP Products, England) and all images archived as digital graphic files (*.bmp). The genetic relationship between banding patterns was interpreted in accordance with the criteria as defined previously by Tenover *et al.* [11] and individual ERIC2 genotypes were assigned an arbitrary code based on ascending alphabetic order. Genotypes were assigned a novel status when differing by a one band shift. RAPD-PCR reproducibility was examined with *P. aeruginosa* isolates on at least two further occasions.

RESULTS

All *P. aeruginosa* isolates examined in this study generated an ERIC2 RAPD banding pattern ranging in size, from approximately 200 bp to >1,000 bp, with between three and eight bands per isolate, with a mean band number of 5.3 bands per isolate examined (Figure 1). Overall, there was a high degree of clonal heterogeneity between all isolates examined, as 24 distinct RAPD genotypes were recorded (Genotype A-Genotype X) from the 63 isolates examined. Clustering of isolates was observed, where Genotype E was the most frequent cluster with 10 (10/63; 15.9% of total isolates) members, followed by Genotype R (8/63; 12.7%), Genotype C (6/63; 9.5%), Genotype B (5/63; 7.9%) and Genotype V (5/63; 7.0%). These five largest clusters accounted for 34/63 isolates (69.8%) examined in the study. Of the remaining genotypes observed, there were three genotypes each containing three isolates, five genotypes each containing two members and 11 genotypes represented by a single isolate. Certain genotypes with larger numbers of isolates had isolates from several years, i.e., for genotype E with 10 members, these isolates were representative from 2005-2007, whilst for genotype R, isolates here represented 2003-2006. Thus, this study demonstrated the persistence and recurrence of certain genotypes over a four year period within the equine population.

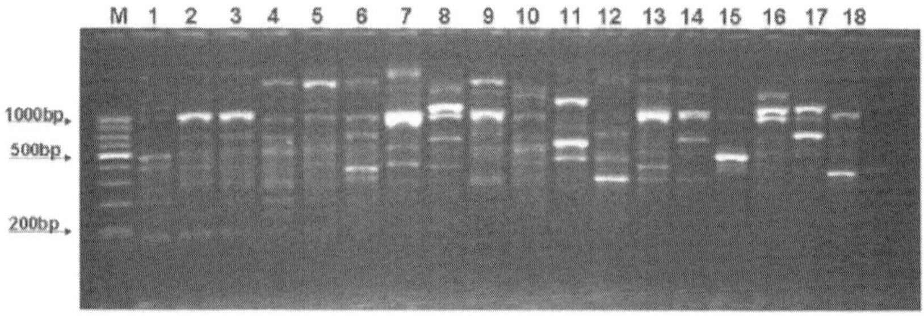

Figure 1. Representative banding profiles associated with ERIC2 RAPD PCR in *P. aeruginosa* isolated from horses. Lane M: Molecular weight marker (100 bp); lane 1: genotype A; lanes 2 and 3, genotype B; lanes 4, 5 and 9, genotype C; lane 6, genotype D; lanes 7 and 13, genotype E; lane 8, genotype F; lane 10, genotype G; lane 11, genotype H; lane 12, genotype H; lane 14, genotype J; lane 15, genotype K; lane 16, genotype L; lane 17, genotype M; lane 18, genotype N.

DISCUSSION

In this study, we attempted to examine the genetic relatedness of a collection of 63 clinical isolates of *P. aeruginosa*, obtained from an equal number of horses throughout Ireland during the five year period, 2003-2007, employing an ERIC RAPD PCR molecular technique. Performing molecular epidemiological studies with such isolate collections is important to help understand acquisition of risk factors for *P. aeruginosa* and to help trace epidemic strains within such patient populations, as well as to help identify common virulent and/or transmissible strains. Previously, several molecular typing schemes have been applied to help elucidate the molecular epidemiology of populations of *P. aeruginosa* in human clinical medicine and these methodologies have been reviewed comprehensively by Speert [10], including pulsed-field gel electrophoresis (PFGE). In the current study, we employed a random amplification of polymorphic DNA (RAPD) method for the sub-species differentiation of a population of equine *P. aeruginosa*, with ERIC2 as the arbitrary primer used, as previously described by Clarke *et al.* [2], who applied this method to help differentiate *P. aeruginosa*, which originated from a human clinical population of patients with cystic fibrosis (CF). Our choice of method and primer was largely driven by the previous publication by Clarke *et al.* [2], as well as Renders *et al.* [8], who reported positively on the value of this method against pulsed-field gel electrophoresis (PFGE) in a CF study of clinical *P. aeruginosa*. In their study, these workers demonstrated that RAPD, using the ERIC2 primer, could expect full concordance with typing data obtained by PFGE, when single band differences are neglected. In addition, these workers showed a higher degree of genetic relatedness amongst the 19 CF patients examined, as demonstrated by highly similar banding profiles.

In this study, we have employed the Tenover criteria [11] to help interpret the banding/clustering patterns and hence the interstrain comparison. Although originally designed to help interprete banding patterns associated with PFGE analysis, many workers have adopted the clustering algorithms associated with the Tenover criteria, with other molecular typing methods, such as RAPD analysis. This may not be the most ideal scenario, but at present, RAPD techniques have not got robust clustering algorithms to help with strain comparison. With such an amount of development of methods to aid in the characterisation of *P. aeruginosa*, it is surprising that similar reports, applying these highly discriminatory molecular methods to collections of *P. aeruginosain* horses, are absent from the literature. The description of previous typing studies on equine *P. aeruginosa* is limited to the relatively old report by Atherton and Pitt [1] from 1982, who employed phenotypic serotyping and phage-typing. This study demonstrated that somatic type O3 accounted for 5% of isolates and that all serotypes had the potential to be pathogenic, given favourable culture conditions. Our current manuscript attempts to redress this lack of application of molecular techniques to collections of *P. aeruginosa* from horses, by applying a simple molecular PCR technique, as a suggested method to other equine microbiologists and others in the field, who wish to examine the genetic relatedness within their equine populations. Consequently, further epidemiological studies are needed from equine centres throughout different continents to help elucidate the molecular epidemiology of *P. aeruginosa* and its disease association with horses, so that common virulent stsrains may be identified and standardised protocols for the molecular typing of *P. aeruginosa* adopted internationally, resulting in a common technique, with added-value, thus aiding in the ability to interpret such typing data on a global basis.

Employment of this primer allowed for the generation of banding patterns containing multiple amplicons, which aided in the differentiation of the isolates. In addition, there was good reproducibility when examining isolates with ERIC2, in that similar banding patterns and hence clusters were obtained for the same isolates examined on different days. This consistency in reproducibility was due to the adoption of several control interventions throughout the study, including: (i) standardisation of genomic DNA template; (ii) employment of single batches of reagents and thermal cycler; (iii) optimisation (iv) standardisation of the RAPD protocol; and, (v) standardisation of image-capture protocol.

Although it has been reported by various groups that RAPD typing is reproducible, historically there have been several reports of problems associated with its reproducibility. Grif *et al.* [3] reported that RAPD suffered from the lack of standardisation using different methods in different laboratories and this may affect the quality of results. RAPD-PCR can also suffer from the same factors that affect ordinary PCR, including magnesium concentration, primer and reagent 'batch-to-batch' variation and quality of the thermal cycler. Power [7] suggested that for RAPD to be a definitive typing technique, the reproducible generation and interpretation of RAPD fingerprints needs to be developed. To achieve this criterion, the use of an automated system for DNA preparation combined with the use of manufactured RAPD master mixes, as well as use of the same thermal cycler and standard procedures for visualisation of fingerprints, may aid reproducibility. Furthermore, RAPD offers a nucleic acid (DNA) amplification-based molecular alternative to PFGE genotyping.

In conclusion, our study demonstrated that there was a high degree of clonal diversity between isolates within this collection, characterising 63 clinical isolates into 24 distinct genotypes, with the largest cluster (genotype E) accounting for 10/63 (15.9%) of the isolates. ERIC2 RAPD PCR proved to be a highly discriminatory molecular typing tool of P. aeruginosa in isolates recovered in horses. With the adoption of several controls to aid reproducibility, this technique may be useful as an alternative to PFGE, particularly in epidemiological investigations of outbreaks where speed may be a significant parameter.

ACKNOWLEDGEMENTS

This work was partially funded by the Research and Development Office, HPSS(NI) through ID-RRG Grant 9.3.

REFERENCES

1. Atherton JG, Pitt TL: Types of *Pseudomonas aeruginosa* isolated from horses. Eq Vet J. 1982, 14 (4): 329-332. 10.1111/j.2042-3306.1982.tb02446.x.
2. Clarke L, Moore JE, Millar BC, et al: Molecular epidemiology of *Pseudomonas aeruginosa* in adult patients with cystic fibrosis in Northern Ireland. Brit J Biomed Sci. 2008, 65 (1): 18-21.
3. Grif K, Karch H, Schneider C, et al: Comparative study of five different techniques for epidemiological typing of *Escherichia coli* O157. Diag Microbiol Infect Dis. 1998, 32 (3): 165-176. 10.1016/S0732-8893(98)00103-5.
4. Gürtler V, Mayall BC: Genomic approaches to typing, taxonomy and evolution of bacterial isolates. Int J Sys Evol Microbiol. 2001, 51 (1): 3-16.
5. Kurlenda J, Grinholc M, Jasek K, et al: RAPD typing of methicillin-resistant *Staphylococcus aureus*: a seven-year experience in a Polish hospital. Med Sci Monitor. 2007, 13 (6): MT13-18.
6. Millar BC, Xu J, Moore JE: Risk assessment models and contamination management: implications for broad-range ribosomal DNA PCR as a diagnostic tool in medical bacteriology. J Clin Microbiol. 2002, 40 (5): 1575-1580. 10.1128/JCM.40.5.1575-1580.2002.
7. Power EG: RAPD typing in microbiology - a technical review. J Hosp Infect. 1996, 34 (4): 247-265. 10.1016/S0195-6701(96)90106-1.
8. Renders N, Romling Y, Verbrugh H, et al: Comparative typing of *Pseudomonas aeruginosa* by random amplification of polymorphic DNA or pulsedfield gel electrophoresis of DNA macrorestriction fragments. J Clin Microbiol. 1996, 34 (12): 3190-3195.
9. Samper JC, Tibary A: Disease transmission in horses. Theriogenology. 2006, 66 (3): 551-559. 10.1016/j.theriogenology.2006.04.019.
10. Speert DP: Molecular epidemiology of *Pseudomonas aeruginosa*. Front Biosci. 2002, 7: e354-361. 10.2741/specrt.
11. Tenover FC, Arbeit RD, Goering RV, et al: Interpreting chromosomal DNA restriction patterns produced by pulsed-field gel electrophoresis: criteria for bacterial strain typing. J Clin Microbiol. 1995, 33 (9): 2233-2239.

CHAPTER 15

Molecular Epidemiology of Campylobacter Isolates from Poultry Production Units in Southern Ireland

Emer O'Mahony[1,2], James F. Buckley[2], Declan Bolton[3], Paul Whyte[1], Se´amus Fanning[1]

[1]UCD Centre for Food Safety, School of Public Health, Physiotherapy & Population Science, UCD Veterinary Sciences Centre, University College Dublin, Belfield, Dublin, Ireland,

[2]Veterinary Food Safety Laboratory, Cork County Council, Inniscarra, County Cork, Ireland,

[3]Food Safety Department, Teagasc Food Research Centre, Ashtown, Dublin, Ireland

ABSTRACT

This study aimed to identify the sources and routes of transmission of Campylobacter in intensively reared poultry farms in the Republic of Ireland.

Citation: O'Mahony E, Buckley JF, Bolton D, Whyte P, Fanning S (2011) Molecular Epidemiology of Campylobacter Isolates from Poultry Production Units in Southern Ireland. PLoS ONE 6(12): e28490. https://doi.org/10.1371/journal.pone.0028490

Breeder flocks and their corresponding broilers housed in three growing facilities were screened for the presence of Campylobacter species from November 2006 through September 2007. All breeder flocks tested positive for Campylobacter species (with C. jejuni and C. coli being identified). Similarly, all broiler flocks also tested positive for Campylobacter by the end of the rearing period. Faecal and environmental samples were analyzed at regular intervals throughout the rearing period of each broiler flock. Campylobacter was not detected in the disinfected house, or in one-day old broiler chicks. Campylobacter jejuni was isolated from environmental samples including air, water puddles, adjacent broiler flocks and soil. A representative subset of isolates from each farm was selected for further characterization using flaA-SVR sub-typing and multi-locus sequence typing (MLST) to determine if same-species isolates from different sources were indistinguishable or not. Results obtained suggest that no evidence of vertical transmission existed and that adequate cleaning/disinfection of broiler houses contributed to the prevention of carryover and cross-contamination. Nonetheless, the environment appears to be a potential source of Campylobacter. The population structure of Campylobacter isolates from broiler farms in Southern Ireland was diverse and weakly clonal.

INTRODUCTION

Campylobacter continues to be the most commonly reported cause of bacterial gastroenteritis in the European Union (EU). In total there were 198,252 confirmed cases of campylobacteriosis in 2009, giving an overall crude incidence rate (CIR) of 45.6 per 100,000 population [1].

While a range of risk factors for infection with Campylobacter have been identified, the most common is the handling and/or consumption of undercooked poultry, in particular chicken. According to a recent opinion of the European Food Safety Authority›s (EFSA) Biological Hazards (BIOHAZ) Panel, 50 to 80% of human cases of campylobacteriosis may be attributed to the chicken reservoir [2]. This report also outlined the considerable underreporting of cases of campylobacteriosis and suggested that no less than 2 million and up to 20 million cases of clinical campylobacteriosis occur per year in the EU.

In 2008, a baseline survey on Campylobacter in broiler batches and carcasses in the EU was undertaken [3]. The prevalence at community level of Campylobacter-colonized broiler batches was reported to be 71.2% and the prevalence of Campylobacter-contaminated broiler carcasses was found to be 75.8%. The colonization of broiler flocks with Campylobacter is therefore a significant food safety issue and a reduction in the number of contaminated poultry products entering the food chain would reduce the negative impact on public health.

It is thought that the reduction of Campylobacter-contaminated poultry meat can be achieved most effectively by implementing on-farm control measures [4]. Extensive research into the most important source of Campylobacter in poultry production units has been carried out. Vertical transmission has previously been implicated [5], [6], [7], [8]. Carryover to subsequent flocks as a result of inadequate disinfection of broiler houses has also been identified as a risk factor [9]. Horizontal transmission from the surrounding environment to the broiler house either via the farm workers or other vectors such as wild birds, vermin and house flies, is considered to be the most likely source of contamination [10], [11]. Risk factors for infection may vary from country to country because of differing farming practices and associated climatic conditions. Therefore, efforts to understand the relative importance of each potential source and transmission route of Campylobacter infection on-farm, continues to have an important role in extending our understanding of the epidemiology of this important pathogen.

In 2009, there were 1,808 confirmed cases of campylobacteriosis in Ireland [12]. The prevalence of Campylobacter-colonized broiler batches in Ireland in 2008 was 83.1% and of that Campylobacter-contaminated broiler carcasses accounted for 98.3% [3]. In order to implement effective control measures and reduce the prevalence of Campylobacter in Irish poultry products, the most significant transmission routes must be identified. The aim of this study was to investigate the occurrence of Campylobacter in a subset of intensively reared Irish poultry flocks and to identify sources of Campylobacter in each farm environment. Molecular sub-typing methods were used to identify the genotypes present in Irish broiler farms and to shed further light on possible transmission routes of Campylobacter in poultry farms.

MATERIALS AND METHODS

Description of Farms in this Study

Three housed broiler flocks (denoted as broiler flocks 1, 2 and 3), located on three different farms (denoted as broiler farms 1, 2 and 3) in different geographical locations of Ireland, were studied throughout their 6 7 week life span. The flocks were screened between November 2006 and September 2007. Farm 1 consisted of three poultry house units sampled from November 2006 to January 2007. Farm 2 consisted of two poultry house units that were sampled between April 2007 and May 2007. Farm 3 consisted of one poultry house unit and was sampled between July 2007 and September 2007. Broiler flock sizes ranged from approximately 18,000 to 34,000 birds per house. Three breeder flocks, located on breeder farm 1 (denoted as breeder farm 1, flocks 1–3), supplied chicks for broiler farm 1. Broiler farm 2 was supplied by breeder farms 2 and 3. Broiler farm 3 was supplied by two flocks from breeder farm 1 (denoted as breeder farm 1, flocks 4 and 5), and one flock from both breeder farms 4 and 5. Each breeder flock comprised approximately 5,000 birds per house.

Sample Collection

Samples were collected approximately every 14 days from (i) the cleaned and disinfected broiler houses prior to chick placement, (ii) the chickens and (iii) the environments inside and outside the broiler houses. The breeder flocks supplying each broiler house were also tested. Samples were transported to the laboratory at 4°C in a cool box and processed on the same day.

Breeder flocks

Sixty fresh faecal samples (5 pooled faecal samples each containing 12 fresh faecal droppings) were collected from the floor of the house of the corresponding breeder farms supplying broiler hatching eggs for each flock in this study.

Broiler flocks

One-day old broiler chicks were tested by enriching the faeces-soiled paper that lined the crates used to transport the birds from the hatchery to the broiler house (100 birds per crate). Faecal samples were then taken at regular intervals throughout the rearing period of each flock (at days 14, 28 and 42 approximately). Sixty individual fresh faecal samples were collected from the broiler house floor and combined (as outlined previously).

Environmental samples

Samples from walls, floors, structural support columns, feed and water dispensers along with concrete aprons (the concreted area outside the front of the house) were taken using sterile swabs pre-moistened with 10 ml Maximum Recovery Diluent (MRD, Lab M Ltd., Bury, UK). An area of 0.1 m^2 of the object›s surface was chosen for sampling, and swabbing continued outside this area until either the entire surface was sampled or the swab was dry. Two litres of water supplying the broiler house drinkers were also taken. Aspiration of air samples onto one *Campylobacter* blood-free selective agar base (CCDA, CM0739, Oxoid, Cambridge, UK), supplemented with cefoperazone and amphotericin B (CCDA selective supplement) (Oxoid), and one Tryptone Soya Agar (TSA, Oxoid Ltd., Cambridge, UK) plate was performed using a Sampl'air MK2 double agar plate sampler (AES Laboratoire Groupe, Combourg, France). Two air samples (500 l×2) were taken from inside and one directly outside each broiler house on each sampling occasion (at days -1, 14, 28 and 42). Occasionally, samples were taken opportunistically inside and outside the broiler house environment (Table 1).

Table 1. ***Campylobacter*** **sources during the sampling period across farms.**

Flock	Sample	Number of positive samples/number of samples tested				
		Day -1*	Day 1	Day 14*	Day 28*	Day 42*
1	Broilers	-	0/48[c]	**3/5**[c]	**5/5**[c]	**5/5**[c]
	Adjacent Broiler 1	-	-	0/2[c]	**2/2**[c]	**2/2**[c]
	Adjacent Broiler 2	-	-	**2/2**[c]	**2/2**[c]	-
	Air	0/6	-	**1/6**	0/6	0/6
	Puddle	0/1	-	**1/2**	0/1	0/2
	Soil	0/2	-	0/2	0/2	**1/2**
	Other	0/20[ap, cl co, d, f, fe, fl, h, w, wa]	-	0/7[ap, h, l, wa]	0/9[ap, h, i, wa]	0/11[ap, h, l, wa]
2	Broilers	-	0/48[c]	0/5[c]	0/5[c]	**5/5**[c]
	Adjacent Broiler	-	-	0/2[c]	0/2[c]	-
	Soil	0/2	-	0/2	**1/2**	0/2
	Other	0/22[a, ap, co, d, fe, fl, p, w, wa]	-	0/10[a, ap, wa]	0/13[a, ap, l, wa]	0/12[a, ap, p, r, wa]
3	Broilers	-	0/48[c]	0/5[c]	**5/5**[c]	**5/5**[c]
	Soil	0/2	-	0/2	0/2	**1/2**
	Other	0/24[a, ap, cl co, d, fe, fl, p, w, wa]	-	0/12[a, ap, p, wa]	0/12[a, ap, p, wa]	0/10[a, ap, wa]

https://doi.org/10.1371/journal.pone.0028490.t001

Isolation of *Campylobacter*

To determine the presence/absence of *Campylobacter*, samples were examined by direct plating and/or enrichment culture methodologies based on the Horizontal Method for Detection and Enumeration of *Campylobacter* spp. (ISO 10272-1:2006) and the Detection and Semi-quantitative Enumeration of Thermotolerant *Campylobacter* spp. (ISO 17995:2005). Isolation of emerging *Campylobacter* species, based on the method previously described [13], was also performed for faecal samples. Following incubation, five suspect colonies were randomly selected from plates and subcultured to obtain pure colonies.

For the lined transport crates, 60 papers from each flock were collected at random and divided into six piles consisting of ten papers on top of one another. Each pile of ten papers was then aseptically cut into 8 strips. Each strip was enriched in 200 ml Campylobacter Enrichment Broth (CEB, Lab M Ltd., Bury, UK) supplemented with 5% (v/v) lysed horse blood (TCS Biosciences, Buckingham, UK) and cefoperazone, vancomycin, trimethoprim and cyclohexamide (CVTC supplement, Lab M Ltd., Bury, UK). Forty-eight composite samples, each consisting of ten strips, from each flock were tested in this way. Swabs were enriched in 100 ml of CEB supplemented with 5% (v/v) lysed horse blood and CVTC. Water samples were filtered using 0.45 μm filters (Millipore, Billerica, MA., USA), which were then enriched in sterile 30 ml containers containing 20 ml of CEB supplemented with 5% (v/v) lysed horse blood and CVTC. For the air samples, the agar from each TSA plate was aseptically removed and enriched in 100 ml CEB supplemented with 5% (v/v) lysed horse blood and CVTC, while each CCDA plate was incubated directly. Faecal samples from horses, crushed flies and beetles were enriched with CEB supplemented with 5% (v/v) lysed horse blood and CVTC using a 1:10 ratio of sample to broth. Water from puddles was enriched with an equivalent volume of double-strength CEB supplemented with 5% (v/v) lysed horse blood and CVTC.

Biochemical Confirmation

Presumptive Campylobacter isolates were confirmed using standard biochemical procedures including a Gram lysis test (3% [w/v] KOH, Sparks Lab Supplies, Dublin, Ireland), oxidase test (Oxoid, Cambridge, UK), catalase test (3% H_2O_2, Sigma Aldrich, St. Louis, MO, USA) and l-alanine aminopeptidase test (Aminopeptidase Test, Fluka Biochemika, Buchs, Switzerland). A latex agglutination test was also used (DrySpot Campylobacter Test Kit, Oxoid, Cambridge, UK). Following biochemical confirmation, isolates were stored at −20 and −80°C on Protect cryobeads (Technical Service Consultants Ltd., Heywood, Lancashire, UK) containing 80% [v/v] glycerol, for further analysis.

Genotyping

Multiplex Polymerase Chain Reaction (mPCR)

DNA purification was carried out using a commercial kit (Wizard Genomic DNA Purification Kit, Promega, Madison, WI, USA). Isolates were identified to species level as described previously [14]. Amplicon sizes were determined by comparison with a molecular weight marker (50 bp ladder, Promega) following migration on a 1.5% [w/v] agarose gel.

flaA-SVR sequencing

Sequencing of an internal 321-bp fragment of the flagellin A short variable region (SVR), was performed using the primer pair Fla242FU [5′-CTA TGG ATG AGC AAT T (AT) A AAAT-3′] and Fla625RU [5′-CAA G (AT) C CTG TTC C (AT)A CTG AAG-3′], as previously described by Meinersmann et al. [15]. All flaA-derived amplicons were purified using a QIAquick PCR Purification kit (Qiagen GmbH, Hilden, Germany) and sequenced commercially by MWG Biotech (Ebersberg, Germany). The nucleotide sequences were deposited in the internet accessible flaA-SVR sequence database (http://hercules.medawar.ox.ac.uk/flaA/) and SVR allele numbers were assigned by sequence comparisons against the existing flaA-SVR sequences. The DNA trace files were submitted for confirmation of novel flaA-SVR alleles.

Multi Locus Sequence Typing (MLST)

Internal fragments of seven gene targets were amplified by PCR and their nucleotide sequences determined with primers and reaction conditions in accordance with the published MLST scheme for *C. jejuni* and *C. coli*. Allele identification followed by sequence type (ST) and clonal complex (CC) assignments were done by interrogation of the *Campylobacter* MLST database for each isolate (http://pubmlst.org/campylobacter).

Data Analysis

Trimmed *flaA*-SVR sequences (including STs) were imported into BioNumerics, Version 6.1 (Applied Maths, Sint-Martens-Latem, Belgium). A dendrogram was created using the unweighted-pair group method with arithmetic mean (UPGMA).

RESULTS

Presence of Campylobacter on poultry farms

Broiler breeders

All broiler breeder flocks supplying each rearing farm were found to be colonized with both *C. jejuni* and *C. coli* (Table 2). A total of 27 *C. jejuni* and 19 *C. coli* were isolated from breeder farm 1 (containing flocks 1, 2 and 3). Two *C. jejuni* and 42 *C. coli* were isolated from breeder farm 2. One *C. coli* isolate and 22 *C. jejuni* isolates were recovered from breeder farm 3. There were 6 *C. jejuni* and 14 *C. coli* isolates recovered from breeder farm 1, flock 4, along with 7 *C. jejuni* and 12 *C. coli* from breeder farm 1, flock 5. A total of 13 *C. jejuni* and 7 *C. coli* were isolated from breeder farm 4, while 26 *C. jejuni* and 25 *C. coli* were recovered from breeder farm 5.

Table 2. ***Campylobacter*** **species recovered from all farms tested.**

Farm	Day	Source	Sample	Species (no.)
1	-	Breeder 1 Flock 1	Faeces	*C. jejuni* (6), *C. coli* (2)
	-	Breeder 1 Flock 2	Faeces	*C. jejuni* (6), *C. coli* (9)
	-	Breeder 1 Flock 3	Faeces	*C. jejuni* (15), *C. coli* (8)
	13	Broiler	Faeces	*C. jejuni* (10)
	13	Adjacent Broiler 2	Faeces	*C. jejuni* (3)
	13	Environment Inside	Air	*C. jejuni* (2)
	13	Environment Outside	Puddle	*C. jejuni* (1)
	32	Broiler	Faeces	*C. jejuni* (9)
	32	Adjacent Broiler 1	Faeces	*C. jejuni* (1)
	32	Adjacent Broiler 2	Faeces	*C. jejuni* (8)
	42	Broiler	Faeces	*C. jejuni* (26)
	42	Adjacent Broiler 1	Faeces	*C. jejuni* (5)
	42	Environment Outside	Soil	*C. jejuni* (1)
Total				*C. jejuni* (93) *C. coli* (19)
2	-	Breeder 2	Faeces	*C. jejuni* (2), *C. coli* (42)
	-	Breeder 3	Faeces	*C. jejuni* (22), *C. coli* (1)
	27	Environment Outside	Soil	*C. jejuni* (5)
	41	Broiler	Faeces	*C. jejuni* (24)
Total				*C. jejuni* (53) *C. coli* (43)
3	-	Breeder 4	Faeces	*C. jejuni* (13), *C. coli* (7)
	-	Breeder 5	Faeces	*C. jejuni* (26), *C. coli* (25)
	-	Breeder 1 Flock 4	Faeces	*C. jejuni* (6), *C. coli* (14)
	-	Breeder 1 Flock 5	Faeces	*C. jejuni* (7), *C. coli* (12)
	27	Broiler	Faeces	*C. jejuni* (23)
	41	Broiler	Faeces	*C. jejuni* (26)
	41	Environment Outside	Soil	*C. jejuni* (1)
Total				*C. jejuni* (102) *C. coli* (58)

https://doi.org/10.1371/journal.pone.0028490.t002

Broiler chickens

Campylobacter could not be cultured from the transport crate paper liners on the day the chicks arrived at the rearing house. All three broiler flocks were found to be contaminated with *Campylobacter* by the end of the rearing period. *Campylobacter jejuni* was the only species isolated from all three poultry flocks under investigation (Table 2). Faecal samples from broiler flock 1 were contaminated with *Campylobacter* on days 13, 32 and 42. Faecal samples from broiler flock 2 were negative until the final sampling day (day 41). Broiler flock 3 was found to be contaminated with *Campylobacter* on days 27 and 41.

Presence of Campylobacter in the poultry farm environment

On all three farms, *Campylobacter* could not be detected in the empty poultry house structure/environment after the cleaning and disinfection procedure had been carried out. All samples taken from the external environment of the cleaned house, including soil, air, the concrete apron and horse faeces were negative for *Campylobacter* (Table 1).

During the rearing period of each flock, the environment was found to be contaminated with *Campylobacter* (Table 1). In the case of broiler farm 1 (day 13), an air sample taken inside the rearing house, an adjacent broiler flock and a puddle outside the house were sampled and found to be contaminated with *Campylobacter*. On day 32, two adjacent broiler flocks tested positive for *C. jejuni*. On day 42, one adjacent broiler house had been depopulated. *Campylobacter jejuni* was isolated from the remaining adjacent flock and was also recovered from an environmental soil sample. In the case of broiler farm 2, *C. jejuni* was isolated from a soil sample taken on day 27. On broiler farm 3, *C. jejuni* was recovered from a soil sample taken on day 41.

Molecular sub-Typing And Spatio-Temporal Tracking of Campylobacter isolates

A representative subset of *Campylobacter* isolates was chosen from each farm and characterized by DNA sequence analysis of the SVR containing region of the *flaA* gene (*flaA*-SVR), to determine if same-species isolates from different sources could be distinguished. Figure 1 shows the distribution of allele types discovered in relation to the source of each isolate. A total of 22 *flaA*-SVR alleles were identified in the 101 isolates investigated.

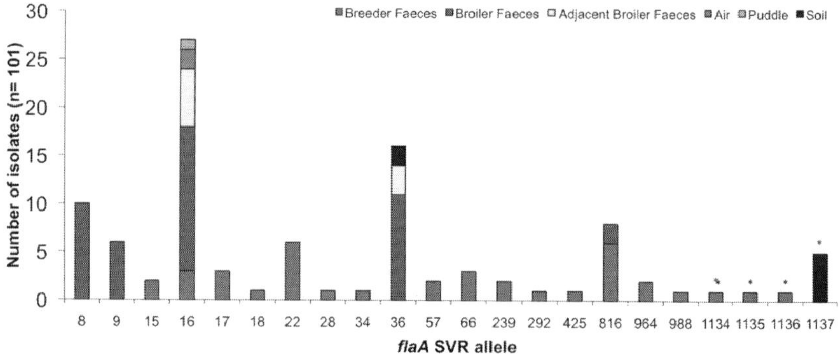

Figure 1. Distribution of *flaA*-SVR alleles detected according to source, (asterisks signify new *flaA*-SVR alleles).

https://doi.org/10.1371/journal.pone.0028490.g001

Figures 2, 3 and 4 show the spatio-temporal distribution of *flaA*-SVR alleles identified on each sampling day across each of the three broiler farms. This sub-typing analysis revealed a diverse and weakly clonal population structure of *C. jejuni*, with multiple subtypes present throughout the lifecycle of each flock. *Campylobacter* isolates originating from each set of breeders and the faeces from their respective progeny presented with non-identical *flaA*-SVR DNA sequences.

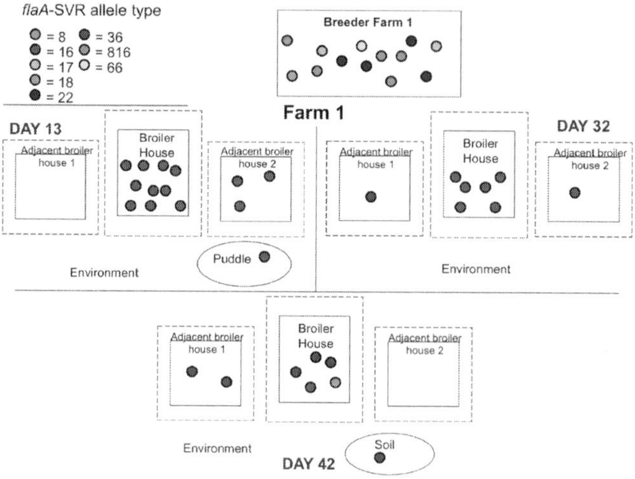

Figure 2. Schematic diagram of the *flaA*-SVR alleles detected on farm 1.

https://doi.org/10.1371/journal.pone.0028490.g002

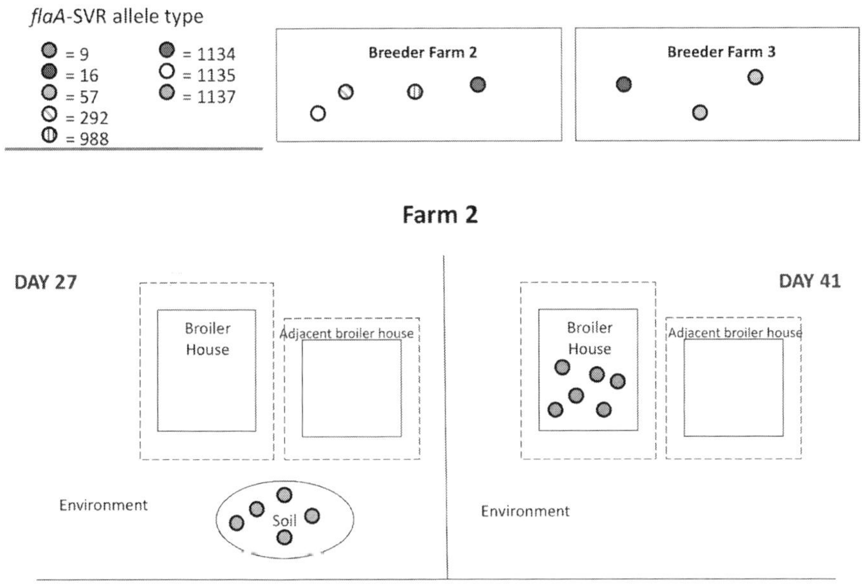

Figure 3. Schematic diagram of the *flaA*-SVR alleles detected on farm 2.

https://doi.org/10.1371/journal.pone.0028490.g003

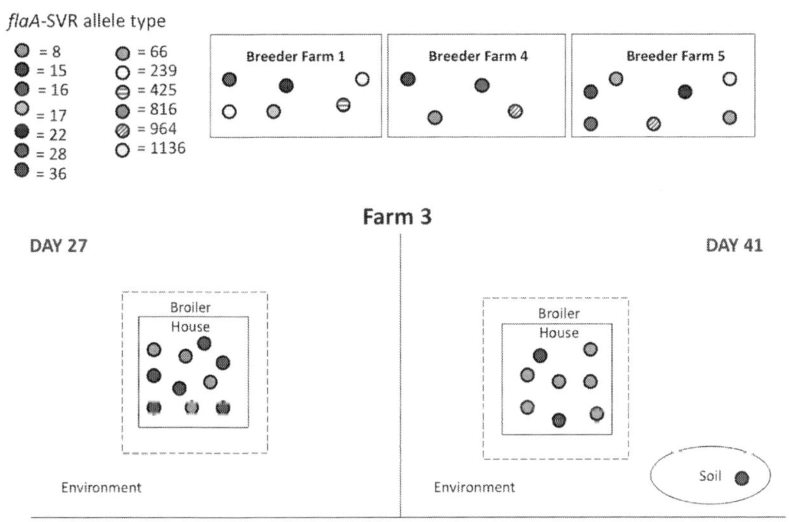

Figure 4. Schematic diagram of the *flaA*-SVR alleles detected on farm 3.

https://doi.org/10.1371/journal.pone.0028490.g004

In the case of farm 1, *flaA*-SVR allele number 16 was detected on day 13 in faeces from the broiler flock, faeces from adjacent broiler flock 2, and in a puddle located outside the house (Figure 2). On day 32, allele number 16 continued to be identified in the broiler and adjacent broiler flock 2 faeces. On the same day, allele number 36 was detected in faeces from adjacent broiler flock 1. On day 42, allele number 36 could be detected in faeces from adjacent broiler flock 1, and was also now identified in broiler flock 1 and in an environmental soil sample. Allele numbers 8 and 16 were also detected in the broiler faeces.

In the case of farm 2, allele number 1137 was identified in an environmental soil sample taken on day 27 (Figure 3). *Campylobacter* was not isolated from broiler flock 2 faeces until the final sampling day, when allele number 9 was detected.

In the case of farm 3, allele numbers 8, 36 and 816 were detected in the broiler faeces on day 27 (Figure 4). On day 41, alleles 8 and 36 were again identified. Allele 26 was also found to be present in a soil sample outside the house on this day.

Diversity of MLST Sequence Types (STs) and Clonal Complexes (CCs)

In order to substantiate *flaA*-SVR findings, a subset of isolates from each farm was chosen for MLST analysis.Eleven STs were identified among the 35 isolates (34 *C. jejuni*, 1 *C. coli*) chosen. Five STs were assigned to more than one isolate while six STs were assigned to single isolates.

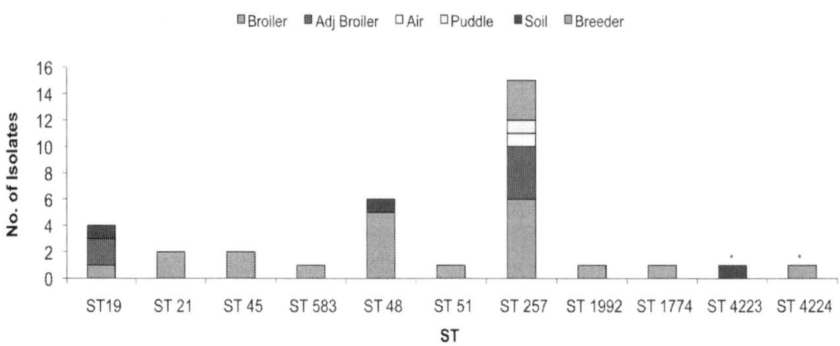

Figure 5. Distribution of STs detected according to source (asterisks signify new STs).

https://doi.org/10.1371/journal.pone.0028490.g005

Figure 5 shows that ST 257 was the most frequently detected subtype (in 15/35 isolates), followed by ST 48 (in 6/35 isolates) and ST19 (in 4/35 isolates). ST 21 and ST 45 were assigned to two isolates each. ST 583, ST 51, ST 1922, ST 1744, ST 4223 and ST 4224 were assigned to single isolates. Two novel sequence types (ST 4223 and ST 4224) were identified and submitted to the *C. jejuni* MLST database (http://pubmlst.org/campylobacter/).

All 11 STs were grouped into 6 CCs. The two novel sequences could not be grouped into a defined CC. The largest CC was found to be CC 257 (consisting of 16 isolates), followed by CC 21 and CC 48 (6 isolates each). In total, 80% (28/35) of isolates analyzed by MLST were grouped into one of these three CCs. Isolates grouped in CC 257 were found to have originated from a variety of sources (Figure 6).

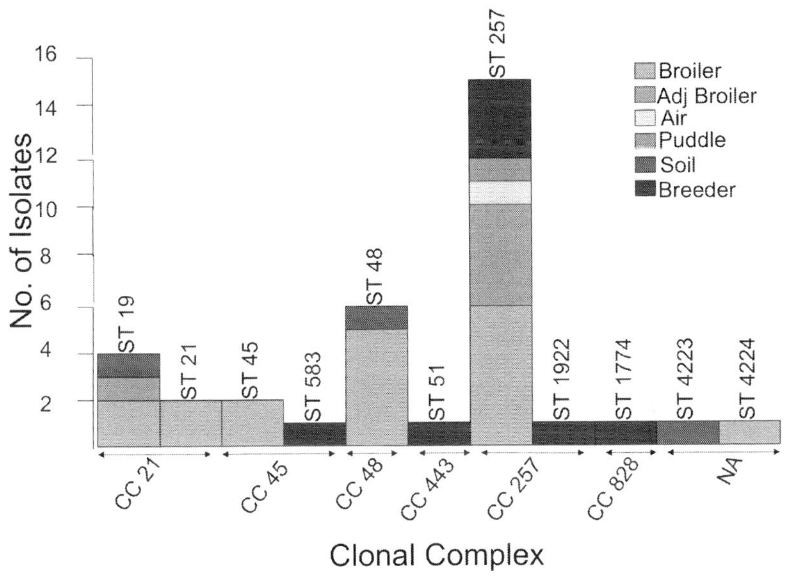

Figure 6. Distribution of CCs and STs according to source.

https://doi.org/10.1371/journal.pone.0028490.g006

Figure 7 depicts the phylogenetic relationship between the 22 *flaA*-SVR types and the 11 STs identified during the study. Nine cluster genotypes were observed using a threshold genetic similarity of 98% as a cut-off coefficient value.

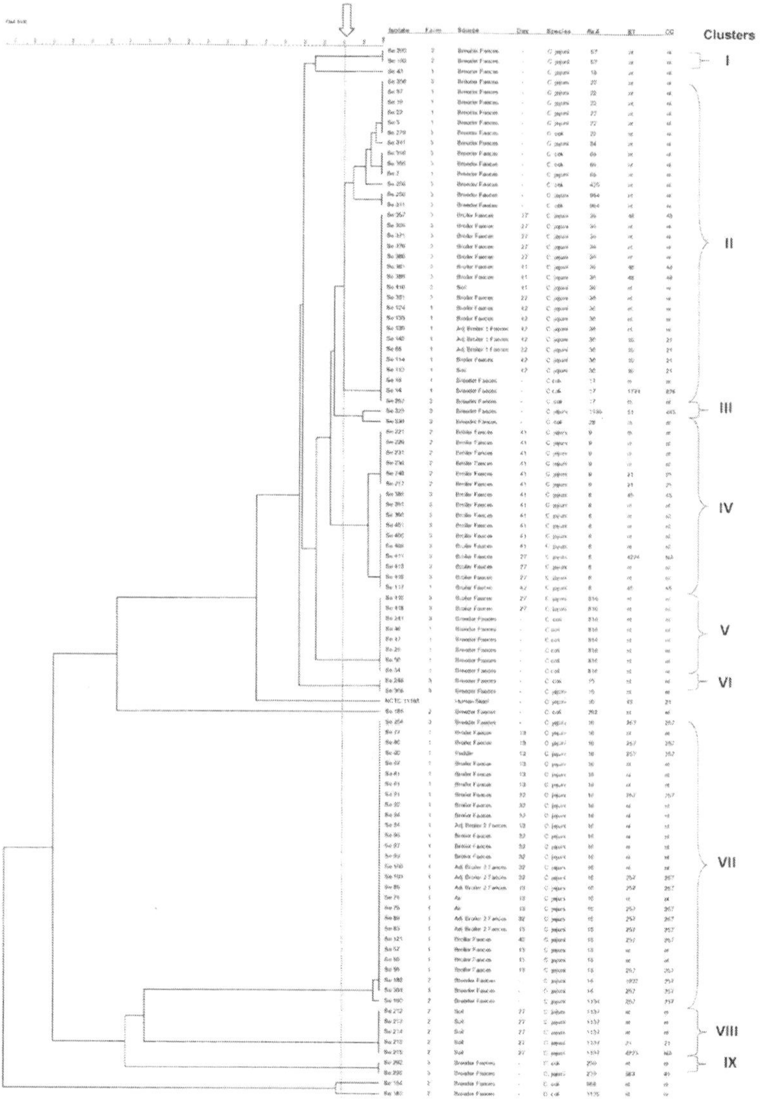

Figure 7. Comparison of *flaA*-SVR sequences of *Campylobacter* **isolates from breeder and broiler farms and farm environments.** A total of 101 isolates were included. Strain *Campylobacter jejuni* NCTC11168 was included as a control strain in the pairwise analysis. The arrow indicates the 98% similarity cut-off point. *flaA*, *flaA*-SVR allele number; ST, sequence type; CC, clonal complex; Nt, not tested; NA, not assigned; -, not applicable.

https://doi.org/10.1371/journal.pone.0028490.g007

DISCUSSION

Despite extensive research, the definitive sources of infection and routes of transmission of *C. jejuni* in the poultry reservoir remain to be fully elucidated.

The purpose of this study was to identify sources of *Campylobacter* in intensively reared broiler flocks in Ireland.

Molecular sub-typing techniques have previously shown the population structure of *C. jejuni* to be highly diverse and weakly clonal [16], [17]. This study confirmed that genetic diversity also exists among Irish *C. jejuni* strains isolated from broiler flocks, adjacent flocks and the farm environment.

The occurrence of vertical transmission of *Campylobacter* in poultry has been a controversial issue. The isolation of *Campylobacter* species from the reproductive tract of broiler breeders has been reported [6], [7] and transmission from breeder hens to broiler chickens has been suggested [18]. Conflicting reports have also been published suggesting that vertical transmission is unlikely or of little importance [19], [20]. In this study, while breeder flocks were found to be colonized with *Campylobacter*, molecular characterization confirmed the isolates to be of different *flaA*-SVR and ST types, suggesting that vertical transmission was not involved in the colonization of these broiler flocks with *Campylobacter*. This work is in agreement with a related study conducted by Patriarchi *et al.*, where none of the genotypes identified in breeder flocks were subsequently identified on any of the broiler farms [21]. Colonization of chickens with *Campylobacter* usually occurs between 3 and 5 weeks of age, and once infected, prevalence in a flock can often be close to 100% [22]. Interestingly, *Campylobacter* was not detected in any of the three broiler flocks in this study before day 13 of the rearing cycle. A possible factor contributing to the delay in colonization of chickens with *Campylobacter* is the presence of protective maternal antibodies in young chicks [23], [24].

In this study, *Campylobacter* was not isolated from broiler flock 2 until after the process of partial depopulation or thinning had been carried out. The practice of thinning has previously been reported as an important risk factor for *Campylobacter* colonization of residual birds [25], [26], [27]. It has been shown *in vitro* that the presence of the neurotransmitter noradrenaline stimulates the growth and motility of *C. jejuni* [28]. As a result of triggering the release of noradrenaline, the stressful thinning process could be expected to contribute to rapid growth of the bacterium in the avian gastrointestinal tract leading to increased shedding of *Campylobacter* by birds, and the subsequent rapid spread of the bacteria.

Challenges in maintaining biosecurity during the thinning process can result in cross-contamination from environmental sources. Allen *et al.* reported the isolation of *Campylobacter* from transport vehicles, equipment, personnel and the farm driveways prior to the thinning process [25]. Using *flaA*-SVR and MLST subtyping methods, Patriarchi *et al.* also identified the practice of partial depopulation as a potential source and route of flock contamination on Irish broiler farms. Molecular evidence of the role of transport crates in introducing *Campylobacter* spp. into the broiler house was also reported [21]. In the case of farm 1 in this study, a *C. jejuni* isolate identified as *flaA*-SVR allele type 36 was isolated from an adjacent broiler flock prior to thinning. On the final sampling day, this *flaA*-SVR allele type was again isolated in the adjacent broiler flock, and was also identified in broiler flock 1 and in an environmental soil sample. Improved biosecurity measures in relation to the depopulation process may contribute to the prevention or delayed colonization of chickens with *Campylobacter*.

Previous studies have reported that poultry strains are frequently found to be genetically distinct from environmental isolates [29], [30]. In the case of broiler farm 2, *C. jejuni flaA*-SVR allele types isolated from broilers and environmental soil samples were confirmed as non-identical. However, identical *C. jejuni* strains were identified from air, soil, water puddles and chickens on broiler farms 1 and 3, suggesting that transfer of campylobacters between these environments may be occurring. However, it is not possible based on the epidemiological data presented here to establish the direction of a given exchange. Bi-directional movement of *Campylobacter* between sources cannot be ruled out and has been implicated previously [31]. In the case of two broiler flocks studied here, contamination of the farm environment was not detected until after the chickens had become infected, highlighting the broilers as a possible source of environmental contamination. MLST data from flock 1 further demonstrate that genetically identical strains can be isolated from broiler faeces and environmental samples. On day 13, ST 257 was found to be present in 2 broiler houses and in a water puddle outside the house. This ST was identified in the broiler flock on all sampling days. On day 42, ST 19 and ST 45 were also detected. These results reflect the findings of previous spatio-temporal studies on broiler farms, where different STs were identified in chicken faeces as the rearing period progressed [11].

During 2007, New Zealand experienced a 50% decline in the rate of campylobacteriosis notifications and hospitalisations [32]. This decline followed the introduction of voluntary and regulatory interventions to reduce contamination of poultry with *Campylobacter* species and was sustained in 2008 and 2009. A number of other countries have reported a reduced incidence of campylobacteriosis infections following the implementation of poultry-focussed control strategies [33], [34], [35], [36]. Various interventions were employed in each country however all strategies included strengthening on-farm biosecurity and monitoring the prevalence of *Campylobacter*-positive flocks. The implementation of similar measures in poultry farms in Ireland could contribute to a reduction in human campylobacteriosis infection rates and lead to improved public health protection.

The population structure of Campylobacter isolates from broiler farms in Southern Ireland was determined, (based on these data), to be weakly clonal. Such genetic diversity complicates the challenge of managing Campylobacter species population dynamics within the poultry farm environment. Nevertheless, our data point to the importance of applying more than one sub-typing method as part of our epidemiological studies to carefully describe this dynamic process.

It is reasonable to conclude that there are multiple sources from which *Campylobacter* can be transmitted on broiler farms. Following their introduction into broiler flocks, the spread of *Campylobacter* is influenced by various host and environmental factors, such as biosecurity measures in place, farming practices, the immune status of the chickens, and the presence of other animals on the farm. A combined protocol individually targeting each potential source and route of transmission is required as a logical approach to effectively reduce the colonization of broiler chickens with *Campylobacter* in a step-wise fashion.

ACKNOWLEDGMENTS

The authors wish to acknowledge the cooperation of the poultry farmers who participated in the study.

REFERENCES

1. **1.**EFSA (2011) The European Union Summary Report on Trends and Sources of Zoonoses, Zoonotic Agents and Food-borne Outbreaks in 2009. EFSA Journal 2011 9(3): 2090. [378pp.] doi:10.2903/j.efsa.2011.2090. Available online: www.efsa.europa.eu/efsajournal [last accessed 02/07/2011].
2. EFSA (2010) EFSA Panel on Biological Hazards (BIOHAZ); Scientific Opinion on Quantification of the risk posed by broiler meat to human campylobacteriosis in the EU. EFSA Journal 2010 8(1): 1437. [89 pp.]. doi:10. 2903/j. efsa. 2010. 1437. Available online: www. efsa. europa. eu [last accessed 02/07/2011].
3. EFSA (2010) Analysis of the baseline survey on the prevalence of *Campylobacter* in broiler batches and of *Campylobacter* and *Salmonella* on broiler carcasses in the EU, 2008 Part A: *Campylobacter* and *Salmonella* prevalence estimates. EFSA Journal 2010 8(03): 1503. [99 pp.]. doi:10. 2903/j. efsa. 2010. 1503. Available online: www. efsa. europa. eu [last accessed 02/07/2011].
4. ACMSF (2005) The Advisory Committee on the Microbiological Safety of Food. Second report on *Campylobacter*.
5. Buhr RJ, Musgrove MT, Richardson LJ, Cox NA, Wilson JL, et al. (2005) Recovery of *Campylobacter jejuni* in feces and semen of caged broiler breeder roosters following three routes of inoculation. Avian Dis 49: 577–581.
6. Buhr RJ, Cox NA, Stern NJ, Musgrove MT, Wilson JL, et al. (2002) Recovery of *Campylobacter* from segments of the reproductive tract of broiler breeder hens. Avian Dis 46: 919–924.
7. Cox NA, Stern NJ, Wilson JL, Musgrove MT, Buhr RJ, et al. (2002) Isolation of *Campylobacter* spp. from semen samples of commercial broiler breeder roosters. Avian Dis 46: 717–720.
8. Pearson AD, Greenwood MH, Feltham RK, Healing TD, Donaldson J, et al. (1996) Microbial ecology of *Campylobacter jejuni* in a United Kingdom chicken supply chain: intermittent common source, vertical transmission, and amplification by flock propagation. Appl Environ Microbiol 62: 4614–4620.
9. Petersen L, Wedderkopp A (2001) Evidence that certain clones of *Campylobacter* jejuni persist during successive broiler flock rotations. Appl Environ Microbiol 67: 2739–2745.

10. Newell DG, Fearnley C (2003) Sources of *Campylobacter* colonization in broiler chickens. Appl Environ Microbiol 69: 4343–4351.
11. Bull SA, Allen VM, Domingue G, Jorgensen F, Frost JA, et al. (2006) Sources of *Campylobacter* spp. colonizing housed broiler flocks during rearing. Appl Environ Microbiol 72: 645–652.
12. HPSC (2010) Health Protection Surveillance Centre. Infectious Disease Notifications in Ireland, 2004–2009. http://www.hpsc.ie/hpsc/NotifiableDiseases/AnnualIDStatistics/ [last accessed 02/07/2010].
13. Lynch OA, Cagney C, McDowell DA, Duffy G (2010) A method for the growth and recovery of 17 species of *Campylobacter* and its subsequent application to inoculated beef. J Microbiol Methods.
14. Wang G, Clark CG, Taylor TM, Pucknell C, Barton C, et al. (2002) Colony multiplex PCR assay for identification and differentiation of *Campylobacter jejuni*, *C. coli*, *C. lari*, *C. upsaliensis*, and *C. fetus* subsp. *fetus*. J Clin Microbiol 40: 4744–4747.
15. Meinersmann RJ, Helsel LO, Fields PI, Hiett KL (1997) Discrimination of *Campylobacter jejuni* isolates by *fla* gene sequencing. J Clin Microbiol 35: 2810–2814.
16. Dingle KE, Colles FM, Wareing DRA, Ure R, Fox AJ, et al. (2001) Multilocus sequence typing system for *Campylobacter jejuni*. J Clin Microbiol 39: 14–23.
17. Meinersmann RJ, Phillips RW, Hiett KL, Fedorka-Cray P (2005) Differentiation of *Campylobacter* populations as demonstrated by flagellin short variable region sequences. Appl Environ Microbiol 71: 6368–6374.
18. Cox NA, Stern NJ, Hiett KL, Berrang ME (2002) Identification of a new source of *Campylobacter* contamination in poultry: transmission from breeder hens to broiler chickens. Avian Dis 46: 535–541.
19. Jacobs-Reitsma WF (1995) *Campylobacter* bacteria in breeder flocks. Avian Dis 39: 355–359.
20. Sahin O, Kobalka P, Zhang Q (2003) Detection and survival of *Campylobacter* in chicken eggs. J Appl Microbiol 95: 1070–1079.
21. Patriarchi A, Fox A, Maunsell B, Fanning S, Bolton D (2011) Molecular characterization and environmental mapping of *Campylobacter* isolates in a subset of intensive poultry flocks in Ireland. Foodborne Pathog Dis 8: 99–108.

22. Jacobs-Reitsma WF (1997) Aspects of epidemiology of *Campylobacter* in poultry. Vet Q 19: 113–117.
23. Sahin O, Luo N, Huang S, Zhang Q (2003) Effect of *Campylobacter*-specific maternal antibodies on *Campylobacter jejuni* colonization in young chickens. Appl Environ Microbiol 69: 5372–5379.
24. Cawthraw SA, Newell DG (2010) Investigation of the presence and protective effects of maternal antibodies against *Campylobacter* jejuni in chickens. Avian Dis 54: 86–93.
25. Allen VM, Weaver H, Ridley AM, Harris JA, Sharma M, et al. (2008) Sources and spread of thermophilic *Campylobacter* spp. during partial depopulation of broiler chicken flocks. J Food Prot 71: 264–270.
26. Hue O, Le Bouquin S, Laisney MJ, Allain V, Lalande F, et al. (2010) Prevalence of and risk factors for *Campylobacter* spp. contamination of broiler chicken carcasses at the slaughterhouse. Food Microbiol 27: 992–999.
27. Hald B, Rattenborg E, Madsen M (2001) Role of batch depletion of broiler houses on the occurrence of *Campylobacter* spp. in chicken flocks. Letters in Applied Microbiology 32: 253–256.
28. Cogan TA, Thomas AO, Rees LE, Taylor AH, Jepson MA, et al. (2006) Norepinephrine increases the pathogenic potential of *Campylobacter* jejuni. Gut.
29. Nesbit EG, Gibbs P, Dreesen DW, Lee MD (2001) Epidemiologic features of *Campylobacter* jejuni isolated from poultry broiler houses and surrounding environments as determined by use of molecular strain typing. American Journal of Veterinary Research 62: 190–194.
30. Petersen L, Nielsen EM, Engberg J, On SL, Dietz HH (2001) Comparison of genotypes and serotypes of *Campylobacter* jejuni isolated from Danish wild mammals and birds and from broiler flocks and humans. Appl Environ Microbiol 67: 3115–3121.
31. Stern NJ, Fedorka-Cray P, Bailey JS, Cox NA, Craven SE, et al. (2001) Distribution of *Campylobacter* spp. in selected U. S. poultry production and processing operations. J Food Prot 64: 1705–1710.
32. Sears A, Baker MG, Wilson N, Marshall J, Muellner P, et al. (2011) Marked campylobacteriosis decline after interventions aimed at poultry, New Zealand. Emerg Infect Dis Jun. [Epub ahead of print].
33. Hofshagen M, Kruse H (2005) Reduction in flock prevalence of *Campylobacter* spp. in broilers in Norway after implementation of an action plan. J Food Prot 68: 2220–2223.

34. Stern NJ, Hiett KL, Alfredsson GA, Kristinsson KG, Reiersen J, et al. (2003) *Campylobacter* spp. in Icelandic poultry operations and human disease. Epidemiol Infect 130: 23–32.
35. Hansson I, Forshell LP, Gustafsson P, Boqvist S, Lindblad J, et al. (2007) Summary of the Swedish *Campylobacter* program in broilers, 2001 through 2005. J Food Prot 70: 2008–2014.
36. Rosenquist H, Boysen L, Galliano C, Nordentoft S, Ethelberg S, et al. (2009) Danish strategies to control *Campylobacter* in broilers and broiler meat: facts and effects. Epidemiol Infect 1–9.

CHAPTER 16

The Epidemiology of Clostridium perfringens type a on Ontario Swine Farms, with Special Reference to cpb2-Positive Isolates

Gloria Chan[1], Abdolvahab Farzan[1], Glenn Soltes[2], Vivian M Nicholson[2], Yanlong Pei[2], Robert Friendship[1] and John F Prescott[2]

[1] Department of Population Medicine, University of Guelph, Guelph, ON N1G 2W1, Canada.

[2] Department of Pathobiology, University of Guelph, Guelph, ON N1G 2W1, Canada.

ABSTRACT

Background

There is poor understanding of most aspects of *Clostridium perfringens* type A as a possible cause of neonatal diarrhea in piglets, and the prevalence

---

Citation: Chan, G., Farzan, A., Soltes, G. *et al.* The epidemiology of *Clostridium perfringens* type A on Ontario swine farms, with special reference to *cpb2*-positive isolates. *BMC Vet Res* **8**, 156 (2012). https://doi.org/10.1186/1746-6148-8-156

and types of *C. perfringens* present on Ontario swine farms is unknown. To study the prevalence of fecal *C. perfringens* and selected toxin genes, 48 Ontario swine farms were visited between August 2010 and May 2011, and 354 fecal samples were collected from suckling pigs, lactating sows, weanling pigs, grower-finisher pigs, and gestating sows, as well as from manure pits. The fecal samples were cultured quantitatively, and toxin genes were detected by real-time multiplex polymerase chain reaction (PCR).

Results

In mixed multivariable linear analysis, $\log_{10}$*C. perfringens* in fecal samples from suckling pigs were higher than that of weanling pigs, grower-finisher pigs, and manure pit samples ($P < 0.05$). In mixed multivariable logistic analysis, the *C. perfringens* isolates recovered from lactating sows (OR=0.069, P <0.001), gestating sows (OR=0.020, P <0.001), grower-finishers (OR=0.017, P <0.001), and manure pits (OR=0.11, P <0.001) were less likely to be positive for the consensus beta2 toxin gene *cpb2* compared to the isolates from suckling pigs. The prevalence of *cpb2* in the isolates recovered from weanlings did not differ significantly from suckling pigs. *C. perfringens* isolates that were positive for *cpb2* were more likely to carry the atypical *cpb2* gene (*atyp-cpb2*) (OR=19, $P < 0.001$) compared to isolates that were negative for *cpb2*. Multivariable analysis did not identify farm factors affecting the presence of consensus *cpb2* and *atyp-cpb2* genes.

Conclusions

This study provides baseline data on the prevalence of *C. perfringens* and associated toxin genes in healthy pigs at different stages of production on Ontario swine farms. The study suggests that if *C. perfringens* type A are involved in neonatal enteritis, there may be strains with specific characteristics that cannot be identified by the existing genotyping system.

BACKGROUND

Clostridium perfringens type A is considered by some to be one of the most common causes of diarrhea in neonatal pigs [1, 2]. However, the pathogenic basis of *C. perfringens* type A diarrhea is unclear, and the current diagnostic methods for this disease are not specific.

Clostridium perfringens are ubiquitous Gram-positive anaerobes that can be isolated from many environments, and their spore-forming ability allows them to persist in the swine ecosystem [2]. Currently, *C. perfringens* are divided into five toxinotypes (A to E) depending on their production of four major toxins. Isolates of any toxinotype may also produce enterotoxin (CPE) and beta2 toxin (CPB2) [3].

In the past decade, diagnosis of neonatal piglet diarrhea due to *C. perfringens* type A has increased, and has been associated with increased pre-weaning mortality [1]. Disease can develop in suckling pigs in the first week of life, and typical clinical signs include non-hemorrhagic, mucoid diarrhea seen within 48 h of birth, and lasting for 5 d [2]. The current method of diagnosis is based on isolation of large numbers of *C. perfringens* type A possessing the consensus CPB2 gene (*cpb2*) detected by polymerase chain reaction (PCR) from the feces or intestinal contents, and the exclusion of other known causes of neonatal diarrhea [3]. There is currently no commercial test available for detecting CPB2 in isolates [1, 3]. Most porcine isolates carry and express *cpb2* [4]. There is poor understanding of most aspects of *C. perfringens* type A as a possible cause of neonatal diarrhea in piglets, and the prevalence and types of *C. perfringens* present on Ontario swine farms is unknown. The objectives of this study were to compare the prevalence of *C. perfringens* and selected toxin genes among pigs at different stages of production, and to identify farm factors affecting this prevalence.

METHODS

Farm selection

Forty-eight farms in Ontario were conveniently selected and visited once or twice between two periods of sample collection from August to December 2010, and from January 2011 and May 2011. For the first period of sample collections, 28 farms were visited. One farm was visited twice in the first period, and visited again in the second period. In the second period, a total of 11 farms from the first period were visited again, and 20 farms were visited for the first time. Some farms that were visited from the first period were not re-visited because they did not have suckling pigs or they were no longer in business.

Sample collection

In both periods of sample collection, for farrowing operations, pooled-fecal samples were collected from gestating sows, weanling pigs, and grower-finisher pigs. In addition, a pooled sample was taken from the manure pit. In the second period of sample collection, fecal samples were also collected from lactating sows and their litters. A lactating sow fecal sample was collected by randomly selecting a sow, and manually obtaining a rectal sample with a gloved hand. A pooled-fecal sample of the lactating sow's litter was also obtained by selecting three suckling pigs in the litter and collecting the feces from the piglet's rectum. If suckling pigs did not defecate sufficient material during the sampling process, fresh piglet feces were obtained from the farrowing area of the litter. Three fecal samples were collected from lactating sows, and three pooled fecal samples were collected from their respective litters from each farrowing operation.

A pooled-fecal sample of gestating sows was obtained by selecting 6 sows from different areas of the barn, and obtaining a rectal sample. A pooled-fecal sample of weanling pigs or grower-finisher pigs was obtained from randomly selecting 6 pens, observing the pigs defecating, and then immediately collecting the fresh feces from the floor of the pens. On farrow-to-feeder farms sampling was similar to farrow-to-finish farms except two pooled samples were collected from weaning pigs. On grower-finisher farms, 3 pooled-fecal samples were taken from the pigs present, as well as a pooled manure pit sample. All farms on the study used a liquid manure system and a sample from the holding tank or manure pit was obtained by filling a plastic bottle from the manure pit. Liquid manure from 3 locations in the pit and at 3 different depths was combined to create the pooled sample.

Questionnaire

A survey was administered to collect information on each farm for different management factors including the following: type of operation (farrowing-to-finish, farrowing-to-feeder, or grower-finisher operation), total number of barns, herd size (number of sows, gilts, weanling pigs, and grower-finisher pigs), sow flow, weanling flow, grower-finisher flow (all-in/all-out or continuous), number of farrowing rooms and crates, the presence of other agricultural species (other species), drug usage and vaccination, history of diarrhea outbreak in suckling pigs or post-weaning pigs in the past year, and treatment of the diarrhea problem. A total of 46 farms were surveyed; two farms were not surveyed because they were no longer in business.

Bacterial Isolation

For bacterial culture, samples were weighed and 4–5 serial 10-fold dilutions were performed in phosphate buffered saline, pH 7.2. Fecal dilutions were plated onto selective SFP (Shahadi Ferguson Perfringens) medium (Difco,

Detroit, MI, USA). The SFP media contains 5% egg yolk emulsion (Oxoid, Nepean, ON), 12 µg/ml kanamycin sulphate, and 30 IU/ml polymyxin B sulphate (SFP Selective Supplement, Oxoid). The plates were transferred to an anaerobic jar, with anaerobic atmosphere provided by GasPak (BD, Sparks, MD, USA) and incubated at 37°C overnight. Colonies with characteristic *C. perfringens* colonial and microscopic morphologies, as well as lecithinase activity on the egg yolk agar, were counted and counts converted to colony-forming units (CFU) per gram.

Genotyping of Isolates

Clostridium perfringens colony lysates were examined by multiplex polymerase chain reaction (PCR) for toxin genes using the method described by Albini and colleagues [5]. Five isolates per sample were genotyped for cpa, consensus cpb2 (cpb2) and atypical cpb2 (atyp-cbp2). The isolates recovered during the first period of the study were also genotyped for the major toxin genes beta toxin (cpb), enterotoxin (cpe), epsilon toxin (etx), and iota toxin (itx), NetB toxin (netB) and large clostridial cytotoxin (tpcL). Since these were all found to be absent they were not examined in the second period. The primer sets for the major toxin genes and cpb2 used were designed using the reported nucleotide sequences [5]. The primer sets for netB, tpeL, and atyp-cpb2 were designed for this study using standard approaches and appropriate bacterial control strains. The Roche LightCycler® 480 SW 1.5 software package was used to analyze the data. Relevant positive control strains were included in all batched PCR tests.

Data Ansalysis

The data were entered in a spreadsheet (Microsoft Excel 2007; Microsoft Corp., Redmond, WA) and imported to Stata 10 Intercooled for Windows XP (StataCorp LP, College Station, TX) for statistical analysis.

Three outcomes were measured, which included *C. perfringens* type A (*Cp*A) count in fecal samples, the presence of *cpb2*-carrying *C. perfringens* type A (*cpb2-Cp*A) isolates in fecal samples, and the presence of *atyp-cpb2*-carrying *C. perfringens* type A (atyp-cpb2-*Cp*A) isolates in fecal samples. The independent variables included in multivariable analysis were stage of production (main effect), type of operation, total number of barns, herd size (small or <1000, medium or ≥1000–3000, and large or >3000), and the presence of other species.

A mixed linear regression modeling method with farm as a random effect was used to analyze the association between *CpA* count in fecal samples and independent variables. Intra-class correlation coefficient was calculated as ICC = σ 2 farm / σ 2 farm + σ 2 sample . In addition, two separate logistic regression models were fitted to investigate the association between the presence of *cpb2-Cp*A and *atyp-cpb2-Cp*A in fecal samples and independent variables. Intra-cluster coefficient was calculated as ICC = σ 2 farm / σ 2 farm + π 2 / 3 by assuming that sample level variance on the logit scale was $\pi^2/3$, π=3.1416.

Univariable analysis of independent variables and their association with *Cp*A count, and presence of *cpb2-Cp*A and *atyp-cpb2-Cp*A isolates in fecal samples was performed using a single linear or logistic regression. Variables with P <0.20 were selected for inclusion in multivariable analyses which was modeled at the sample level with farm as random variable.

A manual stepwise procedure was used to build the multivariable models. Pair-wise correlation coefficients were calculated between independent variables, and coefficients with an absolute value greater than 0.8 were considered colinear. If two independent variables were significantly correlated, the more informative variable was used in the final model. A variable was identified as a confounder if it changed the coefficient of the main effects by 20% or more when the potential confounder variable was removed. If a variable was determined to be a confounder, it was included in the final model regardless of its statistical significance. Interaction was evaluated between all independent variables and the main effects model. Each interaction term was assessed for statistical significance with the main effects, and the interaction terms with *P* <0.05 were included in the final model. Interaction terms that were not significant in the final model were removed if removal of the interaction term did not result in a significant change in the likelihood ratio test (logistic regression) and partial F-test (linear regression). For linear multivariable analysis, a $\log_{10}$ transformation was used on the C. perfringens count (CFU/g) in order to meet the normal distribution and homoscedasticity assumptions for the outcome of interest. The Cook-Weisberg test was performed for the linear multivariable model with farm as a fixed effect, and the assumption of homoscedasticity was met if P >0.05. The Shapiro-Wilk normality test was used to test the normal distribution of $\log_{10}$ CFU/g with farm as a fixed effect, and the assumption of normality was met if P >0.05. The standardized residuals were graphically assessed for outliers, and the leverage values were graphically assessed. The logistic regression multivariable model with farm as a fixed effect for the presence of cpb2 in C. perfringens type A isolates was assessed for goodness-of-fit, which was indicated by P >0.05 in the Hosmer-Lemeshow test. Pearson residuals were graphically assessed for outliers, and the leverage values were graphically assessed. The Spearman's rank correlation test was used to analyze the association between the toxin genes in lactating sows and their litters.

RESULTS

A total of 354 fecal samples were collected, distributed between the sources shown in Table 1. Overall, Clostridium perfringens was isolated from 225 (64%) of 354 fecal samples (98% of suckling piglets, 34% of weanling pigs, 18% of grower-finisher pigs, 89% of gestating sows, 96% of lactating sows, and 75% of manure pit samples) (Table 1). A farm was considered positive for C. perfringens if cpa was detected in the isolate of at least one fecal sample, and a farm was considered positive for consensus cpb2 if it was detected in the C. perfringens isolate of at least one fecal sample. A total of 42 farms (87.5%) were positive for C. perfringens, and 25 farms (52%) were positive for cpb2. C. perfringens was not recovered from samples collected on 6 (12.5%) of 48 farms, all of which were grower-finisher operations. Within-herd prevalence of C. perfringens type A ranged between 0 and 100% with a mean of 53.5±32%.

Table 1. Log $_{10}$ CFU/g of *C. perfringens* **in fecal samples and the positive proportion of toxin genes**

	Log_{10}(CFU/g)	**Percentage of positive samples**		**Total No. of Samples**
	Mean ± SD	***Cpb2***	***Atyp-cpb2***	
Suckling pigs	5.0 ± 1.70	77	44	57
Lactating Sows	4.0 ± 1.86	26	16	57
Gestating Sows	4.3 ± 1.85	5	0	37
Weanling pigs	1.3 ± 1.87	18	7	44
Grower-finisher pigs	0.6 ± 1.24	2	5	99
Manure Pit	2.7 ± 1.82	27	17	60
Total	2.7 ± 2.38	25	15	354

Clostridium Perfringens count (log_{10}CFU/g)

The mean of log_{10} CFU/g of *Clostridium perfringens* in fecal samples is shown in Table 1. The total mean count was 2.7±2.4 log_{10} CFU/g. The distribution of independent variables on 46 surveyed farms is shown in Table 2. Independent variables with $P < 0.20$ that were initially included in the multivariable linear regression analysis were the stage of production, type of operation, total herd size, presence of other species, and sampling period. The type of operation and total herd size did not significantly affect the *C. perfringens* count of fecal samples in the final model ($P > 0.05$). The log_{10}*C. perfringens* count for fecal samples collected from the first and second sampling period was not significantly different ($P > 0.05$); the sampling period was included in the final model because *C. perfringens* count in the samples collected from different stage of production was confounded by sampling period. In multivariable analysis, the log_{10}*C. perfringens* count was higher in suckling pig fecal samples compared to that of weanling pigs, grower-finisher pigs, and the manure pit ($P < 0.05$) (Table 3). The log_{10}*C. perfringens* count was higher in gestating sow samples, lactating sow samples, or the manure pits, respectively, compared to that of weanling pigs and grower-finisher pigs ($P < 0.05$). In the final model, *C. perfringens* count in fecal samples collected from sows (both gestating and lactating) was interacted by the presence of other species on farms ($P < 0.05$) (Table 3). Overall, the log_{10}*C. perfringens* count was higher in the fecal samples collected from sows on farms where other species were absent. The log_{10}*C. perfringens* in gestating sows versus piglets was 4.1=[(−1.35)+(−0.93)+1.5+ 4.9 (constant)] and 5.0=[(−1.35)+1.5+ 4.9 (constant)] in the presence and absence of other species, respectively. Similarly, the log_{10}*C. perfringens* in lactating sows versus piglets was 3.88=[(−1.93)+(−0.93)+1.84+ 4.9 (constant)] and 4.81=[(−1.93)+1.84+ 4.9(constant)] in the presence and absence of other species, respectively.

Table 2. The distribution of independent variables among 46 surveyed farms

Independent Variable		**Percentage of *Cpb2*-positive farms**	**Univariable analysis*P*-value (sample level)**
Type of operation (sample n=220)	Farrowing (farm n=30)	73	0.012
	Grower-finisher (farm n=16)	25	
Other species (sample n=212)	Not present (farm n=23)	57	0.004
	Present (farm n=23)	57	
Herd Size (sample n=217)	Small (<1000) (farm n=9)	33	Referent
	Medium (≥1000-3000) (farm n=28)	61	0.272
	Large (>3000) (farm n=9)	67	0.018
Sow Flow (sample n=84)	All-in/all-out (farm n=16)	75	0.417
	Continuous (farm n=14)	71	
Weanling Flow (sample n=15)	All-in/all-out (farm n=23)	70	0.326
	Continuous (farm n=8)	87.5	
Grower-Finisher Flow (sample n=20)	All-in/all-out (farm n=27)	48	0.327
	Continuous (farm n=13)	54	
Antibiotic usage (sample n=66)	Yes (farm n=7)	100	0.304*
	No (farm n=23)	65	
Diarrhea (sample n=58)	Yes (farm n=11)	82	0.792*
	No (farm n=19)	68	

*Exact logistic regression.

Table 3. Multivariable linear regression model of factors associated with fecal *C. perfringens* **count (log** $_{10}$ CFU/g)

Variables		Coefficient	Standard Error	95% Confidence Interval	*P*-value
Stage	Suckling pig	Referent			
	Lactating sow	−1.9	0.367	(−2.7)-(−1.2)	<0.001
	Gestating sow	−1.4	0.438	(−2.2)-(−0.50)	0.002
	Weanling	−3.8	0.398	(−4.6)-(−3.1)	<0.001
	Grower-finisher	−4.4	0.413	(−5.2)-(−3.6)	<0.001
	Manure pit	−1.7	0.403	(−2.5)-(−0.93)	<0.001
Other species	Absent	Referent			
	Present	−0.93	0.484	(−1.9)-(−0.019)	0.055
	Present x Gestating sow	1.5	0.626	0.30-2.75	0.015
	Present x Lactating sow	1.8	0.533	0.79-2.9	0.001
Sampling Period	First	Referent			
	Second	0.34	0.234	(−0.12)-(−0.80)	0.144

The Intraclass Correlation Coefficient (ICC) was 0.24, indicating a low level of clustering in which the within-farm variation of the *C. perfringens* count accounted for almost a quarter of the total variation of the count.

Potential outliers were identified for two farms, of which the weanling fecal samples had a *C. perfringens* count of 6.0 and 6.7 $\log_{10}$ CFU/g, respectively. Potential extreme observations were identified for two farms, of which *C. perfringens* was not isolated from weanling fecal samples.

Clostridium Perfringens-Associated Toxin Genes

The distribution of *cpb2* and *atyp-cbp2* genes among *Clostridium perfringens* isolates cultured from fecal samples is presented in Table 1.

The independent variables initially included in the multivariable logistic analysis were the stage of production, presence of *atyp-cpb2*, type of operation, presence of other species, total herd size, and sampling period. The type of operation, presence of other species, and total herd size did not significantly affect the likelihood of detecting *cpb2*-positive *C. perfringens* isolates in the final model ($P > 0.05$). The variable of vaccination was not included in the statistical analysis because none of the farms surveyed vaccinated sows or pigs for *C. perfringens*.

Cpb2 was more likely found in the isolates of suckling pigs compared to those of lactating sows, gestating sows, and grower-finisher pigs, and manure pits ($P < 0.05$) (Table 4). The likelihood of finding *cpb2* in *C. perfringens* isolates recovered from weanling pigs was higher compared to that of gestating sows (OR=0.077, P=0.011) and grower-finisher pigs (OR=0.065, P=0.027).

Table 4. Multivariable logistic regression model of factors associated with *cpb2* **-positive** *C. perfringens* **isolates**

Variables		**Odds Ratio**	**Standard Error**	**95% Confidence Interval**	***P*-value**
Stage	Suckling pig	Referent			
	Lactating sow	0.069	0.0437	0.020-0.24	<0.001
	Gestating sow	0.020	0.0189	3.2×10^{-3}-0.13	<0.001
	Weanling	0.26	0.216	0.052-1.3	0.103
	Grower-finisher	0.017	0.0185	2.0×10^{-3}-0.14	<0.001
	Manure pit	0.11	0.0677	3.2×10^{-2}-0.37	<0.001
Atyp-cpb2	Negative	Referent			
	Positive	19	12.0	5.7-65	<0.001

The Intraclass Correlation Coefficient (ICC) that compared the within-farm and between-farm variance was 0.24 and 0.25, respectively for linear regression and logistic regression, indicating a low level of clustering in which the within-farm variation of *C. perfringens* count and the presence of *cpb2* accounted for almost a quarter of the total variation.

The likelihood of detecting *atyp-cpb2* was higher in *cpb2*-positive isolates compared to *cpb2*-negative isolates (P <0.001) (Table 4). The percentages of samples that had *atyp-cpb2* were 51% in *cpb2*-positive samples, and 6% in *cpb2*-negative samples.

Clostridium perfringens isolates recovered from lactating sows (OR=0.20, P=0.001), weanlings (OR=0.20, P=0.039), and the manure pits (OR=0.32, P=0.022) were less likely to be positive for *atyp-cpb2* than isolates recovered from suckling pigs. The correlation of *cpb2* or *atyp-cpb2* in isolates of lactating sows and in isolates from their litters was not statistically significant.

Potential outliers were identified for two farms. On one farm, *cpb2* nor *atyp-cpb2* were detected in the *C. perfringens* isolates of a suckling pig fecal sample. On the other farm, *cpb2* was not detected in the *atyp-cpb2* positive *C. perfringens* isolates of the manure pit fecal sample. Potential extreme observations were identified for one farm, of which the *cpb2*-positive *C. perfringens* isolates were only recovered from two of three suckling pig fecal samples.

The isolates recovered during the first period of the study were all negative for *cpb*, *cpe*, *etx*, *itx*, *netB* and *tpeL*. For this reason, they were not examined in the second period.

DISCUSSION

The current study of enteric *Clostridium perfringens* in pigs identified novel findings that may increase the understanding of the epidemiology of infection and possible diagnosis of the suckling pig diarrhea thought to be caused by this organism. *Clostridium perfringens* type A is speculated to be an emerging pathogen associated with neonatal piglet diarrhea [6–9], although not all prevalence studies conducted on infectious causes of piglet diarrhea have included *C. perfringens* type A in the analysis [6, 10, 11]. Isolates recovered from diarrheic piglets could not be differentiated genetically from isolates recovered from non-diarrheic piglets in the United States [12] or on genotype in Canada [13].

This study is the first to examine systematically the quantitative prevalence of *C. perfringens,* as well as an expanded range of toxin genes, in pigs of different ages on multiple farms. Our results showed that the number of *C. perfringens* type A isolates was higher in suckling pigs compared to pigs in other stages of production. High numbers of *C. perfringens* were also isolated from gestating and lactating sows. The pattern of decline in colonization by *C. perfringens* with age has been noted previously in the intestine of individual pigs but not at the farm level [14]. The decrease in *C. perfringens* count for fecal samples in weanling and grower-finisher pigs was generally expected, since the population of *C. perfringens* decreases in the pig intestine as other bacterial species establish their populations [2]. Host-related factors that affect the decline in colonization may include diet, changes in large intestinal microflora with time, age-related changes in host physiology, or the development of intestinal immunity. The decline in numbers of *C. perfringens* in pigs in the grower-finisher stage may also be partly due to changes in management factors including the use of antimicrobials during the nursery and grower-finisher stages. The effect of other species on the farm in increasing *C. perfringens* counts in sows (Table 3) was an unexpected observation that factors other than "host" can affect colonization by this organism.

Many *C. perfringens* isolated from pigs are relatively unusual among isolates from different animal sources in that they belong to a clonal population [15]. Unusually, these isolates are characterized both by being consensus *cpb2* positive and by expressing CPB2 [4]. There thus appears to be an adaptation of this lineage to swine. Although we did not however examine the isolates made here for their phylogenetic relationships, on the basis of the present study, the adaptation of this lineage appears to be to the young pig, and notably to the suckling piglet. This relationship of *cpb2*-positive *C. perfringens* to the young piglet (Tables 2, 4) has not previously been demonstrated at the farm level. Most of the suckling pig isolates in this study were positive for consensus *cpb2* as noted in other studies [4, 7–9]. The gestating sow isolates were usually not positive for *cpb2* and lacked the *atyp-cpb2* gene that was found in lactating sow and suckling pig isolates, which appear to be specifically colonized by *cpb2*-positive strains [3]. It is possible that the sows in the lactating stage acquired *cpb2*-positive *C. perfringens* strains from suckling pigs in the litter, but that strains were not always maintained when the suckling pigs were weaned and sows returned to the gestating stage. However, there was no significant correlation of the presence of *cpb2* and *atyp-cpb2* in *C. perfringens* type A isolates from sows and their litters, suggesting that the source of the strains in suckling pigs may be different from sows, or that *cpb2* and/or *atyp-cpb2* positive isolates are more likely to proliferate in suckling pigs. Interestingly, there was a dominance of *atyp-cpb2* isolates in *C. perfringens* isolated from grower-finisher pigs. It is possible that other factors contributed to the decline in consensus *cpb2*-positive *C. perfringens* isolates seen in grower-finisher pigs, such as antimicrobial usage; a negative association between *cpb2*-positive *C. perfringens* isolates and resistance to erythromycin and clindamycin was documented in an Ontario study [16]. The isolates from the manure pit were unexpectedly similar to isolates from suckling pigs, reasons for which are unclear.

This study identified genes in the most abundant isolate of *C. perfringens*, so that the less abundant genotypes may not have been detected. *C. perfringens* isolates positive for *cpb2* were more likely to be positive for *atyp-cpb2*, suggesting a linkage between these genes. Further research on the genetic diversity of *C. perfringens* type A isolates in clinically healthy and diarrheic suckling pigs, and the conditions of which *cpb2* and *atyp-cpb2* toxin genes are expressed, may help clarify the role of this agent in neonatal piglet diarrhea.

Since the farms in our study were conveniently selected, the results may not apply to the general population of Ontario swine farms. The difference in bacterial count of *C. perfringens* in fecal samples from pigs at different stages of production might have been affected by the sampling procedure, since samples for weanling pigs and grower-finisher pigs were collected from feces on the pen floor whereas the samples for sows and suckling pigs were collected from the rectum. The fecal samples collected from gestating sows, weanling pigs, grower-finisher pigs, and the manure pits were pooled, so it is conceivable that some samples were misclassified. Fecal samples collected from farrowing operations in the first period of the study did not include those of suckling pigs or lactating sows, which could affect the farm-level sensitivity for the detection of *cpb2*-positive *C. perfringens* isolates from these farms.

This is the first time that *C. perfringens* type A from swine has been examined for *netB* and *tpeL*, genes that are common in isolates from chickens with necrotic enteritis [17]. Their absence further supports the idea of host adaptation of strains of *C. perfringens*, as apparent here in the association of *cpb*2-positive strains with piglets. The absence of *cpb* is consistent with the absence of *C. perfringens* type C in enteric disease in swine and other animals in Ontario. The lack of *cpe* in *C. perfringens* isolates confirms earlier studies [18, 19].

CONCLUSIONS

To our best knowledge, this is the first study on the epidemiology of *C. perfringens* in the swine population. This study provides baseline data from healthy pigs at different stages of production as well as in the manure pits for future studies to investigate the role of *C. perfringens* in swine enteric diseases. The study suggests that if C. *perfringens* type A are involved in neonatal enteritis, there may be strains with specific characteristics that cannot be identified by the existing genotyping system. Further work is required to determine whether this is the case.

ACKNOWLEDGMENTS

The research was funded by the Ontario Ministry of Agriculture Food and Rural Affairs. We also thank the Natural Sciences and Engineering Research Council for support. We thank Bryan Bloomfield for technical assistance and the swine producers for participating in the research.

REFERENCES

1. Lehe K: Successful control of clostridial enteritis: A report from the back forty. Proceedings of American Association of Swine Veterinarians. Perry, IA, USA: American Association of Swine Veterinarians, Dallas; 2009.
2. Schwartz KJ: *Clostridium*-associated diseases in swine. Proceedings of American Association of Swine Veterinarians. Perry, IA, USA: American Association of Swine Veterinarians, Dallas; 2009.
3. Songer JG, Uzal FA: Clostridial enteric infections in pigs. J Vet Diag Invest. 2005, 17: 528-536. 10.1177/104063870501700602.
4. Jost BH, Billington SJ, Trinh HT, Bueschel DM, Songer JG: Atypical *cpb2* genes, encoding beta2-toxin in *Clostridium perfringens* isolates of nonporcine origin. Infect Immun. 2005, 73: 652-656. 10.1128/IAI.73.1.652-656.2005.
5. Albini S, Brodard I, Jaussi A, Wollschlaeger N, Frey J, Miserez R, Abril C: Real-time multiplex PCR assays for reliable detection of *Clostridium perfringens* toxin genes in animal isolates. Vet Microbiol. 2008, 127: 179-185. 10.1016/j.vetmic.2007.07.024.
6. Yaeger M, Funk N, Hoffman L: A survey of agents associated with neonatal diarrhea in Iowa swine including *Clostridium difficile* and porcine reproductive and respiratory syndrome virus. J Vet Diagn Invest. 2002, 14: 281-287. 10.1177/104063870201400402.
7. Garmory HS, Chanter N, French NP, Bueschel D, Songer JG, Titball RW: Occurrence of *Clostridium perfringens* beta2-toxin amongst animals, determined using genotyping and subtyping PCR assays. Epidemiol Infect. 2000, 124: 61-67. 10.1017/S0950268899003295.
8. Bueschel DM, Jost BH, Billington SJ, Trinh HT, Songer JG: Prevalence of cbp2, encoding beta2 toxin, in Clostridium perfringens field isolates: correlation of genotype with phenotype. Vet Microbiol. 2002, 94: 121-129.
9. Waters M, Savoie A, Garmory HS, Bueschel D, Popoff MR, Songer GJ, Titball RW, McClane BA, Sarker MR: Genotyping and phenotyping of beta2-toxigenic *Clostridium perfringens* fecal isolates associated with gastrointestinal diseases in piglets. J Clin Microbiol. 2003, 41: 3584-3591. 10.1128/JCM.41.8.3584-3591.2003.
10. Wieler LH, Ilieff A, Herbst W, Bauer C, Vieler E, Bauerfeind R, Failing K, Klos H, Wengert D, Baljer G, Zahner H: Prevalence of enteropathogens in suckling and weanling suckling pigs with diarrhea

in southern Germany. J Vet Med B. 2001, 48: 151-159. 10.1046/j.1439-0450.2001.00431.x.

11. Katsuda K, Kohmoto M, Kawashima K, Tsunemitsu H: Frequency of enteropathogen detection in suckling and weanling pigs with diarrhea in Japan. J Vet Diagn Invest. 2006, 18: 350-354. 10.1177/104063870601800405.
12. Baker AA, Davis E, Rehberger T, Rosener D: Prevalence and diversity of toxigenic *Clostridium perfringens* and *Clostridium difficile* among swine herds in the midwest. Appl Environ Microbiol. 2010, 76: 2961-2967. 10.1128/AEM.02459-09.
13. Farzan A, Kircanki J, DeLay J, Soltes G, Songer GJ, Friendship R, Prescott JF: An investigation into the association between *cpb2*-encoding *Clostridium perfringens* type A and diarrhea in neonatal piglets. Can J Vet Res. in press
14. Smith WH, Jones JET: Observations on the alimentary tract and its bacterial flora in healthy and diseased pigs. J Path Bact. 1963, 84: 387-412.
15. Jost HB, Trinh TH, Songer GJ: Clonal relationships among *Clostridium perfringens* of porcine origin as determined by multilocus sequence typing. Vet Microbiol. 2006, 116: 158-165. 10.1016/j.vetmic.2006.03.025.
16. Slavic D, Boerlin P, Fabri M, Klotins KC, Zoethout JK, Weir PE, Bateman D: Antimicrobial susceptibility of *Clostridium perfringens* isolates of bovine, chicken, porcine, and turkey origin from Ontario. Can J Vet Res. 2011, 75: 89-97.
17. Chalmers G, Bruce HL, Hunter DB, Parreira VR, Kulkarni RR, Jiang YF, Prescott JF, Boerlin P: Multilocus sequence typing analysis of *Clostridium perfringens* isolates from necrotic enteritis outbreaks in broiler chicken populations. J Clin Microbiol. 2008, 46: 3957-3964. 10.1128/JCM.01548-08.
18. Kanakaraj R, Harris DL, Songer GJ, Bosworth B: Multiplex PCR assay for detection of *Clostridium perfringens* in feces and intestinal contents of pigs and in swine feed. Vet Microbiol. 1998, 63: 29-38. 10.1016/S0378-1135(98)00229-6.
19. Klaasen HL, Molkenboer MJ, Bakker J, Miserez R, Häni H, Frey J, Popoff MR, van den Bosch JF: Detection of the beta2 toxin gene of Clostridium perfringens in diarrhoeic piglets in The Netherlands and Switzerland. FEMS Immunol Med Microbiol. 1999, 24 (3): 325-332.

SECTION 5: PARASITES

CHAPTER 17

Epidemiology of Theileria Bicornis Among Black and White Rhinoceros Metapopulation in Kenya

Moses Y Otiende[1], Mary W Kivata[1] , Joseph N Makumi[4] , Mathew N Mutinda[1] , Daniel Okun[4] , Linus Kariuki[1] , Vincent Obanda[1] , Francis Gakuya[1] , Dominic Mijele[1] , Ramón C Soriguer[2] and Samer Alasaad[2,3]

[1]Veterinary Services Department, Forensic and Genetics Laboratory Kenya Wildlife Service, Nairobi, Kenya.

[2]Estación Biológica de Doñana, Consejo Superior de Investigaciones Científicas (CSIC), Avda. Américo Vespucio s/n 41092, Sevilla, Spain.

[3]Institute of Evolutionary Biology and Environmental Studies (IEU), University of Zürich, Winterthurerstrasse 190, 8057 Zürich, Switzerland.

[4]Department of Biochemistry and Biotechnology, Kenyatta University, Nairobi, Kenya.

Citation: Otiende, M.Y., Kivata, M.W., Makumi, J.N. et al. Epidemiology of Theileria bicornis among black and white rhinoceros metapopulation in Kenya. BMC Vet Res 11, 4 (2015). https://doi.org/10.1186/s12917-014-0316-2

ABSTRACT

Background

A huge effort in rhinoceros conservation has focused on poaching and habitat loss as factors leading to the dramatic declines in the endangered eastern black rhinoceros (*Diceros bicornis michaeli*) and the southern white rhinoceros (*Ceratotherium simum simum*). Nevertheless, the role disease and parasite infections play in the mortality of protected populations has largely received limited attention. Infections with piroplasmosis caused by *Babesia bicornis* and *Theileria bicornis* has been shown to be fatal especially in small and isolated populations in Tanzania and South Africa. However, the occurrence and epidemiology of these parasites in Kenyan rhinoceros is not known.

Results

Utilizing 18S rRNA gene as genetic marker to detect rhinoceros infection with *Babesia* and *Theileria*, we examined blood samples collected from seven rhinoceros populations consisting of 114 individuals of black and white rhinoceros. The goal was to determine the prevalence in Kenyan populations, and to assess the association of *Babesia* and *Theileria* infection with host species, age, sex, location, season and population mix (only black rhinoceros comparing to black and white rhinoceros populations). We did not detect any infection with *Babesia* in the sequenced samples, while the prevalence of *T. bicornis* in the Kenyan rhinoceros population was 49.12% (56/114). White rhinoceros had significantly higher prevalence of infection (66%) compared to black rhinoceros (43%). The infection of rhinoceros with *Theileria* was not associated with animal age, sex or location. The risk of infection with *Theileria* was not higher in mixed species populations compared to populations of pure black rhinoceros.

Conclusion

In the rhinoceros studied, we did not detect the presence of *Babesia bicornis*, while *Theileria bicornis* was found to have a 49.12% prevalence with white rhinoceros showing a higher prevalence (66%) comparing with black rhinoceros (43%). Other factors such as age, sex, location, and population mix were not found to play a significant role.

Keywords: Ixodid, Ticks, Piroplasms, Diceros bicornis michaeli, Ceratotherium simum simum

BACKGROUND

The populations and distribution ranges of the black rhinoceros (*Diceros bicornis*) and the white rhinoceros (*Ceratotherium simum*) have declined in the whole of Africa. The rate of their population decline is faster than any other large terrestrial mammal in recent times [1], a fact that supports their endangered status and calls for robust international efforts towards their recovery. These rhinoceros have been exterminated in the majority of African countries, while their range among the remaining principal countries; Kenya, Tanzania, Namibia, Zimbabwe and South Africa [1] is greatly reduced and currently restricted in artificially created sanctuaries. Habitat loss and vicious poaching are the leading twin drivers of population decline of the rhinoceros [2]. However infectious diseases are also an incipient threat to endangered species [3] having been classified among the top five causes of species extinctions [4].

Piroplasms, which are blood-borne protozoan parasites in the genera *Babesia* and *Theileria* (Order Piroplasmidae), are globally distributed and transmitted by a diverse species of Ixodid ticks. These parasites infect a wide range of domesticated and wild mammals as well as humans. Infections may lead to severe disease and death or it may remain latent depending on virulence of the species and host immune status. Piroplasms have historically been known to infect rhinoceros with some infections associated with fatalities [5-7]. However the causal species were unknown until 10 years ago when *Babesia bicornis* and *Theileria bicornis* were independently associated with stress-induced mortality [8]. The first genetic work on piroplasms by Otiende et al., [9] has shown the existence of infection by piroplasm and the occurrence of three new haplotypes of *Theileria bicornis* circulating in both black and white rhinoceros in Kenya. The factors influencing piroplasms prevalence among Kenyan populations were previously unknown. Piroplasms have coevolved with rhinoceros and they coexist with the host without signs of clinical disease. However, stress induced by translocation has been linked to immune suppression and is a major cause of post translocation morbidity and/or fatality. Translocation is at the core of *in situ* management of rhinoceros metapopulation and yet it is incriminated as a disease inducer besides its inherent role in the spread of pathogens. The link between translocation and piroplasmosis is intricate because it is based on the modulatory effects of stress hormones on the immune system. Translocation elicits stress hormones, which allow uninhibited proliferation of piroplasms in the host resulting in disease and death. However, effects of stress hormones are not predictable or homogenous in the population as underlying individuals' conditions, such as injury, pregnancy, co-infection, vary and may elicit different immune response [10].

The goal of this study is to determine the epidemiology of *Theileria biconis* in Kenya. Specifically, we intended to (a) determine their prevalence in both species of rhinoceros and among sub-populations then (b) test the association of infection prevalence with host species, age, sex, location, season and population mix (black rhinoceros vs black and white rhinoceros populations). Information generated will be useful in guiding management and veterinary options such as translocation, differential diagnosis and chemotherapy.

METHODS

Study area

Lake Nakuru National Park (LNNP) central coordinates are 0°22′S 36°05′E and is 4 km from Nakuru town center. The park covers an area of 188 km^2 completely fenced, of which 44 km^2 lies in the shallow alkaline soda lake, thereby leaving 144 km^2 for wildlife use. The area around the lake is flat bare lowland of 1200 m altitude surrounded by hills and gentle cliffs that rise to 1750 m above sea level. The park receives mean annual rainfall of 850 mm with rainfall in the months of April to May and again in October – November. The park consists of open grassland with elevated areas occupied by dry forests of Acacia xanthophloea, Olea capensis sub sp. Macrocarpa and Croton dichogamus. Marshland along the river inflows and springs are covered by Cyperus laevigatus and Typha spp. Other striking plant species include the invasive Tarchonanthus spp. bush land, the deciduous (Teclea & Olive) forest and the Euphorbia candelabrum forest. The park has 33 white and 69 black rhinoceros that freely interact with other diverse species.

Nairobi National Park (NNP) central coordinates are 1°16′S, 36°49′E and is about 10 km from the city of Nairobi and covers an area of 117 km^2. A large section of the park is fenced with only 20 km left open for wildlife dispersal. Average annual rainfall is 800 mm with rainy season between April-May and October-November. The vegetation consists of mosaic grassland, thickets and Acacia and deciduous forests as well as woodlands especially along River Mbagathi that crosses it. The park has 77 black and 13 white rhinoceros besides many other wildlife species.

Ngulia rhino Sanctuary (NgRS) is within Tsavo West National Park occupying a fenced area of 90 km^2 at 3^0 01′S to 3^0 06′S and 38^0 06E to 38^0

10′E. Altitude ranges from 600 m of lowlands to 1800 m of craggy hills with average annual rainfall of 600 mm. Dry period is between December to March, while rains occur in the months of April to June and again in October and November. The vegetation is thickly wooded by Commiphora-Acacia woodland, dotted with baobab trees. This sanctuary contains 77 black rhinoceros without white rhinoceros, though other small-medium sized wildlife occurs in small density.

Meru Rhino Sanctuary (MRS) is 48Km^2 (central coordinates, N 00^0 15.125, E 038^{0}06.481) located within the Meru National Park has 44 black and 45 white rhinoceros. Altitude ranges from 1000 to 3400 m above the sea level. Average rainfall is 635-762 mm with the wet season occurring in late March to April, while the dry season begins from October. Major rivers such as Makutano, Kanjoo, Kathithi, Rujuwero and Kindani traverse MRS, which contributes to mosaic vegetation types that include thickets, bushland and grassland as well as a thick forest on its southern edge.

Solio and Mugie Rhino Sanctuaries are in the Laikipia-Samburu ecosystem, which is characterized by savannah-type grassland dominated by Euclea divinorum, Acacia spp and Euphorbia woodland while annual rainfall averages 300 to 700 mm. Rainfall level varies annually with intermittent patterns that peaks in April-May, July–August and October–November. The rest of the months are dry. Solio Sanctuary, 0°16′S 37°00′E / 0.27°S 37°E/ -0.27; 37, is 68.9 km^2 and has altitude of 1932 m, located at the base of the Aberdare ranges. The sanctuary holds 50 black and 110 white rhinoceros. Mugie rhino sanctuary (ceased to be a rhino sanctuary at the time of this study, as the entire population was translocated) was 90 km^2 with an altitude 1990 m.

Sampling Design

This study was carried out between 2011 and 2012 in various sub populations of black and white rhinoceros in Kenya. Sampling was cross-sectional whereby samples were collected from apparently healthy rhinos immobilized for translocation or tagging. The biodata of each rhinoceros was obtained from the KWS rhinoceros database. Rhinoceros < 2 years of age were not immobilized for translocation. Juvenile rhinoceros are <3.5 years, sub-adults are (3.6 – 7 years) and adults are >7 years. Rhinoceros were immobilized using a combination of etorphine hydrochloride (M99®), Hyaluronidase (Kyron Laboratories, Benrose 2011, South Africa) and xylazine (Norvatis, [PTY] Ltd, South Africa). Venous blood was drawn from the front limb of the rhinoceros and then collected in EDTA tubes. Blood in EDTA tubes were gently mixed by turning the tubes up and down and then transferring aliquots in labeled cryovials followed by quick freezing in liquid nitrogen. The samples were transported in liquid nitrogen to Forensic and Genetics Laboratory of Kenya Wildlife Service in Nairobi for analysis.

DNA Isolation and PCR Amplification

Genomic DNA was extracted from blood using a genomic DNA extraction kit (DNeasy blood and Tissue Kit, QIAGEN, Southern Cross Biotechnologies, South Africa) following the manufacturers' protocol. A nested PCR amplification specific for the 18S rRNA gene of Babesia and Theileria was performed. A primary amplification was carried out in 50 μl reaction containing 3 μl of the genomic DNA, 45 μl of Platinum blue supermix, 1 μl (10 mM) each forward and reverse primers. The forward primer was ILO-9029, (5′-CGGTAATTCCAGCTCCAATAGCGT-3′) and reverse, ILO-9030 (5′-TTTCTCTCAAAGGTGCTGAAGGAGT-3′) primer [11]. The amplification (Thermocycler, Veriti, ABI) was preceded by a 30 sec polymerase activation step at 95°C followed by 30 cycles of 1 min each at 94°C, annealing at 53°C for 30 sec, extension for 1 min at 72°C. Amplification was terminated by a final extension step 72°C for 9 min. The secondary amplification was in a 50 μl reaction containing 2 μl of the primary amplification product, 45 μl of Platinum blue supermix, 1.5 μl (10 mM) each of forward and reverse primers. The forward primer was MWG4/70, (5′-AGCTCGTAGTTGAATTTCTGCTGC-3′) and the reverse was ILO-7782 (5′-AACTGACGACCTCCAATCTCTAGTC-3′) [11]. The secondary PCR (Thermocycler, Veriti, ABI) was initiated with an initial denaturation at 95°C for 30 sec, followed by 30 cycles of 1 min each at 94°C, annealing at 55°C for 30 s and extension at 72°C for 1 min. The PCR was completed with a final extension step of 72°C for 9 min. PCR products showing successful amplification on agarose gel analysis were directly sequenced for both strands. PCR products were purified for direct sequencing by enzymatic treatment using exonuclease I and shrimp alkaline phosphatase (PCR Product Presequencing Kit, Amersham). All DNA sequencing was carried out by direct cycle sequencing on both strands of purified PCR DNA products from PCR amplification. Sequencing reactions were carried out with the ABI PRISM DigDye Terminator v3.1 cycle sequencing kit and analyzed on an ABI310 DNA sequencer (Applied Biosystems, CA).

Statistical Analysis

Statistical analyses were performed using Fisher Exact test for count data and Chi Square test to determine the relationship between infection of rhinoceros with Theileria and the following variables: rhinoceros age, location, sex, and rhinoceros species. To confirm our results, we also used a Generalized Linear Model with a binomial error, and a complementary log-log link function. All possible interactions were included in the first model. As rhinoceros species are essentially clustered with locality (game park or sanctuary) a generalized linear mixed effect model was also applied, considering locality (game park or sanctuary) as random effect: Theileria Infection ~ Rhinos Species+(1 | Locality). Statistical significance was assessed at $p<0.05$.

Ethic

The Committee of the Department of Veterinary and Capture Services of the Kenya Wildlife Service (KWS) approved the study including animal capture, translocation and sample collection. KWS guidelines on Wildlife Veterinary Practice-2006 were followed. All KWS veterinarians were guided by the Veterinary Surgeons Act Cap 366 Laws of Kenya that regulates veterinary practice in Kenya.

RESULTS

A total of 114 blood samples of black (n=82) and white rhinoceros (n=32) were sampled from seven rhinoceros populations and molecularly examined for infection with Babesia and Theileria. We did not detect any infection with Babesia in the obtained sequences, while the overall Theileria prevalence was 49.1% (56/114). All Theileria sequences belonged to the three haplotypes already described by Otiende et al., [9]. The prevalence of Theileria infection was higher in white rhinoceros (66%) than in black rhinoceros (43%) (Table 1). We confirmed this result using a generalized linear model ($b=-0.652$, $p=0.023$). The simplified glm model was Theileria Infection ~ Rhinos Species, family = binomial (cloglog): Theileria Infection ~ 0.0656±0.2287 Rhinos Species - 0.6516±0.2857. Since prevalence was higher in white rhinoceros, we tested whether presence of white rhinoceros together with black rhinoceros in the same locality was a risk factor for infection with Theileria in black rhinoceros. We found no significant association between infection of black rhinoceros with Theileria and the presence or absence of white rhinoceros in the same locality ($\chi^2=0321$; $p=0.571$). The results were confirmed by applying a generalized lineal mixed effect model, considering locality (game park or sanctuary) as random effect. The last model in this case was Theileria Infection~Rhinos Species, family=binomial (logit): Theileria Infection~0.0647 ± 0.3722 - 0.9414 ± 0.434 (Rhinos Species). Prevalence seemed to increase with age, but the infection by age-groups (juveniles, sub-adults and adults) was not statistically significant (Fisher test, p=0.764, Table 1). Females of both species had higher prevalence 54% (29/54) than males 45% (27/60), this difference however was not statistically significant (Fisher test, p=0.702, Table 1). Inter-population variations in prevalence (Figure 1) were not statistically different (Fisher test, p=0.681, Table 1).

Table 1. Proportion of rhinoceros infected as a function of age, location, sex and species evaluated using Fisher's exact test

Variable	Variable categories	% Negative	% Positive	N	p-value
AGE	Juvenile	58.8	41.2	17	0.764
	Sub-Adult	48.2	51.8	56	
	Adult	51.2	48.8	41	
LOCATION	LNNP	46.7	53.3	35	0.559
	Meru N. P.	50.0	50.0	12	
	Mugie	57.9	42.1	20	
	Ngulia	55.2	44.8	29	
	NNP	55.6	44.4	10	
	Solio	87.5	12.5	8	
SEX	Female	41.7	58.3	59	0.7026
	Male	30.0	70.0	55	
SPECIES	*C. simum*	34.4	65.6	32	**0.037**
	D. bicornis	57.3	42.7	82	

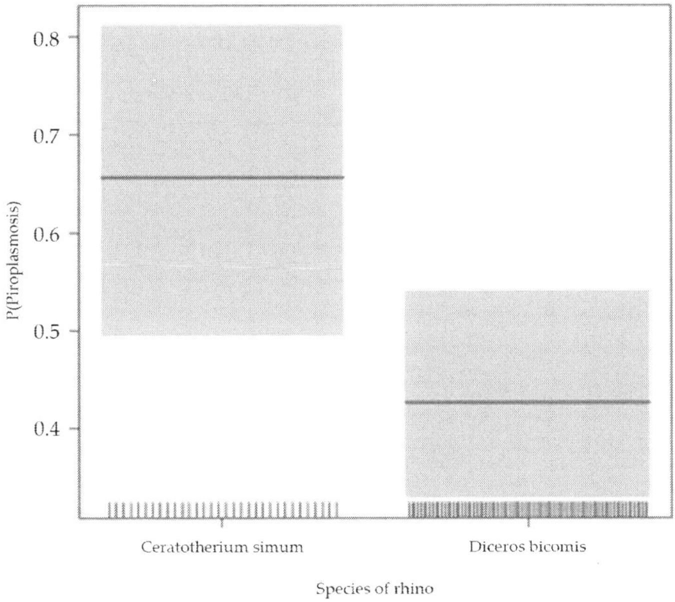

Figure 1. Variation of infection intensity of rhinoceros by species.

DISCUSSION

The six sampled rhinoceros sub-populations in Kenya were infected with piroplasms but we molecularly detected only *Theileria* and not *Babesia* in all studied samples from black and white rhinoceros species.

Ticks and wildlife are the maintenance hosts of piroplasms but efficient transmission fundamentally requires presence of the protozoan, a competent tick species and the host.

Even with *B. bicornis* and *T. bicornis* originally identified from black rhinoceros [8] *T. bicornis* has now been detected in white rhinoceros, Nyala (*Tragelaphus angasii*) and Cattle [12-14]. This means that *T. bicornis* is a multi-host pathogen with possibility of having diverse tick species as vectors. Since *D. rhinocerinus* have been found in other animals such as cattle, sheep, donkey, elephant (*Loxodonta africana*), buffalo (*Syncerus caffer*) and eland (*Taurotragus oryx*) [15,16] this tick could be important in the cross-transmission of rhinocerotid piroplams. Pathogenicity of *T. bicornis* remains unresolved but *T. equi,* which is its close relative [8,17] and recently seen in white rhinoceros [13], has been reported to cause clinical piroplasmosis in translocated equids [18].

Translocation is intimately associated with flare up of latent infections that result in clinical state [18-20]. This is because translocation leads to elevation of glucocorticoids, whose effects are viewed to be obligatorily immunodepressant [10], yet in most cases, especially in transient acute stress, they prepare an animal to survive [10,21,22]. In the present study, sampling was carried out on asymptomatic individuals and despite underlying infection with *T. bicornis*, some of them were subjected to a longer period of stressor condition; during >1000 km road transportation to a new sanctuary in Ruma National Park. Nevertheless, for six months post-release monitoring of this population, they remained asymptomatic. This outcome may support the notion that *T. bicornis* is apathogenic or it may suggest that translocation-stress did not suppress immunity to induce clinical state. Theories behind disease induction by translocation-stress often focus on single parasite infections. However, in nature, wild animals are infected and infested simultaneously with a plethora of parasites that elicits complex immune response that may promote one parasite over the other. For instance, in concomitant infection involving African trypanosome superimposed with piroplasm leads to inhibition of the piroplasm in spite of the trypanosome immunodepressant effect [23]. In reference to fatal piroplasmosis [8], the deceased rhinoceros were subject to diverse and combined stressors; two black rhinoceros in Tanzania did not undergo prior capture and translocation event; the third fatal case involved high parasitemia, severe cold and injury while the fourth case was pregnant and developed translocation myopathy. This suggests that stressor factors that trigger clinical disease are many with maximum effect attained under synergistic state.

In the present study, the prevalence of *T. bicornis* was relatively high (49.1%) but clinical disease was absent in the metapopulation, a state that could mimics endemic stability [24]. This state, which was initially coined for bovine babesiosis and now widely applied in many diseases and hosts, is based on the premise that (1) severity of clinical disease increases with age and (2) that after one infection, the probability that subsequent infections result in disease is reduced [25].

We noted that higher prevalence of *T. bicornis* (odds ratio, 2.502) being detected in white rhinoceros than in black species (Table 1) indicating a species effect. However, we did not find significant effect associating species with prevalence, suggesting that white rhinoceros, even though more susceptible, is not a risk factor to black rhinoceros prevalence. Our result show that the Kenyan white rhino has higher prevalence of *T. bicornis*

(66%) compared to 32.1% - 46.6% in the South African populations [13,26]. The high prevalence of *T. bicornis* in white rhinoceros suggests they are important hosts in the epidemiology of this piroplasm. On the contrary, according to a theory postulated by Schmidt & Ostfeld, [27] we suggest that white rhinoceros could benefit black rhinoceros by acting as 'sinks' for rhinocerotid piroplasms.

Further, our results show that prevalence among the age-groups of rhinoceros did not differ significantly (Table 1) contrary to the infection pattern in white rhinoceros population in South Africa in which female sub-adults had significantly higher prevalence [13]. Nevertheless, the age-associated inclination in our result is comparable with that of Govender *et al.*, [13] in that peak infections were observed among sub-adults (Table 1). It is postulated that sub-adult rhinoceros of both sexes are subject to numerous stress-related changes such as, reproductive maturity, courtship, mating and territorial fights [13,28] that may suppress immunity and enhance susceptibility [29,30].

Sex-biased prevalence is observed in many parasitic infections with males having higher prevalence and intensity of infections than their conspecific females [31]. In our results, there was no significant sex-biased difference in prevalence (Table 1) though females (56%) seemed to have higher prevalence than males (45%) an observation that concurs with that of Govender *et al.,* [13]. Factors predisposing females to piroplasm are likely to be comparable with those affecting sub-adult rhinoceros.

The occurrence of *T. bicornis* in all the sampled rhinoceros sub-populations could have been facilitated by the regular translocations of individuals. Translocation assists spread of both tick vector and haemoparasites among habitat patches/populations. We observed apparent variations in prevalence among the sub-populations (Figure 2) but there was no significant difference (Table 1). This means that local factors in these habitats, such as ecological and weather differences, mammalian diversity, sanctuary size, were not sufficient to cause significant disparity in prevalence. According to Lopez et al., [32], frequent introduction of parasites in to a patch/habitat via host migration contributes to local patch prevalence. This implies that rhino sanctuaries' that frequently receive new individuals are likely to harbor higher parasite prevalence. Okita-Ouma et al., [33] points out that LNNP is more of a source population that has supported 41 outward translocations and received one inward translocation, whereas Ngulia RS is a recipient population having received 16 inward translocations and only one outward translocation. However, lack of significant association between location and prevalence (Table 1) does not concur with the postulation of Lopez et al., [32].

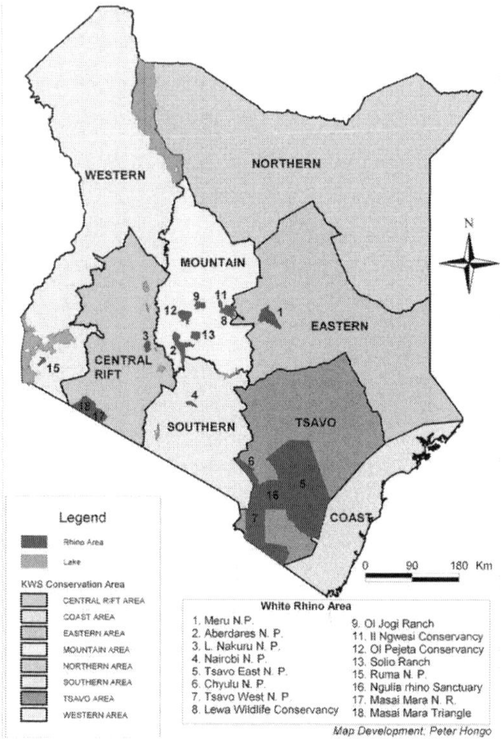

Figure 2. Locations of black rhinoceros conservation areas in Kenya.

CONCLUSION

In the analyzed samples we did not detect the presence of *Babesia bicornis*, while *Theileria bicornis* was found to have 49.12% prevalence with white rhinoceros showing a higher prevalence than black rhinoceros. Other factors such as age, sex, location, and population mix were not found to play a significant role.

REFERENCES

1. Kelly JD, Blyde DJ, Denney IS. The importation of the black rhinoceros (*Diceros bicornis*) from Zimbabwe into Australia. Aus Vet J. 1995;72:369–74.
2. Altizer S, Harvell D, Friedle E. Rapid evolutionary dynamics and disease threats to biodiversity. Trends Ecol Evol. 2003;589:596.
3. Daszak P, Cunningham AA, Hyatt AD. Emerging Threats to Infectious Diseases Wildlife- Health Biodiversity and infectious. Adv Sci. 2000;287:443–9.
4. Wilcove DS, Rothstein D, Dubow J, Phillips A, Losos E. Quantifying Threats to Imperiled Species in the United States. Bioscience. 1998;48:607–15.
5. MucCullogh B, Achard PL. Mortalities associated with capture, translocation, trade and exhibition of black rhinoceroses. Int Zoo. 1960;9:184–95.
6. Brocklesby DW. A *Babesia* species of the black rhinoceros. Vet Rec. 1967;80:484.
7. Mugera GM, Wandera JG. Degenerative polymyopathies in East African domestic and wild animals. Vet Rec. 1967;80:410–3.
8. Nijhof AM, Penzhorn BL, Lynen G, Mollel JO, Morkel P, Bekker CPJ, et al. *Babesia bicornis* sp. nov. and *Theileria bicornis* sp. nov.: tick-borne parasites associated with mortality in the black rhinoceros (*Diceros bicornis*). J Clin Microbiol. 2003;41:2249–54.
9. Otiende MY, Kivata MW, Jowers MJ, Makumi JN, Runo S, Obanda V, et al. Three Novel Haplotypes of *Theileria bicornis* in Black and White Rhinoceros in Kenya. Transbound Emerg Dis. 2014; [Epub ahead of print].
10. Martin LB. Stress and immunity in wild vertebrates: Timing is everything. Gen Com Endocr. 2009;163:70–6.
11. Maamun JM, Suleman MA, Akinyi M, Ozwara H, Kariuki T, Carlsson H-E. Prevalence of Babesia microti in free-ranging baboons and African green monkeys. J Parasitol. 2011;97:63–7.
12. Muhanguzi D, Matovu E, Waiswa C. Prevalence and characterization of *Theileria* and *Babesia* species in cattle under different husbandry systems in western Uganda. Int J Anim Veter Adv. 2010;2:51–8.

13. Govender D, Oosthuisen MC, Penzhorn BL. Piroplasm parasites of white rhinoceroses (*Ceratotherium simum*) in the Kruger National Park, and their relation to anaemia. J S Afr Vet Assoc. 2011;82:36–40.
14. Pfitzer S, Oosthuizen MC, Bosman AM, Vorster I, Penzhorn BL. Tick-borne blood parasites in nyala (*Tragelaphus angasii*, Gray 1849) from KwaZulu-Natal, South Africa. Vet Parasitol. 2011;176:126–31.
15. Arthur DR. Ticks. A Monograph of the Ixodoidea. Part V. The Genera Dermacentor, Anocentor, Cosmiomma, Boophilus and Margaropus. London: Cambridge University Press; 1960.
16. Keirans JE. *Dermacentor rhinocerinus* (Denny 1843) (Acari: lxodida: Ixodidae): redescription of the male, female and nymph and first description of the larva. J Vet Res. 1993;60:59–68.
17. Katzer F, McKellar S, Kirvar E, Shiels B. Phylogenetic analysis of *Theileria* and *Babesia equi* in relation to the establishment of parasite populations within novel host species and the development of diagnostic tests. Mol Biochem Parasit. 1998;95:33–44.
18. Dennig HK. The isolation of *Babesia* species from wild animals. P ICP Rome. 1965;24–26:262–3.
19. Nijhof AM, Pillay V, Steyl J, Prozesky L, Stoltsz WH, Lawrence A, et al. Molecular Characterization of *Theileria* Species Associated with Mortality in Four Species of African Antelopes Molecular Characterization of *Theileria* Species Associated with Mortality in Four Species of African Antelopes. J Clin Microbiol. 2005;43:5907–11.
20. Obanda V, Kagira JM, Chege S, Okita-Ouma B, Gakuya F. Trypanosomosis and other co-infections in translocated black (*Diceros bicornis michaeli*) and white (*Ceratotherium simum simum*) rhinoceroses in Kenya. Sci Parasitol. 2011;12:103–7.
21. Wingfield JC, Maney DL, Breuner CW, Jacobs JD, Lynn S, Ramenofsky M, et al. Ecological bases of hormone—behavior interactions: the "emergency life history stage.". Am Zool. 1998;38:191–206.
22. Dhabhar FS. Stress-induced augmentation of immune function- The role of stress hormones, leukocyte trafficking, and cytokines. Brain Behav Immun. 2000;16:785–98.
23. Millott SM, Cox FEG. Interactions between *Trypanosoma brucei* and *Babesia* spp. and Plasmodium spp. in mice. Parasitology. 1985;90:241–54.

24. Penzhorn BL, Oosthuizen MC, Bosman A-M, Kilian JW, Horak IG. Black rhinoceros (*Diceros bicornis*) populations in northwestern Namibia are apparently not infected with piroplasms. J Wildl Dis. 2008;44:1032–5.
25. Coleman PG, Perry BD, Woolhouse MEJ. Endemic stability—a veterinary idea applied to human public health. Lancet. 2001;357:1284.
26. Bigalke RD, Keep ME, Keep PJ, Schoeman JH. A large *Babesia* sp. and a *Theileria* like piroplasm of the square-lipped rhinoceros. J S Afr Vet Assoc. 1970;41:292–4.
27. Schmidt K, Ostfeld R. Biodiversity and the dilution effect in disease ecology. Ecology. 2001;82:609–19.
28. Brett R. Mortality factors and breeding performance of translocated black rhinos in Kenya: 1984–1995. Pachyderm. 1998;69:82.
29. Glaser R, Kiecolt-Glaser JK. Stress-induced immune dysfunction: implications for health. Nat Rev Immunol. 2005;5:243–51.
30. Muehlenbein MP. Intestinal parasite infections and fecal steroid levels in wild chimpanzees. Am J Phys Anthropol. 2006;130:546–50.
31. Klein SL. Hormonal and Immunological mechanisms mediating sex differences in parasite infection. Parasit Immunol. 2004;26:247–64.
32. Lopez J, Gallinot LP, Wade MJ. Spread of parasites in metapopulations: An experimental study of the effects of host migration rate and local host population size. Parasitology. 2005;130:323–32.
33. Okita-Ouma B, Amin R, Van Langevelde F, Leader-Williams N. Density dependence and population dynamics of black rhinos (*Diceros bicornis michaeli*) in Kenya's rhino sanctuaries. Afr J Ecol. 2010;48:791–9.

CHAPTER 18

Disparities in Beef Tapeworm Identification Rates in the Abattoirs of Gauteng Province, South Africa: A Descriptive Epidemiologic Study

Daniel Nenene Qekwana[1], James Wabwire Oguttu[2], Dries Venter[3], Agricola Odoi[1,4]

[1] Section Veterinary Public Health, Department of Paraclinical Sciences, Faculty of Veterinary Sciences, University of Pretoria, Pretoria, South Africa,

[2] Department of Agriculture and Animal Health, College of Agriculture and Environmental Sciences, University of South Africa, Johannesburg, South Africa,

Citation: Qekwana DN, Oguttu JW, Venter D, Odoi A (2016) Disparities in Beef Tapeworm Identification Rates in the Abattoirs of Gauteng Province, South Africa: A Descriptive Epidemiologic Study. PLoS ONE 11(3): e0151725. https://doi.org/10.1371/journal.pone.0151725

[3] Gauteng Department of Agriculture and Rural Development, Veterinary Services Directory, Pretoria, South Africa,

[4] Biomedical and Diagnostic Sciences, College of Veterinary Medicine, University of Tennessee, Knoxville, Tennessee, United States of America

ABSTRACT

Background

Bovine *Taenia saginata* cysticercus infections (also called bovine cysticercosis or beef measles) is usually diagnosed in cattle only during post-mortem meat inspection. The aim of this study was to investigate the identification rates of these infections in and to identify predictors/determinants of variations in the identification rates in abattoirs in Gauteng province, South Africa.

Methods

Retrospective data for over 1.4 million cattle carcasses inspected in 26 abattoirs between January 2010 and December 2013 were used for the study. The identification rates (proportion of bovine *Taenia saginata* cysticercus positive carcasses) were computed and generalized estimating equations used to identify predictors/determinants of identification rates.

Results

The overall identification rate was 0.70% (95% CI: 0.45, 0.95). Significantly ($p < 0.05$) lower rates were reported during summer (0.55%) than other seasons. Some geographic areas reported significantly ($p<0.05$) higher rates than others. The identification rates in high throughput abattoirs was significantly ($p<0.05$) higher (RR: 9.4; 95% CI: 4.7–19.1) than in low throughput abattoirs. Similarly, the identification rates among animals from feedlots were significantly ($p<0.05$) higher (RR: 1.6; 95% CI: 1.7–3.5) than those from non-feedlot sources. No significant ($p>0.05$) association was identified between identification rates and either the number of meat inspectors per abattoir or the provider of inspection services.

Conclusion

Although no significant association was found between identification rates and provider of inspection services, follow-up studies will need to be done to specifically investigate the potential conflict of interest arising from the fact that abattoir owners hire meat inspection services directly. Capture of abattoir surveillance data needs to include farm address and for each case to be reported separately. Finally, information on the type of identified cysts (alive or calcified) need to be collected to help better estimate risk to consumers. This study provides useful baseline data to guide future studies, surveillance and control efforts.

INTRODUCTION

Cattle are the intermediate hosts for the larval stages of *T. saginata*, while humans act as definitive hosts. Thus, for the life cycle of *T. saginata* to be complete, there must be a link between humans and animals [1]. Therefore, humans acquire *T. saginata* infestation when they consume raw or undercooked beef having *T. saginata* cysts. Grazing contaminated pastures and drinking water contaminated with the intermediate stages of the parasite passed out by humans are risk factors for *Taenia* infestation in cattle. [1, 2].

Following ingestion of *T. saginata* eggs by cattle, the larvae develop into cysticerci primarily in the skeletal and cardiac muscles. These cysts are fully developed in 4–5 months and measure approximately 0.5–1.0 cm. Cysts may vary in appearance depending on the degree of inflammation, necrosis, and mineralization of the resulting lesion [3]. *Taenia saginata* cysts are potentially infective to humans by 10 weeks [1].

Infested cattle do not usually show symptoms and therefore detection of infection usually only occurs during meat inspection [4–6]. Although meat inspection has 100% specificity [7], it has low sensitivity (11.5–15.6%) [8, 9], which increases the risk of potential exposure/infection to consumers due to false negative meat inspection results.

In South Africa, all bovine carcasses have to be individually inspected for the presences of bovine *T. saginata* cysts by visual examination and incision of the masseter, heart, diaphragmatic and triceps muscles. Palpation of the tongue and visual examination of the offal is also prescribed. If a carcass presents a generalised infestation, the carcass and offal have to be declared unfit for human consumption and condemned. However, if the infestation is localised, the carcass could be partially passed on condition that it is treated by freezing at temperatures not exceeding −10°C for more than 14 days [10, 11]. The implication of this, is serious financial losses that occurs due to either carcass condemnation or costs of cold treatment and downgrading of affected carcasses [2, 12, 13].

According to Section 11 of the South African Meat Safety Act 40 of 2000 [14], abattoir owners are required to procure meat inspection services for their abattoirs. Meat inspection services can be provided by either companies or private individuals. Moreover, the Act stipulates that the inspection services have to be independent of the abattoir management. However, because abattoir owners pay the meat inspection service providers directly, a conflict of interest is likely to arise especially where private individuals are involved, which could adversely affect the identification rates of the condition in different abattoirs. This is a problem that needs to be addressed to enhance the meat inspection services in the country. In November of 2012, a proposal by a meat inspection working group was tabled to address the issue [10]. However, the question that arises is whether indeed the provider of the meat inspection services does influence the rate of identification of bovine *T. saginata* cysticercus positive carcasses in abattoirs.

The burden of bovine cysticercosis globally differs by region with high prevalence proporations reported in developing countries. Studies done in Europe have demonstrated a prevalence of 3.09% of bovine cysticercus in Belgium [15] and 1.23% in France [16]. In Africa, a prevalence of 19.7% has been reported in Ethiopia [17] and 0.2% in the North West province of South Africa [18]. However, there is no evidence of any studies that have been done to assess the prevalence of the condition among animals slaughtered in abattoirs in Gauteng province, South Africa.

In light of this, the objectives of this study were to estimate identification rates of bovine *T. saginata* cysticercus positive beef cattle slaughtered in abattoirs located in Gauteng Province of South Africa between 2010 and 2013 and to identify predictors/determinants of the variations in identification rates.

MATERIALS AND METHODS

Ethics approval

This study was approved by the Animal Ethics Committee of the University of Pretoria.

Study area

This study was conducted in Gauteng Province which is South Africa's most populous province with an estimated population of 12.3 million people. The province is approximately 18,178 km^2 and consists of three metropolitan municipalities (City of Johannesburg, City of Tshwane and Ekurhuleni) and two district municipalities (Sedibeng and West Rand). The two district municipalities are further subdivided into seven local municipalities for administrative purposes [19]. The province as a whole has a subtropical climate, but Johannesburg tends to be cooler than Pretoria. Gauteng Province is located in the Highveld region with an annual summer rainfall of approximately 700 mm, with December and January being the wettest months of the year. The province experiences average annual maximum temperatures of about 22°C in the south and about 25°C in the north.

Data source

Retrospective data used in the study were from 26 abattoirs located in Gauteng Province of South Africa and covered the time period January 2010—December 2013. Approval to use the data was obtained from the Gauteng Department of Agriculture and Rural Department (GDARD) that also provided the data to the investigators. The Meat Safety Act 40 of the year 2000 requires that all animals slaughtered in South African abattoirs to be inspected and for details of the animals and findings of meat inspection to be recorded [14]. The Act defines a bovine *T. saginata* cysticercus positive carcass as one whose head, active muscles and red offal is found to have one or more parasitic intermediate stages of the parasite that is either alive or calcified. The information collected during meat inspection is sent to the Gauteng Department of Agriculture and Rural Development (GDARD) where it is entered into an electronic database as part of disease monitoring and surveillance.

The data used in the present study included monthly abattoir reports of the total number of cattle slaughtered, number of carcasses inspected, as well as the number of bovine *T. saginata* cysticercus positive carcasses identified. The study included data from both high and low throughput abattoirs located in Gauteng and registered with GDARD. A low throughput abattoir is defined as one that handles a maximum of 20 units per day [14]. However, if only one species of animals is slaughtered per day, then the maximum accepted limits are: (a) no more than 20 units if handling cattle or horses or sausage pigs larger than 90kgs; (b) no more than 40 units if handling sheep or goats, or (c) no more than 30 units if handling pigs. High throughput abattoirs, on the other hand, are classified at the discretion of the provincial executive officer based on the capacity of the lairages, hourly throughput potential relative to available equipment and facilities, such as hanging space and chiller capacity, as well as handling of rough offal [10].

Data collection and management

The data were collated into monthly numbers of carcasses inspected at each abattoir, and bovine *T. saginata* cysticercus positive carcasses identified during meat inspection. The data were evaluated for missing values and any inconsistencies such as implausible values. The following variables were included in the final dataset: number of animals slaughtered and inspected, number of cysticercus positive carcasses, municipality where the abattoir was located, month of the year when animals were slaughtered and sources of the animals (feedlot or non-feedlot). For purposes of this study, a feedlot was defined as an intensive animal farming system where beef cattle are fattened prior to slaughter. Non-feedlot sources, on the other hand, included all sources of animals that did not fit the above definition.

Data analysis

Descriptive analysis was performed to determine the proportions of bovine *T. saginata* cysticercus positive carcasses (presented as identification rates) by month, year, season and municipality. To assess whether the data were normally distributed or not, Shapiro-Wilks test was used. Exact Wilcoxon test was used to compare the differences in identifications rates when the data were not normally distributed otherwise Chi-square analysis was used. Statistical significance was assessed at $p \leq 0.05$.

To investigate associations between identification rates and the suspected predictors/determinants, a Generalized Estimating Equation (GEE) model was fit to the data. The dependent variable was specified as the number of bovine *T. saginata* cysticercus positive carcasses identified per abattoir per month, and the natural log of the number of carcasses inspected per abattoir per month was used as the offset. The error distribution and link function were specified as Poisson and log, respectively. Using this modelling framework, simple/two-way associations were first investigated to identify variables that had significant simple associations with the outcome. The following variables were assessed for the simple associations with the dependent variable: class of the abattoir (high vs low throughput), source of the animals (feedlot vs non-feedlot), meat inspection service provider (private individuals vs company), and municipality. Variables that had significant ($p<0.05$) simple associations with the outcome were offered for assessment in the multivariable GEE model. A backward elimination procedure with a critical p-value of 0.05 was used to identify significant predictors/determinants. All possible two-way interactions of variables in the final main effects model were assessed for significance. Since GEE is not a likelihood-based method, Quasi-likelihood under the Independence Model Criterion (QIC), which is analogous to the Akaike Information Criterion (AIC), was used to assess model fit and identify the best correlation structure for the data.

RESULTS

Descriptive analysis results

A total of 1,415,005 animals were slaughtered and the same number of carcasses inspected during the study period (January 2010-December 2013). Most (1,373,229) of the carcasses inspected at the abattoirs were of animals from non-feedlot sources (Fig 1). A total of 9,920 out of .... (0.70%, 95% confidence interval [95% CI]: 0.45, 0.95) cattle slaughtered during the study period were positive for bovine cysticercosis.

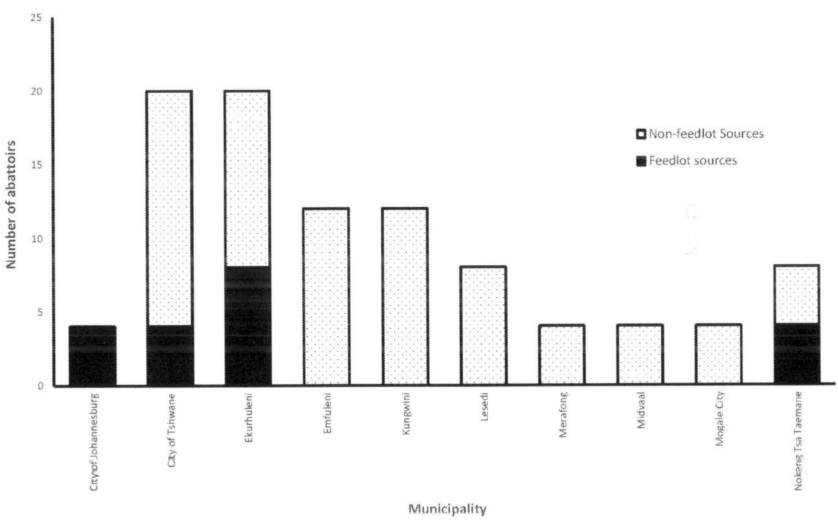

Figure 1. Distribution of sources of slaughtered animals by municipalities in Gauteng Province (South Africa), 2010–2013.

https://doi.org/10.1371/journal.pone.0151725.g001

Simple/Univariable Associations

There were significant (p<0.05) differences in bovine cysticercosis identification rates across years, with the highest identification rates reported in 2013 and the lowest in 2012 (Table 1). There were monthly fluctuations in identification rates over the study period with summer months (November to March) having relatively low identification rates compared to the other months (Fig 2). Similar patterns were revealed by the results of seasonal analysis that also showed significantly (p<0.05) lower bovine cysticercosis identification rates during the summer months (0.55%) than other seasons. This coincided with the time when relatively more animals were slaughtered (Fig 3). Although there seemed to be evidence of an increasing temporal trend ($R^2 = 0.46$) in identification rates of bovine cysticercus positive carcasses over the study period, the observed increase was not statistically significant (p = 0.325).

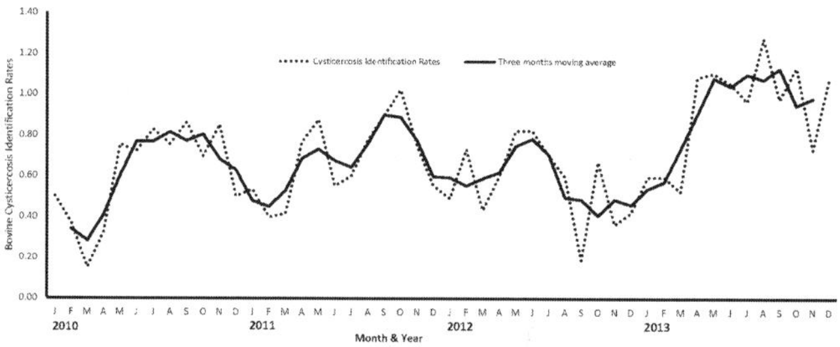

Figure 2. Temporal trends in monthly bovine *T. saginata* cysticercus identification rates in abattoirs across Gauteng Province (South Africa) between 2010 and 2013.

https://doi.org/10.1371/journal.pone.0151725.g002

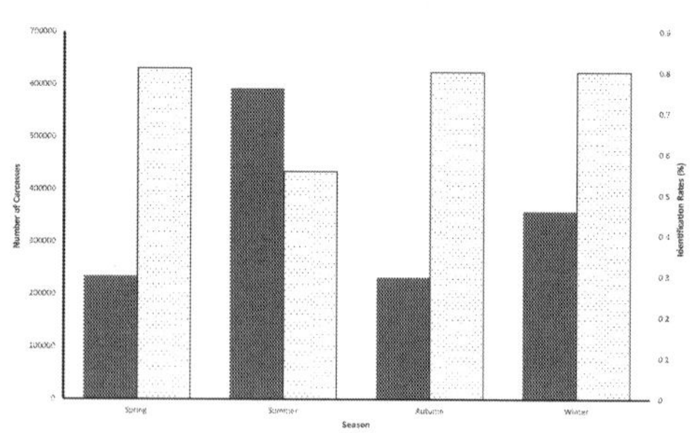

Figure 3. Seasonal patterns in number of carcasses inspected and bovine *T. saginata* cysticercus identification rates in Gauteng Province (South Africa), 2010–2015.

https://doi.org/10.1371/journal.pone.0151725.g003

Table 1. The total number of cattle slaughtered in Gauteng Province (South Africa) between 2010 and 2013 and proportion of bovine *T. saginata* cysticercus positive carcasses identified, treated and condemned.

Year	Number of abattoirs	Number of carcasses inspected	Number of bovine cysticercosis positive carcasses	Percentage (95% CI*) of bovine cysticercosis positive carcasses	Number of treated carcasses	Percentage (95% CI*) of treated carcasses	Number of condemned carcasses	Percentage (95% CI*) of condemned carcasses
2010	25	356,006	2, 169	0.61 (0.58, 0.63)	2, 118	0.59 (0.57, 0.62)	51	2.34 (1.78, 3.1)
2011	25	349,458	2, 389	0.68 (0.66, 0.71)	2, 380	0.68 (0.65, 0.71)	9	0.38 (0.13, 0.62)
2012	25	348,309	1, 980	0.57 (0.54, 0.59)	1, 975	0.57 (0.54, 0.59)	5	0.25 (0.03, 0.47)
2013	26	361,232	3, 382	0.92 (0.90, 0.97)	3, 374	0.93 (0.90,0.97)	8	0.24 (0.07, 0.40)

* CI: Confidence interval

https://doi.org/10.1371/journal.pone.0151725.t001

The bovine cysticercosis identification rates were significantly (p<0.05) higher during the months of May (0.89%), October (0.88%), August (0.85%), and lower during February (0.53%) and March (0.37%) (Table 2).

Table 2. Monthly identification rates of bovine *T. saginata* cysticercus in Gauteng Province (South Africa), 2010–2013.

Month	Number of carcasses inspected	Number of Bovine cysticercosis Positive carcasses	Percentage (95% CI*) of Bovine cysticercosis Positive carcasses
January	100,581	535	0.53 (0.49, 0.58)
February	102,738	546	0.53 (0.49, 0.58)
March	117,741	440	0.37 (0.34, 0.41)
April	111,914	785	0.70 (0.65, 0.75)
May	119,218	1, 067	0.89 (0.90, 1.02)
June	116,059	902	0.78 (0.73, 0.83)
July	122,039	948	0.78 (0.73, 0.83)
August	120,025	1, 020	0.85 (0.80, 0.90)
September	112,709	825	0.73 (0.68, 0.78)
October	121,230	1, 071	0.88 (0.83, 0.94)
November	125,759	847	0.67 (0.63, 0.72)
December	144,992	934	0.64 (0.60, 0.69)

*CI: 95% Confidence interval

https://doi.org/10.1371/journal.pone.0151725.t002

A comparison of the identification rates across municipalities showed significantly (p<0.05) higher rates in Nokeng Tsa Taemane (1.23%) and Mogale City (1.04%) than the rest of the provinces, while the lowest rates were reported in the City of Tshwane (0.01%). No cases were observed in the Midvaal municipality (0%) (Table 3). Identification rates for *T. saginata* cysticercus infection were significantly (p<0.002) higher in the high throughput abattoirs (0.024%) compared to low throughput abattoirs (0.05%). There was no significant (p<0.4226) difference between the identification rates observed among animals sourced from feedlots (0.054%) as compared to those from non-feedlot sources (0.033%). Moreover, significantly (p = 0.0295) higher identification rates were observed in abattoirs where meat inspection was performed by independent service providers (0.05%) compared to those in which the owners (0.0%) were responsible for performing meat inspection (Table 4).

Table 3. Bovine *T. saginata* cysticercus identification rates by municipalities of Gauteng Province (South Africa), 2010–2013.

Metropolitan Municipality	District municipality	Local municipality	Number carcasses inspected	Number of Bovine cysticercosis Positive carcasses	Percentage (95% CI[†]) of Bovine cysticercosis Positive carcasses
City of Johannesburg			707	4	0.57 (0.15, 1.44)
City of Tshwane			31,455	4	0.01 (0.003, 0.03)
Ekurhuleni			54,002	65	0.12 (0.09, 0.15)
	Sedibeng	Emfuleni	110,014	552	0.50 (0.46, 0.54)
		Lesedi	256,868	1,726	0.67 (0.64, 0.70)
		Midvaal	715	0	0.00 (0.0, 0.51)
	West Rand	Kungwini	142,919	197	0.14 (0.12, 0.16)
		Merafong	71,753	275	0.38 (0.34, 0.43)
		Mogale City	238,808	2,495	1.04 (1.00, 1.09)
	*Metwseding	Nokeng Tsa Taemane	525,115	6,479	1.23 (1.20, 1.26)

*The Metsweding District together with Nokeng Tsa Taemane local municipality were closed and merged into the Tshwane Metropolitan Municipality in 2011.

[†]CI: Confidence interval.

doi.10.1371/journal.pone.0151725.t003

https://doi.org/10.1371/journal.pone.0151725.t003

Table 4. Characteristics of abattoirs and animal sources in Gauteng Province (South Africa) between 2010 and 2013.

Variable	Number positive for bovine cysticercosis	Number carcasses inspected	Number of bovine cysticercosis positive carcasses	Percentage (95% CI*) of Bovine cysticercosis positive carcasses
Class of the abattoir				
Low Throughput	34	68,217	34	0.05 (0.03, 0.07)
High Throughput	9,886	1,356,708	9,886	0.73 (0.71, 0.74)
Source of animals				
Non-feedlot	9,920	1,373,229	9,920	0.72 (0.71, 0.74)
Feedlot	5,704	497,412	5,704	1.15 (1.12, 1.18)
Meat inspection service provider				
Independent service provider	9,554	1,292,654	9,554	0.74 (0.72, 0.75)
Service provided by owner	366	122,351	366	0.30 (0.27, 0.33)

*CI: Confidence interval

doi:10.1371/journal.pone.0151725.t004

https://doi.org/10.1371/journal.pone.0151725.t004

Determinants of Bovine Cysticercosis Identification Rates Based on Multivariable GEE Model

Although the municipality in which the abattoir was located and the provider of the inspection services had significant simple/univariable associations with identification rates of bovine cysticercosis, the observed associations were not statistically significant in the multivariable GEE model. In addition, the number of meat inspectors in the abattoir were also not significantly associated with bovine cysticercosis identification rates.

Bovine cysticercosis identification rates of high throughput abattoirs were 9 times higher (RR: 9.4; 95% CI: 4.7–19.1) than those of low throughput abattoirs (Table 5). Similarly, the rate of identification of bovine cysticercosis was 1.6 times higher (RR: 1.6; 95% CI: 1.7–3.5) for carcasses of animals from feedlots compared to those from non-feedlot sources (Table 5).

Table 5. Results of the GEE Poisson model showing identified predictors/determinants of the identification rates of bovine *T. saginata* cysticercus in Gauteng Province (South Africa), 2010–2013.

Predictor	Relative risk	Standard error	P-value	95% Confidence interval
Class of the abattoir				
High Throughput	9.436	3.387	0.0001	4.670, 19.067
Low Throughput	Referent			
Source of Animals				
Feedlot	2.431	0.464	0.001	1.672, 3.533
Non Feedlot	Referent			

doi:10.1371/journal.pone.0151725.t005

https://doi.org/10.1371/journal.pone.0151725.t005

DISCUSSION

This study was designed to estimate cysticercosis identification rates in Gauteng Province of South Africa, and to identify predictors/determinants of variations in the identification rates. This is the first study in South Africa that investigated determinants of cysticercosis identification rates using a multivariable modelling approach. The overall cysticercosis identification rates was generally low (0.7%), although higher than the 0.2% reported in an earlier study done in the North West Province of South Africa in 2011 [18]. However, it was lower than the 1.6% reported in Matabeleland Province of Zimbabwe [13]. Due to the low sensitivity of meat inspection, as has been reported in other studies [8, 9, 20, 21], it is possible that results reported here are an underestimation of the true proportion of the cysticercosis positive carcasses. Therefore, additional studies are needed to further investigate this issue.

The significantly lower proportions of cases of bovine cysticercosis observed during the summer months (November to March) was in contrast to reports by Sungirai et al (2014) and Dzoma et al (2011) who did not observe seasonal differences in the occurrence of bovine cysticercosis cases in Zimbabwe and North West Province of South Africa respectively [13, 18]. The reason for the lower identification rates observed in the summer is unclear and will require further investigations.

The reason for the relatively high identification rates of bovine cysticercus positive carcasses in Nokeng Tsa Taemane and Mogale City is unclear. However, it could be attributed to differences in data capturing practices across abattoirs (some abattoirs might be doing a better job of recording than others) as well as differences in the rigor of meat inspection procedures and abattoir management (some abattoirs might have more experienced inspectors who do a better job of identifying the cysts). Similar reports were made by Terefe et al (2014) who observed that management practices within the abattoir affects the proportion of bovine cysticercosis positive carcasses identified and reported [22]. Moreover, other authors have reported that the accuracy of meat inspection data is dependent not only on the number of cysts present but also on the skill, rigor and number of meat inspectors employed in the abattoir [23].

The Meat Safety Act 40 of the year 2000 and associated regulations require abattoir owners to hire independent meat inspection services for their abattoirs and therefore the owners are responsible for paying for the services directly to the providers [14]. Although this arrangement seems to have a potential to lead to conflict of interest, this study found that the type of service provider was not a significant determinant of bovine cysticercosis identification rates. However, these results should be interpreted with caution, and more detailed primary base studies need to be done to specifically further investigate the role of inspection service providers in the identification and reporting of bovine cysticercosis.

Contrary to the findings of a study by Sungirai et al (2014), who reported a significantly higher prevalence of bovine cysticercosis in cattle from communal farmers compared to feedlots [13], this study found significantly higher identification rates in carcasses of cattle from feedlots as compared to those sourced from non-feedlot sources. The findings of the current study is consistent with those of other studies that have reported that animals raised on feedlots farms are more likely to be exposed to point source contamination such as sewage, feed and water [24, 25]. This has been attributed to large numbers of animals becoming infected at the same time. For example, in the USA, bovine *T. saginata* cysticercus outbreaks have been reported in feedlot cattle [26–28]. In Alberta, contaminated water with human sewage waste [25] and in South Wales, Australia, imported copra meal, which was used as a feed supplement have been implicated in outbreaks of *T. saginata* cysticercus infection among beef cattle [24].

The higher identification rates in high throughput abattoirs as compared to low throughput abattoirs contradicted the findings by Dzoma (2011), who reported a high proportion of cysticercosis in low throughput abattoirs [18]. This could be due to the fact that meat inspectors working in high throughput abattoirs are more likely to be experienced due to the large numbers of carcasses they handle resulting in the higher identification rates observed in the current study. Moreover, high throughput abattoirs have more resources to employ potentially more experienced meat inspectors.

This study used retrospective administrative data and, therefore, the investigators had no control over the quality of the data collection and hence the findings reported here should be interpreted with caution. This is because as reported by Dorny et al (2000), it is difficult to accurately estimate the proportion of cysticercosis positive carcasses based wholly on abattoir data [1]. Moreover, the data used for this study did not include some variables such as age and sex of the animal that have been reported to be associated with the occurrence of bovine cysticercosis [16, 29]. Also lacking in the records was indication of whether the cysts were alive or calcified making it difficult to directly infer risk to the consumer. Furthermore, the records did not include data on addresses of farms of origin of slaughtered animals, making it difficult to trace back the animals to the source. In view of this, there is a need to improve surveillance data collection to include these additional variables. Nonetheless, the findings of this study provide useful information on variations of cysticercosis identification rates and their predictors/determinants to guide future studies, surveillance and control efforts.

CONCLUSIONS

Bovine cysticercosis is frequently observed in bovine carcasses in abattoirs across the Gauteng Province of South Africa, albeit at low levels. The class of the abattoir and the source of the animals were significant predictors/ determinants of bovine cysticercosis identification rates. Interestingly, the number of meat inspectors and the type of service providers were not significant determinants. To better understand the differences in identification rates observed for the different variables, the authors recommend that capture of data on bovine *T. saginata* cysticercus infections in abattoirs need to be improved to include more variables such as the address of the farm of origin of the animal, and for each case to be reported separately to help with traceability. We further recommend that efforts to identify whether cysts are alive or calcified are needed to help better estimate risk to consumers. This study provides useful baseline data to guide future studies, surveillance data collection and control efforts.

ACKNOWLEDGMENTS

The authors would like to thank the Gauteng Department of Agriculture and Rural Development for providing access to the study data and the Language Services of the University of South Africa for copy editing the manuscript.

REFERENCES

1. Dorny P, Praet N. Taenia saginata in Europe. Special section: Food-borne Parasitic Zoonosis Papers presented at the Fourth Annual Meeting of the European Veterinary Parasitology College (EVPC). 2007;149(1–2):22–4.
2. Boone I, Thys E, Marcotty T, de Borchgrave J, Ducheyne E, Dorny P. Distribution and risk factors of bovine cysticercosis in Belgian dairy and mixed herds. Preventive veterinary medicine. 2007;82(1–2):1–11. pmid:17559956
3. Scandrett B, Parker S, Forbes L, Gajadhar A, Dekumyoy P, Waikagul J, et al. Distribution of Taenia saginata cysticerci in tissues of experimentally infected cattle. Veterinary parasitology. 2009;164(2–4):223–31. pmid:19524368
4. Karanikas ID, Sakellaridis TE, Alexiou CP, Siaperas PA, Fotopoulos AC, Antsaklis GI. Taenia saginata: a rare cause of bowel obstruction. Trans R Soc Trop Med Hyg. 2007;101(5):527–8. pmid:16973197
5. Collier L, Cox F, Topley WW. Topley and Wilson's Microbiology and Microbial Infections: Parasitology: London: Edward Arnold Publishers; 1998.
6. Patel NM, Tatar EL. Unusual colonoscopy finding: Taenia saginata proglottid. World J Gastroenterol. 2007;13(41):5540–1. pmid:17907306; PubMed Central PMCID: PMC4171297.
7. Eichenberger RM, Stephan R, Deplazes P. Increased sensitivity for the diagnosis of Taenia saginata cysticercus infection by additional heart examination compared to the EU-approved routine meat inspection. Food Control. 2011;22(6):989–92. pmid:WOS:000288418600027.
8. Eichenberger RM, Lewis F, Gabriel S, Dorny P, Torgerson PR, Deplazes P. Multi-test analysis and model-based estimation of the prevalence of Taenia saginata cysticercus infection in naturally infected dairy cows in the absence of a 'gold standard' reference test. Int J Parasitol. 2013;43(10):853–9. pmid:23831108.
9. Dupuy C, Hendrikx P, Hardstaff J, Lindberg A. Contribution of meat inspection to animal health surveillance in Bovine animals. In: EFSA, editor. European Food Safety Authority2012. p. 53.
10. Department of Agriculture F, Fisheries. Proposal for a meat inspection service in South Africa. 2012.

11. Geysen D, Kanobana K, Victor B, Rodriguez-Hidalgo R, de Borchgrave J, Brandt J, et al. Validation of Meat Inspection Results for Taenia saginata Cysticercosis by PCR-Restriction Fragment Length Polymorphism. Journal of food protection. 2007;70(1):236–40. pmid:17265888
12. Cabaret J, Geerts S, Madeline M, Ballandonne C, Barbier D. The use of urban sewage sludge on pastures: the cysticercosis threat. Veterinary research. 2002;33(5):575–97. pmid:12387491
13. Sungirai M, Masaka L, Mbiba C. The prevalence of Taenia saginata cysticercosis in the Matabeleland Provinces of Zimbabwe. Tropical animal health and production. 2014;46(4):623–7. pmid:24429809
14. Department of Agriculture Forestry and Fisheries. Act No. 40, 2000: Meat Safety Act, 2000. Pretoria, South Africa: Department of Agriculture, Forestry and Fisheries; 2000.
15. Dorny P, Vercammen F, Brandt J, Vansteenkiste W, Berkvens D, Geerts S. Sero-epidemiological study of Taenia saginata cysticercosis in Belgian cattle. Veterinary parasitology. 2000;88(1–2):43–9. pmid:10681021
16. Dupuy C, Morlot C, Gilot-Fromont E, Mas M, Grandmontagne C, Gilli-Dunoyer P, et al. Prevalence of Taenia saginata cysticercosis in French cattle in 2010. Vet Parasitol. 2014;203(1–2):65–72. pmid:24655724.
17. Terefe Y, Redwan F, Zewdu E. Bovine cysticercosis and its food safety implications in Harari People's National Regional State, eastern Ethiopia. Onderstepoort Journal of Veterinary Research. 2014;81(1):676–.
18. Dzoma BM, Setlhodi EK, Molefe MM, Motsei LE, Bakunzi FR, Ndou RV, et al. Prevalence of Bovine Cysticercosis in the North West Province of South Africa from 2000 to 2010. Journal of Human Ecology. 2011;36(1):9–12.
19. Dyson LL. Heavy daily-rainfall characteristics over the Gauteng Province. Water SA. 2009;35(5):627–38.
20. Allepuz A, Napp S, Picado A, Alba A, Panades J, Domingo M, et al. Descriptive and spatial epidemiology of bovine cysticercosis in North-Eastern Spain (Catalonia). Vet Parasitol. 2009;159(1):43–8. pmid:19027236.
21. Dorny P, Vercammen F, Brandt J, Vansteenkiste W, Berkvens D, Geerts S. Sero-epidemiological study of Taenia saginata cysticercosis in Belgian cattle. Vet Parasitol. 2000;88(1–2):43–9. pmid:10681021.

22. Terefe Y, Redwan F, Zewdu E. Bovine cysticercosis and its food safety implications in Harari People's National Regional State, eastern Ethiopia. Onderstepoort J Vet Res. 2014;81(1). pmid:24833237.
23. Garedaghi Y, Saber APR, Khosroshahi MS. Prevalence of bovine cysticercosis of slaughtered cattle in Meshkinshahr Abattoir. American Journal of Animal and Veterinary Sciences. 2011;6(3):121–4.
24. Jenkins DJ, Brown GK, Traub RJ. 'Cysticercosis storm' in feedlot cattle in north-west New South Wales. Aust Vet J. 2013;91(3):89–93. pmid:23438459.
25. Lees W, Nightingale J, Brown D, Scandrett B, Gajadhar A. Outbreak of Cysticercus bovis (Taenia saginata) in feedlot cattle in Alberta. Can Vet J. 2002;43(3):227–8. pmid:11901599; PubMed Central PMCID: PMC339212.
26. Ebel E, Hancock D, Combs B, Yoder D, editors. Investigation of an outbreak of bovine cysticercosis in Idaho. Proceedings of the 8th International Congress on Animal Hygiene; 1994 12–16 September 1994; St.Paul, Minnesota, USA.
27. Slonka GF, Matulich W, Morphet E, Miller CW, Bayer EV. An outbreak of bovine cysticercosis in California. Am J Trop Med Hyg. 1978;27(1 Pt 1):101–5. pmid:626265.
28. Yoder DR, Ebel ED, Hancock DD, Combs BA. Epidemiologic findings from an outbreak of cysticercosis in feedlot cattle. J Am Vet Med Assoc. 1994;205(1):45–50. pmid:7928548.
29. Dupuy C, Morlot C, Demont P, Callait M-P, Ducrot C, Calavas D, et al. Spatial analysis of bovine cysticercosis in France in 2010. Food Control. 2014;47:348–52.

CHAPTER 19

Host Species Heterogeneity in The Epidemiology of Nesopora Caninum

Karla I. Moreno-Torres[1], Laura W. Pomeroy[1], Mark Moritz[2], William Saville[1], Barbara Wolfe[3], Rebecca Garabed[1,4]

[1] The Ohio State University, Department of Veterinary Preventive Medicine, Columbus, Ohio, United States of America,

[2] The Ohio State University, Department of Anthropology, Columbus, Ohio, United States of America,

[3] Morris Animal Foundation, Denver, Colorado, United States of America,

[4] The Ohio State University, Public Health Preparedness for Infectious Diseases Program, Columbus, Ohio, United States of America

ABSTRACT

Pathogen transmission across species drives disease emergence; however, mechanisms by which multi-host pathogens cross species boundaries are not well identified. This knowledge gap prevents integrated and targeted control

Citation: Moreno-Torres KI, Pomeroy LW, Moritz M, Saville W, Wolfe B, Garabed R (2017) Host species heterogeneity in the epidemiology of Nesopora caninum. PLoS ONE 12(8): e0183900. https://doi.org/10.1371/journal.pone.0183900

in an epidemiologically continuous ecosystem. Our goal is to describe the impact of host species heterogeneity on the epidemiology of *Neospora caninum* circulating between livestock and wildlife in southeastern Ohio. We collected biological samples from Père David's deer (*Elaphurus davidianus*) located at an outdoor wildlife conservation center; from cattle raised at farms adjacent to the center; and from wild white-tailed deer that roamed across farm and center boundaries. We designed nested infectious disease models of competing hypotheses about transmission and used collected data to fit the models, thereby estimating important immunological and transmission quantities which describe the species-specific contribution to the persistence of this pathogen in the community. We applied these data and models to suggest appropriate species-specific disease control methods. Results show that immunity in cattle and Pére David's deer wanes over time, while in white-tailed deer immunity appears to be lifelong. Transmission quantities for cattle were estimated at values below the threshold for an outbreak ($R_t < 1$), meaning that chains of transmission are not maintained within this population and infections must occur due to reintroduction from an outside source. Pére David's deer and white-tailed deer both could maintain continuous chains of transmission within their group ($R_t > 1$). Therefore, we propose that control of contact with outside sources will be useful for disease control in cattle; boosting immunity with vaccines might be an avenue to prevent infection in cattle and Père David's deer. White-tailed deer are a potential maintenance host for infection and require further study to determine optimal control methods. Community-level investigations like this allow us to better evaluate heterogeneities in transmission processes that ultimately guide targeted control.

INTRODUCTION

Most pathogens that threaten human and livestock populations persist by infecting multiple species [1]. These are called multi-host pathogens. The enormous impact of multi-host pathogens has been quantified: >61% of human pathogens, >77% of livestock pathogens, and >90% of carnivore pathogens participate in cross-species transmission events [1, 2]. Nevertheless, most research on multi-host pathogens investigates only one part of the system by considering either unidirectional transmission or single-host pathogens [1, 3]. This gap in research may result, in part, from the difficulty in identifying the role of each host species in multi-species disease transmission dynamics. Each host species can be identified as a maintenance population, as a part of a multi-species reservoir population, or as both. Maintenance populations are defined as closed populations that can maintain chains of transmission because they exceed the critical community size [4, 5]; reservoirs are epidemiologically connected populations or environments–including both maintenance and non-maintenance populations–in which pathogens can be maintained and transmitted to populations of interest [4]. Interventions to control and prevent disease, logically, will differ depending on how the reservoir is constituted and which species act as maintenance populations. Thus, understanding host species heterogeneity in the transmission dynamics of multi-host pathogens is essential for targeted control [6–8].

In natural settings, quantifying host species heterogeneity is often confounded by multiple sources of interspecies variation, including immunity, management, behaviors, and age structure of herds or packs. One solution for determining host species heterogeneity–even with species-specific confounders–is to estimate key transmission parameters using infectious disease models statistically fit to serology data collected from multiple species. A key parameter is the *force of infection* (FoI), defined as the per-capita rate at which susceptible individuals acquire infection, which can quantify heterogeneities in transmission [9]. In addition, the FoI can be used to calculate the basic reproductive number (R_0), which is defined as the average number of secondary cases arising from an average primary case in an entirely susceptible population. R_0 is a critical metric that can inform the maximum transmission potential for an infectious disease in a population. Thus by estimating R_0 for each population that constitutes the community, the species-specific transmission of a pathogen in that community can be compared, and targets for disease control can be established [9, 10].

We use the case study of *Neospora caninum*, a protozoan parasite capable of infecting a wide range of canid and ungulate species [11–13] to quantify host species heterogeneity of this understudied, but financially important, multi-host parasite [14–16]. The disease is important due to the substantial economic losses the parasite causes to the national and international cattle industry, mainly derived from premature culling, reduced milk yields and abortions [17, 18]. *N. caninum* is estimated to cost $843 million dollars annually in US dairy farms alone [17].

The life cycle of *N. caninum* involves three known stages. Oocysts are shed to the environment in feces by the definitive host–i.e., dogs and wild canids–and consumed by the intermediate host, which are primarily ruminants such as cattle and deer. Tachyzoites and bradyzoites develop in the intermediate host, causing damage to infected tissues and resulting in abortion, maternal infertility, and other clinical signs [13]. Prevention of *N. caninum* infection in cattle is based on reducing direct and indirect interactions between the definitive host (canids) and the intermediate host (ruminants) [19]. Also, because transplacental transmission from cow to calf seems to be the primary source of maintaining the parasite within the herd as opposed to horizontal transmission via colostrum and/or interactions with canids, maintaining a closed herd [19] and vaccination [20] have been suggested as control strategies to limit infection.

In current practice, however, there is no vaccine nor effective treatment on the market for *N. caninum* in cattle [21]. To our knowledge, current control and prevention strategies are not based on actual studies of transmission across species. Furthermore, neither identification nor inclusion of other plausible hosts, such as rodents and birds, have been explored as part of a control strategy that would target an *N. caninum* reservoir. Disease control currently targeting cattle populations may be targeting part of the reservoir community, but might not be targeting maintenance populations. To move forward with disease control, identification of maintenance populations is crucial.

In our study area in southeastern Ohio, differences in estimated true prevalence for intermediate ruminant host species suggested differences in the epidemiology of *N. caninum* for these interacting populations [22]. To understand why those differences were found, an evaluation of the environmental phase of *N. caninum* shed in wild canid scats was pursued. Results suggested that in our study population, environmental transmission to ruminants is likely rare [23]. Consequently, our next logical step was to evaluate host species heterogeneity in the epidemiology of *N. caninum* in ungulates in the community.

Therefore, our goal is to better understand the role of host species heterogeneity in the epidemiology of *N. caninum* circulating in a community. We evaluated transmission dynamics and immunodynamics in three intermediate ruminant host species by: 1) estimating and comparing their FoI; and 2) estimating the duration of immunity. Also, we estimated the species-specific transmission by calculating the reproductive number of each of the three ruminant host species. Results of this study can identify maintenance populations and inform control strategies at the community level.

MATERIALS AND METHODS

Study area and Population

The ecological system is located in southeastern Ohio and contains our focal species: cattle (*Bos Taurus*), Père David's deer (*Elaphurus davidianus*) and white-tailed deer (*Odocoileus virginianus*). This region is rural and includes the largest wildlife conservation center in North America, the International Center for the Preservation of Wild Animals (DBA, *the Wilds*). Several rare and endangered ruminant species are bred at the *Wilds* as part of their conservation program. Beef cattle farms predominate in the area surrounding the conservation center, but dairy cattle and small ruminant farms also are found in the community. In addition, farms keep domestic dogs primarily as companions or livestock guards. White-tailed deer in this community are free-ranging and capable of crossing fences and comingling with both domestic and non-domestic captive herds. Free-ranging wild carnivores such as coyote and fox in the area are considered pests, tourist attractions, and a recreation source for the local community. This ecosystem allows us to explore mechanisms of the persistence of multi-host pathogens, such as *N. caninum*, across the wildlife-livestock interface in a community of adjacent and intermingling populations.

Sample collection

In a cross-sectional study design, we collected jugular vein blood samples (10 ml per individual) during March 2013 for 37 (23 females, 13 males, 1 unrecorded) Père David's deer managed at *the Wilds*. Birth date records were accessed from *the Wilds* database; thus, individuals' age was calculated from the birth date to the bleeding date. For 137 female cattle, tail vein blood samples (10 ml per individual) were collected at the Muskingum Livestock Auction, from June 2014 to June 2015 (see below for cattle selection methods). In addition, we recorded breed, reproductive status, fetus gestational age and cow age. Reproductive status and gestational-fetal age [24] were estimated via rectal palpation by the auction veterinarian, and cattle age was determined through dentition [25] by an experienced worker at the auction. The same veterinarian and worker evaluated all cattle during the entire study. Forty-two free-ranging white-tailed deer (37 females and 5 males), with approximately 4 (depending on gender and the number of fetuses) samples collected from each animal (blood, brain, placenta and fetal tissues) were obtained from the study area by volunteer hunters during

Ohio's hunting season (October–February) in 2012, 2013 and 2014. We collected 10ml of blood per deer directly from the heart soon after death, to avoid degradation of antibodies. We estimated individuals' ages through dentition [26], recorded reproductive status based on visual inspection of the uterus, ovaries, and estimated gestational-fetal age through crown-rump length measurements [27].

Cattle selection at the auction was conducted systematically, stratified based on township, and weighted based on approximate numbers of cattle farms in each township using estimates provided by the county extension agents in the area (Muskingum, Morgan, Guernsey and Noble counties, OH; Table B in S1 Text). We continued to sample cattle arriving at the auction from the targeted townships until the sample size goal for that township was achieved. Additionally, we limited our sampling to a maximum of three cows selected per owner/seller per auction day. Targeted townships were located within 25 km radius from *the Wild*s. The area selected in this study (25 km radius) was based on the average home range of coyotes (17.5 km^2) and follows previous study design [23]. The auction was used for sampling because adequate management facilities do not exist on most farms in this area to effectively restrain cattle for blood collection. The cattle at auction are not a random sample of all cattle in the area–cattle at auction tend to be young healthy beef cattle and culled breeding/ dairy stock. Therefore, we limited our population to adult cows.

Laboratory Testing

Serum was separated by centrifugation and collected tissues were frozen. Serum and tissues were stored at -20°C. All species' serum was tested for *N. caninum* antibodies using a commercial cELISA kit (VMRD, Inc., Pullman, WA, USA) and white-tailed deer tissues were tested for *N. caninum* DNA by using species-specific primers: Np4-Np7 targeting the *pNC-5* gene of *N. caninum* through PCR [28, 29] at the Ohio Department of Agriculture Animal Disease Diagnostic Laboratory, Reynoldsburg, OH. The cELISA was performed according to manufacturer recommendations. Individual results were reported as percentage inhibition values; a value $\geq$ 30% inhibition was considered a positive result and <30% a negative result, as that cut-off is currently used by the manufacturer and has been validated for cattle [30] and for Père David's deer and white-tailed deer [22]. For each white-tailed deer, three tissue samples were tested (each 20 mg): 1) pooled brain (frontal lobe, cerebellum, cerebral cortex, and spinal cord), 2) pooled placenta, cotyledon and uterus, (multiple areas of each organ), and 3) pooled fetal tissue (brain, lung, heart, kidney, and liver).

Descriptive Data Analysis

Descriptive statistics were calculated regarding breed, reproductive status, and gestational-fetal age for cattle, and reproductive status and gestational-fetal age for white-tailed deer. Data on reproductive status was not available for Père David's deer. Age distribution of each species was also described. In addition, a logistic regression analysis was performed in R [31] to estimate associations between sero-positivity and the potential confounders described above.

Transmission Model Implementation, Selection, and Parameterization

For each species sampled, we considered eight models for *N. caninum* transmission: the catalytic model; the catalytic model with maternal antibodies; the catalytic model with age-specific FoI; the catalytic model with maternal antibodies and age-specific FoI; the reverse catalytic model; the reverse catalytic model with maternal antibodies; the reverse catalytic model with age-specific FoI; and the reverse catalytic model with maternal antibodies and age-specific FoI [32, 33]. For cattle, models with maternal antibodies were eliminated because sampled individuals were at least two years of age, and therefore beyond the lifespan of maternal antibodies.

The catalytic model categorizes animals into two states–seronegative or seropositive–assuming that individuals are born into the seronegative state, convert to the seropositive state at the rate designated by the FoI, and remain in the seropositive state due to the lifelong presence of a protective immunological response. Mathematically, let λ represent the constant FoI and let a represent the age of each individual in years. According to the catalytic model [32, 33], the age-specific seroprevalence is given by

$$P(\alpha) = 1 - e^{-\int_0^{\alpha} \lambda d\alpha}. \quad (1)$$

Let *t* represent the age at which maternal antibodies wane. Then, the age-specific seroprevelenace can be calculated by the catalytic model with maternal antibodies, such that

$$P(\alpha) = 1 - e^{-\int_t^{\alpha} \lambda d\alpha} \quad \textbf{(2)}$$

In the catalytic models with age-specific FoI, we followed Eqs (1) and (2) with the only difference being that instead of a constant FoI, we fit the data using a b-splined FoI with two equally spaced internal knots.

The reverse catalytic model, similarly, categorizes individuals as seronegative or seropositive, but relaxes the lifelong immunity assumption and allows us to evaluate waning immunity [33]. We assume that immunity lasts for an average duration of $1/\omega$ and then wanes [32]. Again, let λ represent the constant FoI and let represent the age of each individual in years. Then, the age-specific seroprevalence is given by

$$P(\alpha) = \frac{\lambda}{\lambda + \omega}\left(1 - e^{-\int_0^{\alpha} (\lambda+\omega) d\alpha}\right). \quad \textbf{(3)}$$

Let t represent the age at which maternal antibodies wane. Then, the age-specific seroprevalenace can be calculated by the reverse catalytic model with maternal antibodies, such that

$$P(\alpha) = \frac{\lambda}{\lambda + \omega}\left(1 - e^{-\int_t^{\alpha} (\lambda+\omega) d\alpha}\right). \quad \textbf{(4)}$$

In the reverse catalytic models with age-specific FoI, we followed Eqs (3) and (4) with the only difference being that instead of a constant FoI fit to the data, we used a b-splined FoI with two equally spaced internal knots.

We estimated the FoI and the rate of waning immunity by maximizing the binomial likelihood of seropositive animals, such that

$$y_a = Bin(N_a, p_a), \quad \textbf{(5)}$$

where N_a is the number of animals sampled at each age, p_a is the number of seropositive animals at each age, given in Table 1. We fit candidate models to serology data independently for each species following the method of Pomeroy *et al.* [33]. It is reasonable to assume independence because *N. caninum* is not transmitted directly (horizontally) between ruminant species. Likelihood estimations were performed using the "optim" function in R [34]. To select the model that best fit the data for each species, we used the Akaike's Information Criterion (AIC) [35]. Therefore, we can assess which immunological and transmission processes best described each species.

Table 1. Number of samples positive using apparent and true prevalence, by species and age.

Species	Age (years)	Apparent Prevalence (AP)	True Prevalence (TP)	Total sampled
Cattle	2	0	1	6
	3	1	1	7
	4	2	1	6
	5	1	1	16
	6	1	1	13
	7	1	1	14
	8	2	2	18
	9	1	1	10
	10	1	1	16
	11	1	2	23
	12	1	1	8
White-tailed deer	0.5	2	2	3
	1.5	3	2	4
	2.5	3	4	10
	3.5	6	7	14
	4.5	0	1	3
	5.5	3	3	5
	6.5	2	2	3
Père David's deer	1	2	1	2
	2	4	2	6
	3	1	1	4
	4	5	2	6
	5	3	1	4
	6	2	1	4
	7	0	0	1
	8	4	1	4
	9	1	0	1
	10	0	0	0
	11	0	0	0
	12	0	0	0
	13	0	0	0
	14	1	1	2
	15	2	1	2
	16	0	0	1

https://doi.org/10.1371/journal.pone.0183900.t001

https://doi.org/10.1371/journal.pone.0183900.t001

To estimate true prevalence, informative prior distributions from models a7, b4 and c4 were used from [22]. The prior prevalence for model a7 was modified for prevalence of cattle in Ohio [36], Beta (alfa = 7.039, beta = 69.479). The mode of the overall cattle population TP was 7.2% (95 PI% 3.9–12%), for Père David's deer population TP was 53% (95 PI% 39.4–66%) and for white-tailed deer population TP was 53.5% (95 PI% 36–78%). The overall cattle population AP was 8.8% (95 PI% 5.1–14.7%), for Père David's deer population AP was 67.6% (95 PI% 51.5–80.4%) and for white-tailed deer population AP was 45.2% (95 PI% 31.2–60.1%).

Positive test results can be used to fit models either as apparent prevalence (AP) or as true prevalence (TP). Commonly, AP is used in investigations, however biased conclusions on epidemiological parameters are plausible [37, 38]. Therefore, to compare the impact of fitting models with AP or TP on epidemiological parameters, first, we fit the AP, which is the proportion of animals that tested seropositive at each age. Second, we fit the TP, which is the true proportion of animals seropositive at each age when accounting for diagnostic error. Thus, each model was run for both AP and TP data. We calculated age-specific AP by using seropositive test results by age as the numerator and the total of animals sampled by age as the denominator. We calculated age-specific TP by using Bayesian inference methods with informative prior distributions explicitly described in Moreno-Torres et al. (2016) [22].

We calculated the reproductive number (R_t), the number of secondary cases produced by a primary infectious case, by multiplying the estimated FoI and the average life expectancy of each species [39]. Finally, we estimated the critical proportion that needs to be vaccinated to prevent an outbreak as a $Pc = 1-(1/Rt)$, assuming that the vaccine is completely protective [39]. As most vaccines are not completely protective, P*c* can be thought of as the proportion of the population that must be immunologically protected to prevent an outbreak, and the proportion to be vaccinated can be adjusted based on vaccine efficacy.

Ethics Statement

Animal use protocols were reviewed and approved by the Ohio State University Institutional Animal Care and Use Committee (Protocol Number: 2012A00000154). Both captive species (cattle and Père David's deer) were physically restrained in a chute and, once restrained, the procedure lasted about 10–15 minutes per animal. Individuals were sampled early in the morning to avoid heat stress and acutely stressed individuals were removed from the study (one cow). A white-tailed deer scientific collection permit was granted by the Division of Wildlife, Ohio Department of Natural Resources (Permit Number: 14–266). The collection of white-tailed deer samples was opportunistic during Ohio's hunting season; thus, the accessibility and availability of sample collection was restricted to the number of hunters that verbally consented to provide samples and their bag limits. *The Wilds* and the Muskingum Livestock Auction gave written consent to participate in this study. Both facilities provided specialized personnel to handle animals and the necessary equipment to physically restrain animals. Cattle owners/sellers verbally consent cattle sampling at the auction and provided the township location where the cattle resided prior to the auction.

RESULTS

Descriptively, the three sampled ruminant populations varied in age distribution, reproductive status, and gestational age (Figs 1–5). For the

137 cattle sampled, 10% were dairy breeds, 25% crossbreed, and 65% beef breeds (Table C in S1 Text). In logistic regression models, there was no association between seropositivity and the covariates breed, reproductive status and breed, or gestational age and breed for cattle or white-tailed deer. As reproductive data were not available for the Père David's deer, similar associations could not be tested in this species.

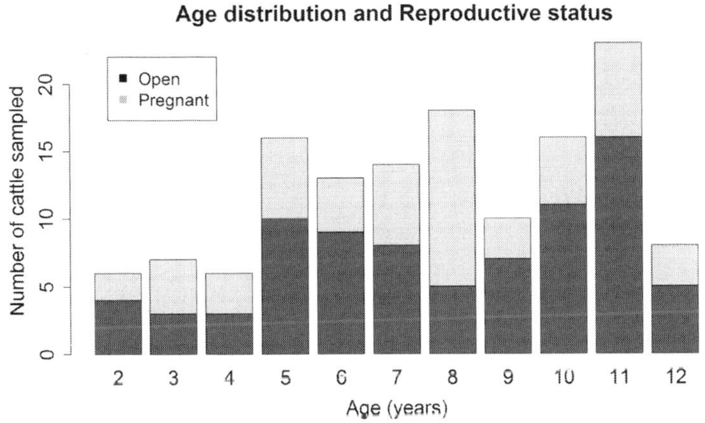

Figure 1. Age distribution and reproductive status of the 137 sampled female cattle. A total of 59% of cattle were open and 41% pregnant.

https://doi.org/10.1371/journal.pone.0183900.g001

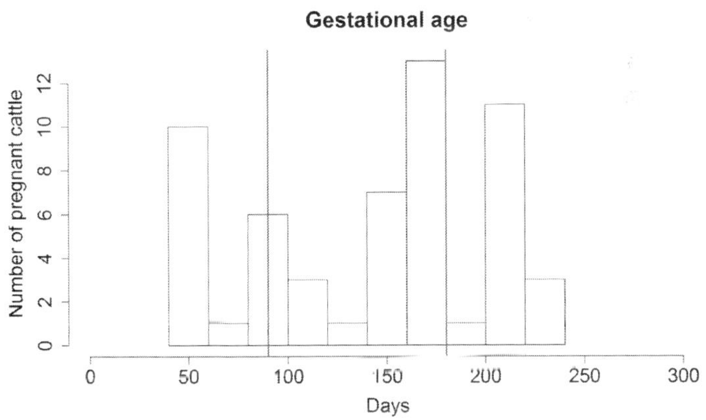

Figure 2. Gestational age of pregnant cattle. Pregnant cattle were distributed in all three trimesters of gestation, indicated by the vertical red lines.

https://doi.org/10.1371/journal.pone.0183900.g002

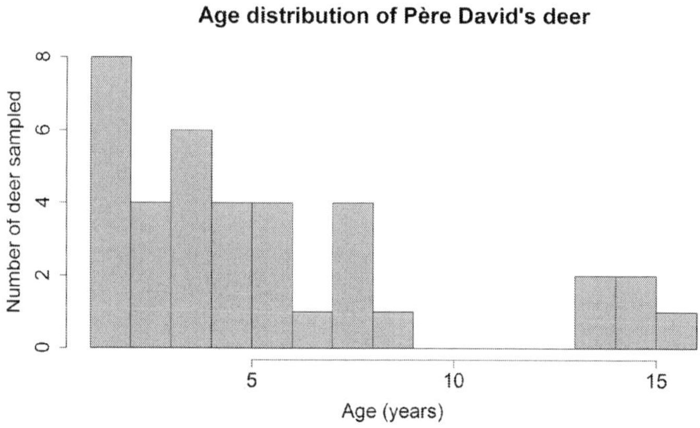

Figure 3. Age distribution of the 37 (23 females, 13 males, 1 unrecorded) sampled Père David's deer.

There were no individuals from 9 to13 year old. Reproductive status was not recorded but pregnant individuals were in their third trimester of gestation, by The *Wilds* records and based on seasonality of breeding in this species.

https://doi.org/10.1371/journal.pone.0183900.g003

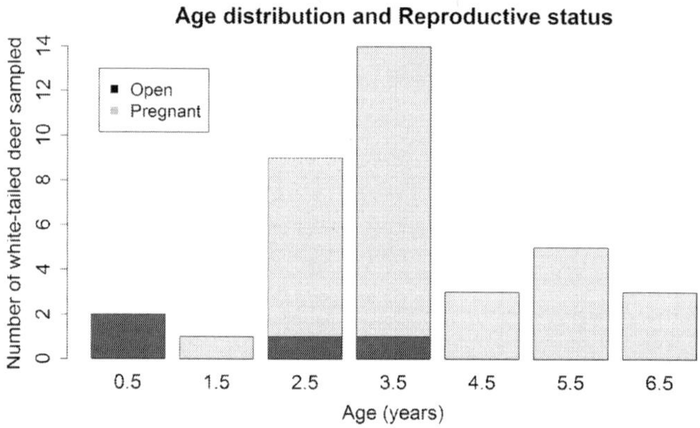

Figure 4. Age distribution and reproductive status of the 37 sampled female white-tailed deer.

There were 5 males sampled: 0.5 year old (n = 1), 1.5 year old (n = 3) and 2.5 year old (n = 1) (not displayed in this figure). A total of 11% of female white-tailed deer were open and 89% pregnant.

https://doi.org/10.1371/journal.pone.0183900.g004

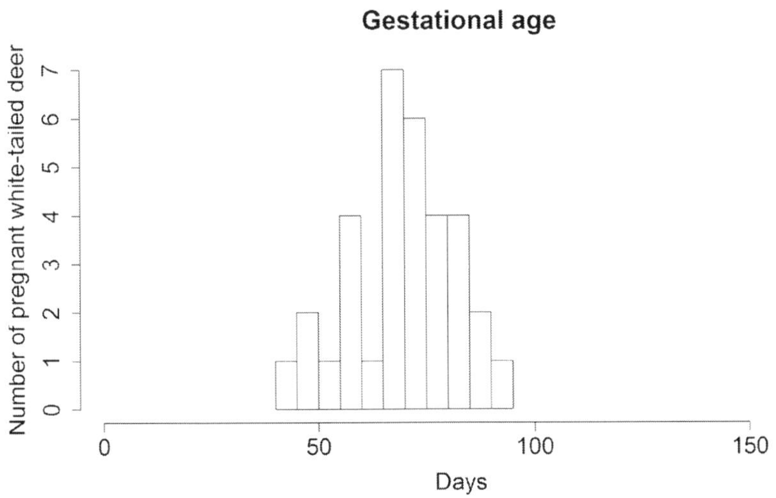

Figure 5. Gestational age of pregnant white-tailed deer. Pregnant white-tailed deer were all in the second trimesters of gestation.

https://doi.org/10.1371/journal.pone.0183900.g005

To determine the prevalence of *N. caninum* in white-tailed deer, both jugular vein blood samples and tissue samples were collected from free-ranging animals. These samples were analyzed using PCR and cELISA. PCR tests for the presence of DNA, giving information about active infection, latent or recently degraded DNA; however, PCR is more likely to give false negative results if a low load of parasites is present and if tissue selection is arbitrary rather than focused in histopathology results. The cELISA tests for the presence of antibodies–with inhibition above a predetermined threshold considered seropositive–indicating if a past exposure occurred but not necessarily providing information on infection. All white-tailed deer tissue samples were negative for *N. caninum* DNA by PCR test. These negative PCR test results contrast with the cELISA test results, of which 45% tested positive (overall AP) The proportion of the white-tailed deer population that tested seropositive by cELISA varied by age (Fig 6). These results suggest that the white-tailed deer sampled were infected with the parasite sometime in the past but were not experiencing active or latent infection at the time of sampling.

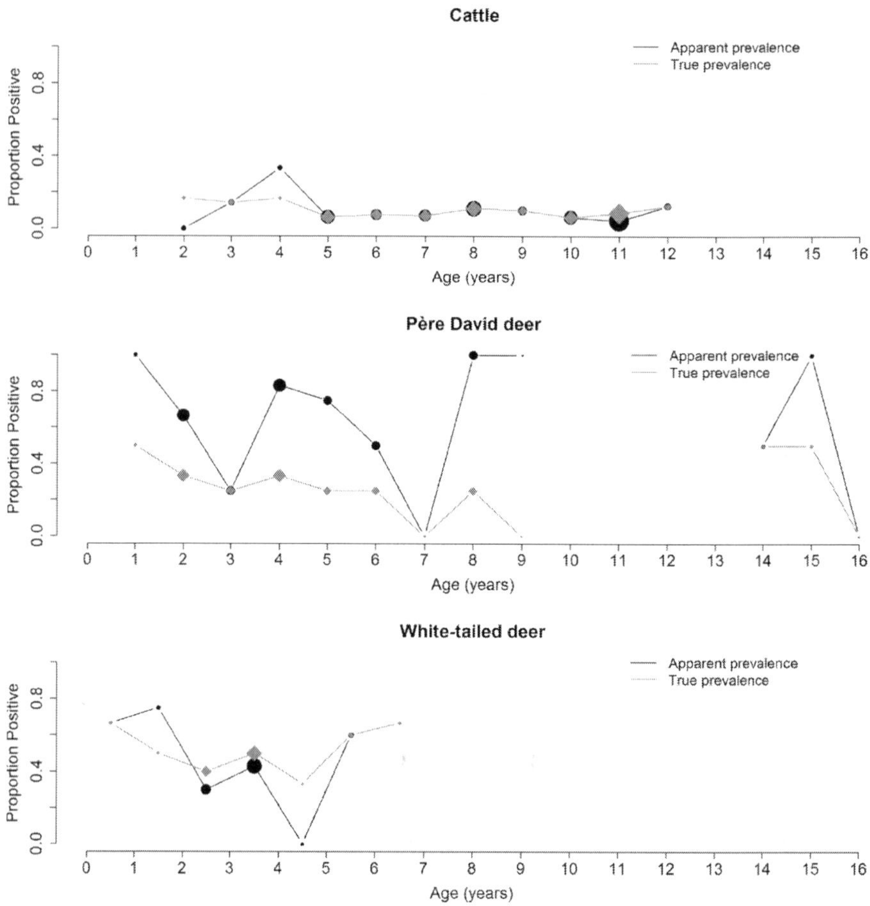

Figure 6. Apparent and true prevalence of proportion positive individuals by species and age. The size of the circles or diamonds is proportional to the number of animals sampled at each age.

https://doi.org/10.1371/journal.pone.0183900.g006

To determine the prevalence of exposure of *N. caninum* in cattle and Père David's deer, cELISA tests were performed. Like the white-tailed deer, both cattle and Père David's deer showed evidence of seropositivity, 8.7% overall AP and 67.5% overall AP, respectively. The proportion of seropositive individuals per age by species is shown in Fig 6. These results suggest that both cattle and Père David's deer were infected with the parasite sometime in the past.

The estimated TP of *N. caninum* in all three species differed from AP. These differences between AP and TP were expected, given that we are accounting for the diagnostic error in the TP. The mode of the overall cattle population TP was 7.2% (95 PI% 3.9–12%), for Père David's deer population TP was 53% (95 PI% 39.4–66%) and for white-tailed deer population TP was 53.5% (95 PI% 36–78%).

To quantify the duration of seropositivity for each host species, we fit age-structured serology data (Table 1) to the catalytic and reverse catalytic models and selected the better-fit model according to AIC value. The catalytic model assumes lifelong seropositivity; the reverse catalytic model assumes a seropositive response that wanes over time. Serologic data from cattle and Père David's deer fit the reverse catalytic models better (Figs 7 and 8). The duration of natural seropositivity was estimated to be 2 years for cattle and 1 year for Père David's deer using TP data, and 4 years for Père David's deer using AP data (Table 2). Serologic data from white-tailed deer fit the catalytic model better, which indicates lifelong seroconversion after infection (Fig 9). While the same model was selected regardless of whether the AP or TP data were fitted, the estimates for the parameters differed (Table 2). These results indicate that using AP data may identify the appropriate biological mechanism for which the parasite persists in the population, but control strategies based on AP may be biased due to the underestimation or overestimation of the parameters. Therefore, seropositivity wanes for cattle and Père David's deer–with the duration of seropositivity varying by species–while seropositivity lasts for the entire lifespan in white-tailed deer.

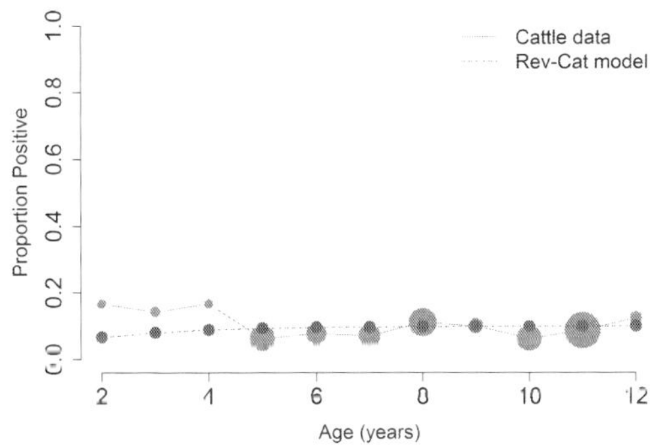

Figure 7. Model selection for cattle. Reverse catalytic model was the best-fit model for cattle. Size of points represents the relative number of animals sampled with that age.

https://doi.org/10.1371/journal.pone.0183900.g007

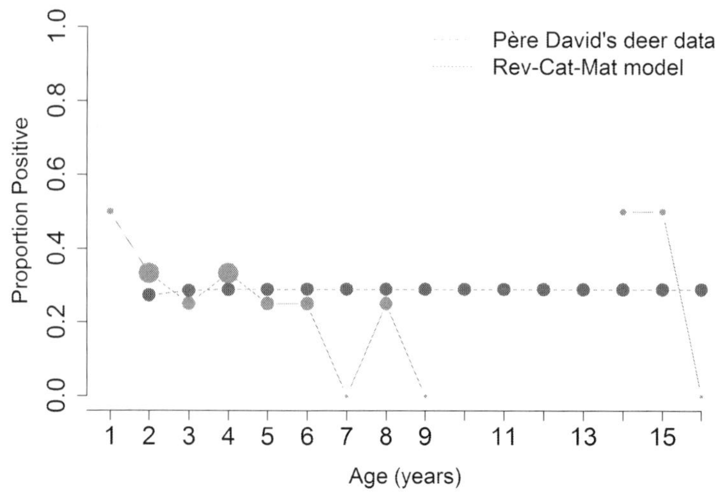

Figure 8. Model selection for Père David's deer. Reverse catalytic model with maternal antibodies was the best-fit model for Père David's deer. Size of points represents the relative number of animals sampled with that age.

https://doi.org/10.1371/journal.pone.0183900.g008

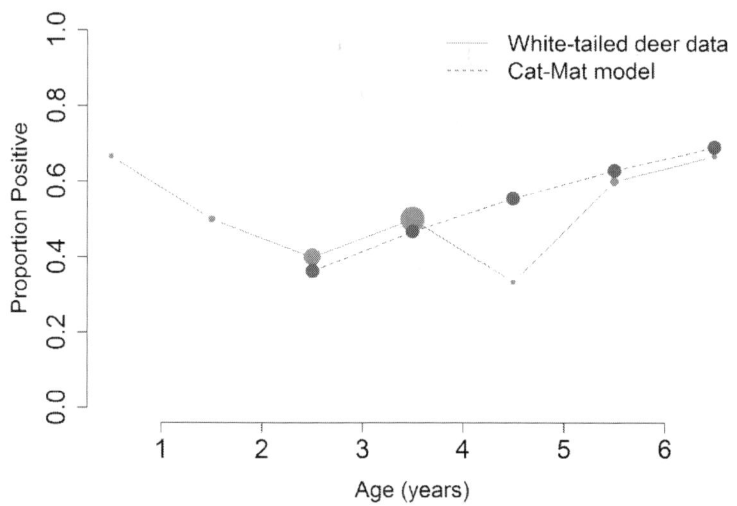

Figure 9. Model selection for white-tailed deer. Catalytic model with maternal antibodies was the best-fit model for white-tailed deer. Size of points represents the relative number of animals sampled with that age.

https://doi.org/10.1371/journal.pone.0183900.g009

Table 2. Parameter estimates from best-fit model with apparent (AP) versus true prevalence (TP).

Species	Data	Model selected	Force of infection (years^{-1})	Duration of natural sero-positivity (years)	Maternal sero-positivity duration (years)	Reproduction number	Critical proportion to vaccinate
Cattle	AP	Reverse Catalytic	0.051	2	n/a	0.61	n/a
	TP	Reverse Catalytic	0.055	2	n/a	0.66	n/a
Pere David's deer	AP	Reverse Catalytic [a]Mat	0.55	4	1.5	8.85	0.89
	TP	Reverse Catalytic [a]Mat	0.41	1	1.5	6.53	0.85
White-tailed deer	AP	Catalytic [a]Mat	0.135	lifelong	1.5	0.88	n/a
	TP	Catalytic [a]Mat	0.170	lifelong	1.5	1.11	0.09

[a]Mat is defined as the antibodies found in less than 2 year old individuals. Those antibodies could be maternal antibodies or fetal antibodies from congenital infection with *N. caninum*.

https://doi.org/10.1371/journal.pone.0183900.t002

To quantify the duration of maternal antibody persistence for both deer species, we fit reverse catalytic models with maternal immunity to the Père David's deer serology data and catalytic models with maternal immunity to the white-tailed deer data to see if they fit the serology data better than their corresponding models without maternal immunity. The models with maternal immunity fit the data better and were selected, by AIC score, over the models did not assume maternal antibody presence (Tables E and F in S1 Text). The duration of maternal antibody persistence was estimated at 1.5 years for both deer species, regardless of whether AP or TP data were used (Table 2). To determine if transmission dynamics are age-structured in all three species, we fit models with b-splined FoI and constant FoI to serologic data for each host species. The models without age-structured transmission were selected for all populations, indicating that the FoI was constant for all ages in all species (Table 2). This indicates, that the main route of transmission, likely, is vertical with a possibility of low rates of horizontal transmission.

To quantify the constant FoI in all host species, we fit both AP and TP serologic data to the best fit model selected by AIC. We found that Père David's deer had the highest FoI, between 0.41 yr^{-1} and 0.55 yr^{-1}. Cattle demonstrated the lowest FoI, between 0.051 yr^{-1} and 0.055 yr^{-1}, and white-tailed deer had intermediate FoI estimates, between 0.135 yr^{-1} and 0.170 yr^{-1}. There are clear differences in FoI estimates for each species.

To determine if each species could sustain chains of transmission only within their species, we calculated R_t. When this value is greater than one, we assume that the species can maintain chains of transmission internally;

when this value is less than one, we assume that the species cannot maintain chains of transmission internally. We found that Père David's deer R_t estimates were 7 to 9, while cattle R_t estimates were <1. White-tailed deer R_t estimates were ~1; however, this species had an estimate of R_t >1 when TP was used. Based on these estimates, the two deer species seem to maintain chains of transmission within their populations, while cattle do not appear to be capable of maintaining transmission without an outside source (Table 2).

DISCUSSION

Our goal was to describe host species heterogeneity in the epidemiology of a multi-host pathogen, N. caninum, circulating in a community. As in many natural settings, our study area supports diverse livestock and wildlife species. This community structure is rarely investigated along an epidemiological continuum with the idea that each species might affect pathogen transmission. These kinds of studies are difficult to achieve, since studies designed to obtain detailed data measuring longitudinal transmission heterogeneities in the field could be prohibitively cost- and time-intensive. However, we have demonstrated that the most common and easily gathered data, serology and age, can be used to test hypotheses regarding transmission heterogeneities using mathematical models. Our results show that immunity in cattle and Père David's deer wanes over time; for white-tailed deer, immunity appears to be lifelong. For white-tailed deer, natural boosting of the immune system might be occurring or natural infections may truly result in lifelong immunity to re-infection.

Our molecular data on white-tailed deer corroborates the model findings: none of the white-tailed deer tissue that was molecularly tested was positive; thus, seropositive individuals were either exposed rather than infected at sampling or detection was limited due to a low load of parasites and the arbitrary selection of tissue tested rather than focused in histopathology results. Another reason for the discrepancy of molecular and serological results of white-tailed deer was the possibility of cross-reaction with antibodies. However, the cELISA test uses the 65-kDa N. caninum tachyzoite antigen and a monoclonal antibody (MAb 4A 4–2) to detect antibodies in sera instead of whole tachyzoite antigens used in indirect ELISA and IFA tests, which largely alleviates the problem of cross-reaction with antibodies produced by antigenically related parasites such as Toxoplasma gondii, Sarcocystis cruzi, Sarcocystis hominis, and Sarcocystis hirsute [30, 40]. The reproductive number, R_t, for cattle was below the threshold for sustaining internal epidemic transmission, suggesting that infections are the result of exposure to an outside source. For cattle, models with maternal antibodies were eliminated because sampled individuals were at least two years of age, and therefore beyond the lifespan of maternal antibodies. Unlike cattle, Père David's deer and white-tailed deer could both maintain continuous chains of transmission (Rt > 1) within their populations, presumably via vertical transmission. For Père David's deer and white-tailed deer the models with maternal antibodies fitted better the data. These results may reflect the extent of colostrum consumption or in utero exposure with fetal antibody production. Our results indicate that transmission and immunodynamics of this pathogen are complex and differ by host species.

Differences in N. caninum seroprevalence, rates of abortion and immune responses between cattle breeds have been suggested [15, 16, 41, 42]. A proportion of each of the three ruminant populations sampled were in various stages of gestation. However, neither breed, reproductive status, nor gestational age was associated with the probability of being seropositive. We were unable to test associations for Père David's deer; however, a previous retrospective study on N. caninum seroprevalence in the Père David's deer herd found no correlation between calving rates and seropositivity [43].

Previously, French et al., 1999 studied transmission and control options of N. caninum in cattle. One of their remarkable findings was that, even with a high vertical transmission probability, some horizontal transmission was needed to maintain the endemism of the parasite in a cattle herd [44], a finding that is supported by our results (R_t <1 in cattle). Furthermore, seroprevalence studies in cattle, buffalo and Père David's' deer have suggested both age-specific and constant N. caninum prevalence with age [43, 45–48]. In our study none of the three ruminant populations showed age-structured FoI. This suggests that transmission within each of these populations is mainly vertical with low rates of horizontal transmission. High rates of horizontal transmission would be suspected if prevalence had increased with age or increased and decreased due to waning immunity and re-infection. Moreover, previous work in our study area found an estimated low environmental prevalence of N. caninum oocysts shed by wild canid hosts, which supports the hypothesis that transmission from canids to ruminants is infrequent [23].

Ecologically informed targeted control has been suggested for other multi-host pathogen systems: for example, targeting canines to control rabies in the Serengeti ecosystem [7] and targeting asymptomatically infected sheep to control peste des petits ruminants in multi-host communities [8]. Because results from our study suggest that vertical transmission is the main route by which N. caninum persists in the community, targeted control should be directed at the species of interest. For example, cattle would be the population of interest for farmers, while Père David's deer would be the population of interest for conservationists. Specifically, our results suggest that, for cattle, controlling outside sources of infection (purchased animals and canid feces) may be helpful in controlling disease. However, control of outside sources would not likely affect disease transmission in either deer species, because these species can maintain chains of transmission within their populations. We also showed that immunity in cattle and Père David's deer wanes over time, suggesting that boosting immunity with vaccines might prevent re-infection within these populations.

Currently, there is no vaccination for N. caninum and the development of a vaccine is not trivial. N. caninum is an obligatory intracellular parasite that seems to avoid host immunity by differentiating from tachyzoites into bradyzoites within cysts (tissue cysts). That differentiation changes the health status of a host from being acutely to chronically infected, and chronically infected individuals appear to be asymptomatic. However, during pregnancy, recrudescence of bradyzoites back to tachyzoites is plausible, resulting in symptomatic individuals. Pregnancy is associated with changes in the immune system, which can allow the parasite to infect the fetus. Therefore, candidate vaccines should prevent tachyzoite proliferation and dissemination to avoid transplacental transmission, and prevent tissue cyst formation [49]. Although various vaccine strategies (live attenuated vaccines, killed parasite lysates, total antigens or antigen fractions from killed parasites, and subunit vaccines) have been investigated to mediate protection in mice and cattle models, there are still disadvantages such as cost of production, safety, and stability that limit their application [20]. However, as one of the most economically important pathogens for the cattle industry with a global distribution, vaccine development for N. caninum should still be a priority.

Our study had a number of limitations. For instance, differences in the magnitude of the FoI per species captured species-specific differences in disease transmission. Despite that, the FoI cannot distinguish individual mechanisms that affect the health status of the individual. In addition, these models relate to the transition from sero-negative to sero-positive and does not assume that sero-positive animals are infectious. Therefore, conclusions about the mechanism behind the transmission cannot be done. To better confirm our findings a further evaluation incorporating a larger sample size for white-tailed deer and cattle is suggested. It would be interesting to study the context of optimal control strategies, and determine if they change regionally. Nevertheless, our conclusions hold for our studied populations, the mathematical models have previously illustrated their functionality with this type of data, and our results agree with previously-published, peer-reviewed literature.

Differentiating reservoirs from maintenance populations along the epidemiological continuum is an important step towards disease control and prevention in both domestic and wildlife populations. The methods we have demonstrated suggest which populations maintain disease (deer) and which require outside introduction of the pathogen (cattle). To our knowledge, this is the first study to suggest control and prevention strategies based on a better understanding of the heterogeneity of transmission across species.

These methods guide targeted control and prevention by suggesting in which species biosecurity (cattle) or vaccination (cattle and Père David's deer) would be helpful, with implications for other multi-host pathogens as well.

ACKNOWLEDGMENTS

We would like to thank the hunters, farmers, volunteer field assistants, the Wilds and Muskingum Livestock Auction for granting research permission and collaboration. In addition, we would like to thank Craig Hicks (USDA Wildlife Services, National Wildlife Disease Program) for providing advice on wildlife collection permits, Dr. Foster Anderson (Auction veterinarian) and Randi (Auction worker) for providing expertise on handling and sampling collection, and Pallavi Oruganti and Erika Helgeson for field and laboratory assistance. Also, we would like to thank the Ohio Department of Agriculture Animal Disease Diagnostic Laboratory, Reynoldsburg, OH for performing the serological and molecular tests. We are grateful for the insightful discussions provided by the Disease Ecology and Computer Modelling Laboratory group.

REFERENCES

1. Woolhouse MEJ, Taylor LH, Haydon DT. Population Biology of Multihost Pathogens. Science. 2001;292(5519):1109–12. pmid:11352066
2. Jones KE, Patel NG, Levy MA, Storeygard A, Balk D, Gittleman JL, et al. Global trends in emerging infectious diseases. Nature. 2008;451(7181):990–4. pmid:18288193
3. Giraudoux P, Raoul F, Pleydell D, Craig PS. Multidisciplinary studies, systems approaches and parasite eco-epidemiology: something old, something new. Parasite. 2008;15(3):469–76. Epub 2008/09/26. pmid:18814725.
4. Haydon DT, Cleaveland S, Taylor LH, Laurenson MK. Identifying reservoirs of infection: a conceptual and practical challenge. Emerg Infect Dis. 2002;8(12):1468–73. Epub 2002/12/25. pmid:12498665; PubMed Central PMCID: PMC2738515.
5. Bartlett MS. The Critical Community Size for Measles in the United States. Journal of the Royal Statistical Society Series A (General). 1960;123(1):37–44.
6. Streicker DG, Fenton A, Pedersen AB. Differential sources of host species heterogeneity influence the transmission and control of multihost parasites. Ecol Lett. 2013;16(8):975–84. Epub 2013/05/30. pmid:23714379; PubMed Central PMCID: PMC4282463.
7. Lembo T, Haydon DT, Velasco-Villa A, Rupprecht CE, Packer C, Brandao PE, et al. Molecular epidemiology identifies only a single rabies virus variant circulating in complex carnivore communities of the Serengeti. Proc Biol Sci. 2007;274(1622):2123–30. Epub 2007/07/05. pmid:17609187; PubMed Central PMCID: PMC2279181.
8. Munir M. Role of Wild Small Ruminants in the Epidemiology of Peste Des Petits Ruminants. Transboundary and Emerging Diseases. 2014;61(5):411–24. pmid:23305511
9. Keeling MJ, Rohani P. Modeling infectious diseases in humans and animals: Princeton University Press; 2008.
10. Dobson A. Population dynamics of pathogens with multiple host species. Am Nat. 2004;164 Suppl 5:S64–78. Epub 2004/11/13. pmid:15540143.
11. Almeria S. Neospora caninum and Wildlife. International Scholarly Research Notice Parasitology. 2013;2013:23. pmid:27335866

12. Gondim LFP. Neospora caninum in wildlife. Trends in Parasitology. 2006;22(6):247–52. pmid:16616642
13. Dubey JP, Schares G. Neosporosis in animals—The last five years. Veterinary Parasitology. 2011;180(1–2):90–108. pmid:21704458
14. Almería S, López-Gatius F, García-Ispierto I, Nogareda C, Bech-Sàbat G, Serrano B, et al. Effects of crossbreed pregnancies on the abortion risk of Neospora caninum-infected dairy cows. Veterinary Parasitology. 2009;163(4):323–9. pmid:19464119
15. Armengol R, Pabon M, Santolaria P, Cabezon O, Adelantado C, Yaniz J, et al. Low seroprevalence of Neospora caninum infection associated with the limousin breed in cow-calf herds in Andorra, Europe. J Parasitol. 2007;93(5):1029–32. Epub 2008/01/01. pmid:18163335.
16. Santolaria P, Almería S, Martínez-Bello D, Nogareda C, Mezo M, Gonzalez-Warleta M, et al. Different humoral mechanisms against Neospora caninum infection in purebreed and crossbreed beef/dairy cattle pregnancies. Veterinary Parasitology. 2011;178(1–2):70–6. pmid:21216103
17. Reichel MP, Alejandra Ayanegui-Alcerreca M, Gondim LFP, Ellis JT. What is the global economic impact of Neospora caninum in cattle—The billion dollar question. Int J Parasit. 2013;43(2):133–42. pmid:23246675
18. Trees AJ, Davison HC, Innes EA, Wastling JM. Towards evaluating the economic impact of bovine neosporosis. Int J Parasit. 1999;29(8):1195–200.
19. Haddad JPA, Dohoo IR, VanLeewen JA. A review of Neospora caninum in dairy and beef cattle—a Canadian perspective. Canadian Veterinary Journal-Revue Veterinaire Canadienne. 2005;46(3):230–43.
20. Monney T, Hemphill A. Vaccines against neosporosis: What can we learn from the past studies? Experimental parasitology. 2014;140:52–70. pmid:24602874
21. Weston JF, Heuer C, Williamson NB. Efficacy of a Neospora caninum killed tachyzoite vaccine in preventing abortion and vertical transmission in dairy cattle. Prev Vet Med. 2012;103(2–3):136–44. pmid:21925752
22. Moreno-Torres K, Wolfe B, Saville W, Garabed R. Estimating Neospora caninum prevalence in wildlife populations using Bayesian inference. Ecol Evol. 2016;6(7):2216–25. Epub 2016/04/22. pmid:27099713; PubMed Central PMCID: PMC4831453.

23. Moreno Torres K. The Wildlife-Livestock Interface of Infectious Disease Dynamics: A One Health Approach. Chapter 3. Spatial distribution of *Neospora caninum* and morphologically similar oocysts shed by wild canid host into the environment. Columbus: The Ohio State Univeristy; 2016.
24. Richardson C, Jones PC, Barnard V, Hebert CN, Terlecki S, Wijeratne WV. Estimation of the developmental age of the bovine fetus and newborn calf. Vet Rec. 1990;126(12):279–84. Epub 1990/03/24. pmid:2343510.
25. England RB. A Guide to age determination of cattle: incisor tooth eruption, development and attrition. Santa Barbara, Ca.: Veterinary Practice Pub. Co.; 1984.
26. Cain AT, Wallace M. A guide to age determination of white-tailed deer: Texas Parks and Wildlife; 2001.
27. Hamilton R, Tobin M, Moore W, editors. Aging fetal white-tailed deer. Proceedings of the Annual Conference of the Southeastern Association of Fish and Wildlife Agencies; 1985.
28. Baszler TV, Gay LJC, Long MT, Mathison BA. Detection by PCR of Neospora caninum in fetal tissues from spontaneous bovine abortions. J Clin Microbiol. 1999;37(12):4059–64. pmid:10565932
29. Ghanem ME, Suzuki T, Akita M, Nishibori M. Neospora caninum and complex vertebral malformation as possible causes of bovine fetal mummification. Canadian Veterinary Journal-Revue Veterinaire Canadienne. 2009;50(4):389–92.
30. Baszler TV, Adams S, Vander-Schalie J, Mathison BA, Kostovic M. Validation of a commercially available monoclonal antibody-based competitive-inhibition enzyme-linked immunosorbent assay for detection of serum antibodies to Neospora caninum in cattle. J Clin Microbiol. 2001;39(11):3851–7. pmid:11682497
31. R Core Team. R: A Language and Environment for Statistical Computing. Version 313 Vienna, Austria. http://www.R-project.org: R Foundation for Statistrical Computing; 2015.
32. Muench H. Catalytic models in epidemiology: Harvard University Press; 1959.
33. Pomeroy LW, Bjornstad ON, Kim H, Jumbo SD, Abdoulkadiri S, Garabed R. Serotype-Specific Transmission and Waning Immunity of Endemic Foot-and-Mouth Disease Virus in Cameroon. PLoS One.

2015;10(9):e0136642. Epub 2015/09/04. pmid:26327324; PubMed Central PMCID: PMC4556668.

34. Nash JC. {optimr}: A Replacement and Extension of the 'optim' Function. 2016.
35. Akaike H. Factor analysis and AIC. Psychometrika. 1987;52(3):317–32.
36. Hinrichs KC. Seroprevalence of antibodies to Neospora Caninum in Ohio Cattle. Columbus: The Ohio State University; 2003.
37. Greiner M, Gardner IA. Epidemiologic issues in the validation of veterinary diagnostic tests. Prev Vet Med. 2000;45(1–2):3–22. pmid:10802331
38. Lewis FI, Torgerson PR. A tutorial in estimating the prevalence of disease in humans and animals in the absence of a gold standard diagnostic. Emerging Themes Epidemiology. 2012;9(1):9. Epub 2012/12/29. pmid:23270542; PubMed Central PMCID: PMC3558341.
39. Anderson RM, May RM, Anderson B. Infectious diseases of humans: dynamics and control: Wiley Online Library; 1992.
40. Baszler TV, Knowles DP, Dubey J, Gay JM, Mathison BA, McElwain TF. Serological diagnosis of bovine neosporosis by Neospora caninum monoclonal antibody-based competitive inhibition enzyme-linked immunosorbent assay. J Clin Microbiol. 1996;34(6):1423–8. pmid:8735092
41. Eiras C, Arnaiz I, Álvarez-García G, Ortega-Mora LM, Sanjuánl ML, Yus E, et al. Neospora caninum seroprevalence in dairy and beef cattle from the northwest region of Spain, Galicia. Prev Vet Med. 2011;98(2–3):128–32. pmid:21145605
42. Bartels CJ, Arnaiz-Seco JI, Ruiz-Santa-Quitera A, Björkman C, Frössling J, von Blumröder D, et al. Supranational comparison of Neospora caninum seroprevalences in cattle in Germany, The Netherlands, Spain and Sweden. Veterinary Parasitology. 2006;137(1–2):1–2.
43. Bapodra P, Wolfe BA. Investigation of Neospora caninum seroprevalence and potential impact on reproductive success in semi-free-ranging Père David's deer (Elaphurus davidianus). Veterinary Record Open. 2015;2(1):e000123. pmid:26392906
44. French NP, Clancy D, Davison HC, Trees AJ. Mathematical models of Neospora caninum infection in dairy cattle: transmission and options for control. Int J Parasit. 1999;29(10):1691–704.

45. Davison HC, Otter A, Trees AJ. Estimation of vertical and horizontal transmission parameters of Neospora caninum infections in dairy cattle. Int J Parasit. 1999;29(10):1683–9.
46. Rinaldi L, Fusco G, Musella V, Veneziano V, Guarino A, Taddei R, et al. Neospora caninum in pastured cattle: determination of climatic, environmental, farm management and individual animal risk factors using remote sensing and geographical information systems. Veterinary Parasitology. 2005;128(3–4):219–30. pmid:15740859
47. Sanderson M, Gay J, Baszler T. *Neospora caninum* seroprevalence and associated risk factors in beef cattle in the northwestern United States. Veterinary Parasitology. 2000;90(1):15–24.
48. Moore DP, Konrad JL, San Martino S, Reichel MP, Cano DB, Mendez S, et al. Neospora caninum serostatus is affected by age and species variables in cohabiting water buffaloes and beef cattle. Vet Parasitol. 2014;203(3–4):259–63. Epub 2014/05/06. pmid:24792747.
49. Monney T, Debache K, Hemphill A. Vaccines against a Major Cause of Abortion in Cattle, Neospora caninum Infection. Animals. 2011;1(3):306–25. pmid:26486502

SECTION 6: FUNGUS

CHAPTER 20

Clinical and Molecular Epidemiology of Veterinary Blastomycosis in Wisconsin

Jennifer L Anderson[1], Brian L Sloss[2] and Jennifer K Meece[1]

[1] Marshfield Clinic Research Foundation, Marshfield WI 54449, USA.

[2] U.S. Geological Survey, Wisconsin Cooperative Fishery Research Unit, College of Natural Resources, University of Wisconsin-Stevens Point, Stevens Point WI 54481, USA.

ABSTRACT

Background

Several studies have shown that *Blastomyces dermatitidis*, the etiologic agent of blastomycosis, is a genetically diverse pathogen. Blastomycosis is a significant health issue in humans and other mammals. Veterinary and human isolates matched with epidemiological case data from the same geographic area and time period were used to determine: (i) if differences in

Citation: Anderson, J.L., Sloss, B.L. & Meece, J.K. Clinical and molecular epidemiology of veterinary blastomycosis in Wisconsin. BMC Vet Res 9, 84 (2013). https://doi.org/10.1186/1746-6148-9-84

genetic diversity and structure exist between clinical veterinary and human isolates of *B. dermatitidis* and (ii) if comparable epidemiologic features differ among veterinary and human blastomycosis cases.

Results

Genetic typing of 301 clinical *B. dermatitidis* isolates produced 196 haplotypes (59 unique to veterinary isolates, 134 unique to human isolates, and 3 shared between canine and human isolates). Private allelic richness was higher in veterinary (median 2.27) compared to human isolates (median 1.14) ($p = 0.005$).

Concordant with previous studies, two distinct genetic groups were identified among all isolates. Genetic group assignment was different between human and veterinary isolates ($p < 0.001$), with more veterinary isolates assigned to Group 2.

The mean age of dogs diagnosed with blastomycosis was 6 years. Thirty cases were in male dogs (52%) and 24 were females (41%). The breed of dog was able to be retrieved in 38 of 58 cases with 19 (50%) being sporting breeds. Three of four felines infected with blastomycosis were domestic shorthair males between ages 6–12, and presented with disseminated disease. The other was a lynx with pulmonary disease. The equine isolate was from an 11-year-old male Halflinger with disseminated disease. Disseminated disease was reported more often in veterinary (62%) than human cases (19%) ($p < 0.001$).

Conclusions

Isolates from all hosts clustered largely into previously identified genetic groups, with 3 haplotypes being shared between human and canine isolates confirming that *B. dermatitidis* isolates capable of infecting both species occur in nature. Allelic diversity measures trended higher in veterinary samples, with a higher number of total alleles and private alleles. Veterinary isolates of *B. dermatitidis* contributed a substantial amount of diversity to the overall population genetic structure demonstrating the importance of including veterinary isolates in genetic studies of evolution and virulence in this organism.

Keywords: Blastomycosis, Canine, Epidemiology, Blastomyces dermatitidis, Genetics

BACKGROUND

Blastomycosis is a significant health issue in humans and other mammals. This fungal infection often presents as a non-specific febrile illness that mimics viral and bacterial pneumonia, causing cough, fever, hospitalization and sometimes death. Moreover, dissemination of the infection to bone, skin, central nervous system and other organs is often observed. In dogs, it is thought to predominantly affect young, male, large sporting breeds, chiefly because of their propensity for outdoor activity [1–4]. Only Minnesota considers canine blastomycosis a reportable disease, therefore the number of cases occurring nationally is grossly underestimated. In Minnesota, canine cases occurred at 2.5 times that of humans from 1999–2009 [5]. Studies in Arkansas and Wisconsin report incidence of canine blastomycosis to be 10–13 times that of humans [6, 7]. In highly endemic areas, incidence rates reach 1-2% annually, and in one 10 year retrospective, epidemiological study, a fatality rate of 41% was observed [3]. Although reported less frequently, blastomycosis infections in felines show similar symptomology and severity [8, 9].

Several studies, have explored the genetic diversity of *Blastomyces dermatitidis*, the etiologic agent of blastomycosis [10–12]. More recently, genetic characterization using microsatellite markers has revealed that *B. dermatitidis* is a genetically diverse pathogen [13] with the 2 principally divergent genetic groups being associated with different clinical phenotypes in humans [14]. Previous studies have focused on the genetics of human clinical isolates as isolates from veterinary sources are less available for research. It is unknown whether canine, or other veterinary, clinical isolates of *B. dermatitidis* exhibit genetic diversity similar to human isolates or whether genetic groups could be associated with specific clinical phenotypes, as has been shown in humans.

Veterinary and human isolates matched with epidemiological case data from the same geographic area and time period were used to determine: (i) if differences in genetic diversity and structure exist between clinical veterinary and human isolates of *B. dermatitidis* and (ii) if comparable epidemiologic features differ among veterinary and human blastomycosis cases. The results of this study will provide a more complete understanding of the population genetic diversity of *B. dermatitidis* from a wider range of host species and assess the existence of host-specific and shared genotypes.

METHODS

Isolates

The majority of isolates used in this study are part of the Marshfield Clinic Research Foundation's *B. dermatitidis* bio-bank, which have been collected through our relationship with Marshfield Labs since 1999. Over this period Marshfield Clinic Research Foundation has obtained an extensive library of human and veterinary clinical isolates including 57 cultures of *B. dermatitidis* from canines, 4 from felines, and 1 equine isolate. In addition, we were gifted an isolate obtained from the stool of a dog diagnosed with acute blastomycosis [15]. The human and veterinary isolates used in this study were collected over the same time period and geographic area. It is unknown whether any isolates in this study were acquired from epidemiologically linked owners and pets, both diagnosed with blastomycosis.

Genetic Analysis

DNA was extracted from all veterinary *B. dermatitidis* isolates (n = 63) and genotyped with 27 microsatellite markers as previously described [13]. Up to 3 attempts were made for isolates that did not amplify initially for a given locus. Veterinary isolate genotyping data were combined with previously genotyped human isolates (n = 238). Rarefacted allelic and private allelic richness [16, 17] were calculated with HPRARE [18] employing a rarefaction method to account for unequal sample sizes in the two isolate sets. Non-parametric Wilcoxon signed rank tests were used to compare genetic diversity estimates between human and veterinary isolates of *B. dermatitidis*.

Genetic analyses focused on delineating genetic structure, predicting clusters of related isolates, and comparing the diversity of strains across the different host species. Briefly, genetically identical isolates were identified using Genetic Analysis in Excel v6.41 [19] and subsumed to representative haplotypes. The program STRUCTURE [20] was used to predict the minimum number of genetic clusters (*K*) within the composite data, with analysis settings as previously described [14]. The method of Evanno *et al.* [21] as employed by Structure Harvester [22] was used to estimate the most likely *K* given the data in conjunction with the mean and variance of the ln probability of *K* across multiple replicates. Principle coordinate analysis (PCoA) of the standardized covariance of the haplotypic genetic distance as performed in Genetic Analysis in Excel v6.41 was used to analyse and graphically plot clusters of haplotypes. Isolates were assigned to genetic group based on haplotypic results from both STRUCTURE and PCoA analysis. A Chi-square test was used to assess difference in genetic group assignment between human and veterinary isolates.

Medical Record Review

The test accession number was retrieved from each clinically confirmed isolate of *B. dermatitidis* and used to obtain the following information on each veterinary patient: city and state of veterinary clinic submitting the sample for testing; age, breed, and gender of the veterinary patient; and sample source. Location of infection was inferred from the sample source submitted for testing and cases were categorized as pulmonary only or disseminated infections (which were presumed to have originated as a respiratory infection). Samples submitted for culture that led to a classification of pulmonary disease included: tracheal or transtracheal wash, bronchial lavage, lung biopsy, and respiratory secretions. Samples classified as obtained from disseminated disease included biopsies or discharge collected from masses or lesions on any part of the body not considered part of the respiratory tract. Clinical information associated with each human isolate was obtained previously for another study [14]. A Chi-square test was used to assess differences in location of infection between human and veterinary cases.

RESULTS

Genetic Analysis

Typing of *B. dermatitidis* veterinary isolates (n = 63) across 27 loci yielded a 97.5% success rate. Forty-three alleles were unable to be determined as a result of either failed PCR or null alleles. Most of these undetermined alleles (n = 35) were from 3 isolates, obtained in 2001 and 2002. After genetic typing of veterinary isolates was completed and combined with previously typed human isolates, a total of 301 isolates represented 196 distinct haplotypes (59 unique to veterinary isolates, 134 unique to human isolates, and 3 shared between canine and human isolates). A summary of haplotypic allele diversity of *B. dermatitidis* isolated from veterinary and human hosts is shown in Table 1. The total number of alleles and private alleles identified in veterinary isolates was higher than those in human isolates (Table 1). The median (adjusted mean # of alleles/locus) rarefacted allelic richness of veterinary isolates and human isolates was 9.26 and 8.13, respectively. Testing the null hypothesis that overall allelic richness is the same in veterinary and human isolates showed no significant difference (Wilcoxon signed ranks 2-tailed, $p = 0.1319$).

Table 1. Haplotypic allele diversity of *B. dermatitidis* **isolated from veterinary and human hosts**

	Veterinary haplotypes (n = 59)		**Human haplotypes (n = 134)**		**Shared haplotypes (n = 3)**
Locus	**# of alleles**	**# of private alleles**	**# of alleles**	**# of private alleles**	**# of alleles**
1	4	1	3	0	2
2	8	0	8	0	1
3	13	3	12	2	2
4	10	3	7	0	2
5	8	4	5	1	1
6	14	3	13	2	3
7	9	3	7	1	2
8	9	2	10	3	1
9	7	2	6	1	2
10	9	1	9	1	1
11	12	3	11	2	1
12	8	0	10	2	2
13	8	0	8	0	1
14	9	0	9	0	1
15	7	0	8	1	1
16	12	3	9	0	1
17	5	1	6	2	1
18	11	1	14	4	1
19	9	3	11	5	2
20	12	5	10	3	1
21	12	3	10	1	1
22	11	3	10	2	1
23	7	2	6	1	1
24	7	3	5	1	1
25	6	0	8	2	1
26	13	5	9	1	2
27	10	1	10	1	2
Totals	**250**	**55**	**234**	**39**	**39**

However, private allelic richness differed significantly between veterinary isolates (median = 2.27) and human isolates (median = 1.14) (Wilcoxon 2-tailed, $p = 0.0050$) with 20 of 27 loci having higher private allelic richness for veterinary versus human isolates. In addition, a subsequent 1-tailed Wilcoxon signed ranks test showed veterinary private allelic richness was greater than human private allelic richness (Wilcoxon 1-tailed, $p = 0.0025$).

Summary of haplotypic allele diversity of *B. dermatitidis* isolated from veterinary and human hosts. The 301 clinical *B. dermatitidis* isolates were subsumed to 196 unique haplotypes; 59 unique to veterinary isolates, 134 unique to human isolates, and 3 shared between veterinary and human isolates.

Structure analysis of the unique haplotypes resolved two genetic clusters ($K = 2$) in the data based on the method of Evanno et al. [21] and the linearity and variance of lnP(D). The individual ancestry of each haplotype (q-value) based on $K = 2$ revealed 191 haplotypes with a majority q-value >80% (Figure 1). Five haplotypes (1, 61, 111, 134, and 190, representing only one isolate each) had a majority q-value <80%. Principal coordinate analysis (Figure 2) confirmed clustering of the haplotypes, primarily, into 2 distinct groups. Fifty percent of variance was accounted for by the first principle coordinate (Coordinate 1) and separated Group 1 and 2 isolates. The secondary axis (Coordinate 2) explained an additional 16% of variance. The 5 haplotypes (numbered in Figure 2) with <80% ancestral membership in STRUCTURE analysis, were observed to be genetic intermediates by PCoA analysis and could not be assigned to either genetic group. After removing genetic intermediates (1 canine isolate and 4 human isolates), Group 1 contained 48 haplotypes (representing 17 veterinary isolates and 135 human isolates). Group 2 contained 143 haplotypes (representing 45 veterinary isolates and 99 human isolates). Testing the null hypothesis that group assignment is the same for veterinary and human isolates showed that significantly more canine isolates were assigned to Group 2 (chi-square = 17.98, 1 df, $p < 0.001$). Although not statistically supported by STRUCTURE analysis as a separate genetic group, visual examination of Figure 2 reveals a small cluster of exclusively human isolates within Group 2.

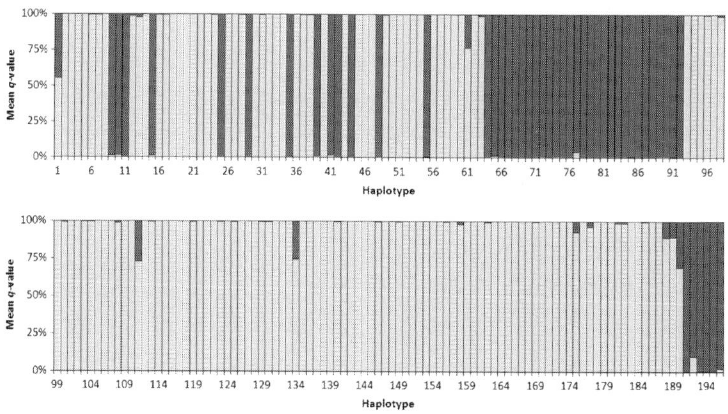

Figure 1. STRUCTURE analysis of *B. dermatitidis* haplotypes**.** Mean *q*-values from STRUCTURE for all 196 unique haplotypes (representing 301 isolates) with K = 2. Group 1 haplotypes are shown in black and Group 2 haplotypes in gray. Haplotypes 1–54 represent exclusively canine isolates. Haplotype 55 represents the single equine isolate. Haplotypes 56–59 represent feline isolates. Haplotypes 61–87, 89–91, 93–194, and 196 represent exclusively human isolates. Haplotypes 88, 92, and 195 contained both human and canine isolates. Five haplotypes (1, 61, 111, 134, and 190, representing only one isolate each) had a majority q-value <80% and could not be assigned to either genetic group.

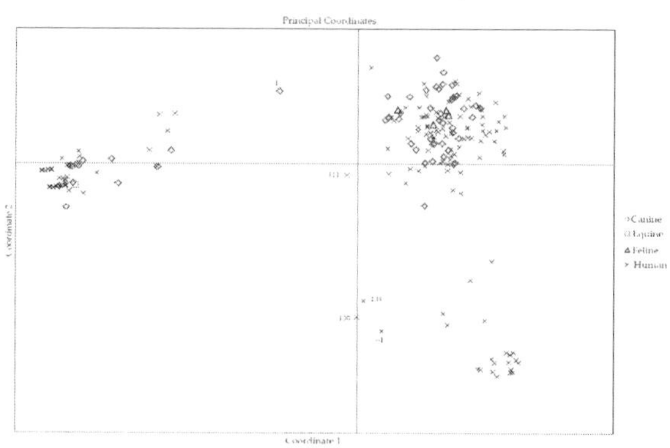

Figure 2. Principle coordinate analysis of *B. dermatitidis* haplotypes. Principle coordinate analysis of the haplotypic pairwise covariance distance matrix for all 196 unique haplotypes (representing 301 isolates). Fifty percent of variance was accounted for by the first principle coordinate (coordinate 1) and separated Group 1 and 2 isolates. The five numbered haplotypes represent single isolates with a mean majority *q*-values <80% in STRUCTURE results and could not be assigned to either genetic group.

Medical Record Review

Samples were submitted for either fungal or aerobic culture testing by 42 veterinary clinics in 34 cities throughout Wisconsin, Illinois, Michigan and Minnesota. The highest numbers of samples submitted for culture over the 13 year period were received from veterinary clinics in Buffalo Grove, IL (n = 6), Plover, WI (n = 5), and Appleton, WI (n = 5). The age range of dogs diagnosed with blastomycosis was 1–17 years (mean 6), with 8 cases having unspecified ages. Thirty cases were male dogs (52%), 24 were female (41%), and 4 had unspecified gender (7%). The highest numbers of isolates were obtained in 2006, 2007, and 2009 with 9, 13, and 8 isolates cultured, respectively, in each of those years. The breed of dog was able to be retrieved in 38 of 58 cases with 19 (50%) being purebred sporting breeds (Table 2). Three of the 4 felines were domestic shorthair (*Felis catus*) neutered males between the ages of 6–12 and presented with disseminated disease. The other was a lynx (*Lynx canadensis*), which presented with pulmonary disease, and its age and gender were not specified. The equine isolate was obtained from an 11-year-old Halflinger neutered male (*Equus ferus caballus*) and was the result of disseminated disease.

Table 2. Breeds of 58 canines diagnosed with blastomycosis

Breed	No. of dogs
Labrador retriever[1]	9
Golden retriever[1]	5
Mixed[2]	4
German shepard	4
Cocker spaniel[1]	3
Shih tzu	2
Other[3]	11
Unspecified	20
Total	**58**

[1]breeds belonging to AKC sporting group.

[2]Mixed breeds included 1 mixed Doberman, 1 mixed Labrador retriever, 1 mixed Golden retriever, and 1 mixed Rottweiler/German shepard.

[3]Other includes one each of the following breeds: Alaskan malamute, Beagle, Australian cattle dog, Chow chow, German shorthaired[1], Siberian husky, Jack russell terrier, Shetland sheepdog, Treeing walker coonhound, Weimaraner[1], and Yorkshire terrier.

Twenty-one (33%) veterinary infections were identified as pulmonary, whereas 39 (62%) were classified as disseminated infections. Three infections were unable to be classified based on available information. The location of infection in human cases, abstracted from medical charts as previous described, showed 190 (80%) were pulmonary, 46 (19%) were disseminated, and 2 (< 1%) infections had undetermined location [14]. Testing the null hypothesis that location of infection is the same in veterinary and human cases showed a significant difference in the number of pulmonary and disseminated infections between human and veterinary cases (chi-square = 48.399, 1 df, $p < 0.001$).

DISCUSSION

Studies exploring the genetic diversity of *B. dermatitidis* have primarily focused on human clinical isolates as clinical isolates obtained from veterinary sources are less available to researchers. It is unknown whether canine, or other veterinary, clinical isolates of *B. dermatitidis* exhibit genetic diversity similar to human isolates or whether genetic groups could be associated with specific clinical phenotypes, as has been shown in humans [14]. Understanding strain level phenotypic variation across host species could help us better anticipate clinical symptoms and outcomes and potentially inform treatment options.

Genotyping of veterinary isolates with these previously published microsatellite markers continued to provide a reliable and informative tool for assessing population level diversity. Compared to human isolates, the veterinary isolates had a slightly lower genotyping success rate. Most of the missing genotype data came from 3 of the oldest isolates in our collection (obtained in 2001 and 2002) and are likely the result of DNA degradation. The remaining missing data points are across isolates and markers and likely represent genuine alleles (null alleles) especially since repeated attempts were made to amplify missing alleles. Furthermore these markers were initially designed from genomic sequence of a human *B. dermatitidis* isolate [13]. Given the larger number of total and private alleles detected in canine isolates of *B. dermatitidis*, it is probable that alleles exist in canine isolates that the primers sets used in this study cannot detect. Missing alleles are not reported in counts of the number of alleles or the number of private alleles shown in Table 1. However isolates with missing alleles were left in the haplotypic analysis.

In the current study, PCoA and STRUCTURE analyses resolved veterinary and human isolates intermixed within previously identified genetic groups with no distinct resolution of veterinary isolates discrete from human isolates. Allelic diversity measures trended higher in veterinary samples, with a higher number of total and private alleles, despite the more than two-fold higher number of haplotypes in the human isolates. Three haplotypes were shared between human and canine isolates confirming that *B. dermatitidis* isolates capable of infecting both humans and canines occur in nature. Despite this, a large number of alleles (94 of 289, 33%) and most haplotypes (193 of 196, 98%) were not shared between host species. In fact, all four feline isolates and the single equine isolate comprise unique haplotypes individually. One notable observational finding is the presence of a small genetic cluster within Group 2 that is exclusively comprised of human isolates. Veterinary isolates of *B. dermatitidis* contributed a substantial amount of diversity to the overall population genetic structure demonstrating the importance of including veterinary isolates in genetic studies of evolution and virulence in this organism.

B. dermatitidis isolates exhibit considerable population genetic diversity by microsatellite typing. This is especially true of Group 2 isolates regardless of the host species from which it is isolated. This may indicate that other factors, such as genetic recombination rates and ecological or environmental partitioning, may impact the population genetic structure of *B. dermatitidis* as has been observed in the closely related fungal pathogens *Coccidioides* and *Penicillium marneffei* [23, 24]. Very little work has been done exploring genetic recombination in *B. dermatitidis*. Although *B. dermatitidis* can reproduce sexually, the mating-type locus was only recently identified [25], and it is unknown how frequently sexual reproduction occurs in nature. The rate of recombination between and among the 2 major genetic groups of *B. dermatitidis* could significantly impact the population genetic structure of this pathogen. Further exploration of ecological partitioning could also provide interesting insight into the genetic structure and evolution of *B. dermatitidis*. Mapping of exposures and the identification of point-source outbreaks have provided clues about the growth requirements and preferred environmental conditions of *B. dermatitidis* [26, 27]. Although no apparent geospatial relationships have been identified in the distribution of *B. dermatitidis*, this work has been previously hindered by the limited number of environmental isolates and cultured clinical isolates with associated exposure data. It is possible that spatial partitioning exists on a more subtle level and is impacting the genetic diversity of *B. dermatitidis*. It is also recognized that *B. dermatitidis* exists in the environment in geographically-restricted microfoci that are apparently short-lived [28]. This may complicate attempts at fully understanding spatial relationships as time and seasonality may also play a role [29].

Another explanation of the varying levels of genetic diversity between host species observed in this study is that they are the result of host sampling. Canines and felines are exposed to a wider range of environments and in more direct environmental contact than humans. For example, the amount of direct contact and exposure of dogs (both surface area covered and time exposed directly to that surface) to the moist soil environments thought to be the primary habitat of *B. dermatitidis* is immeasurably higher than that of even the most-avid outdoorsman. Thus, dogs should be a better representative or sampling vector of the species diversity than the limited exposures available in human isolates. This is especially relevant to researchers as environmental isolates are rare. As seen in previous epidemiologic studies of canine blastomycosis, we observed a high rate of infection in male, sporting breed dogs. Unexpectedly, we found that the mean age of infected dogs in our study was higher than in previous reports [2–4, 7].

Veterinary isolates were obtained more often from cases of disseminated disease versus pulmonary disease. This may be due to delay of diagnosis in dogs as owners may fail to notice the initial symptoms of blastomycosis, which can present as a non-specific febrile illness. Alternatively, it may be the result of intrinsic disease differences in varying host species, unrelated to strain type. One difference in disease manifestation that has been observed between humans and dogs is the higher rate of eye dissemination in dogs, which to date, has remained unexplained. However, the higher rate of dissemination in dogs may be a bias of our study associated with the ease of obtaining a biopsy in a disseminated case versus a sample such as a bronchial wash, which would be necessary in the case of pulmonary only disease presentation. Due to this, we felt confident classifying the disease location in veterinary cases based on the assumption that a veterinarian would not submit a pulmonary sample if another sample could be obtained with a less invasive procedure. Therefore, any isolates for which we had a non-respiratory sample were considered disseminated infections. Samples that were unable to be classified as either disseminated or pulmonary infections included one unidentified sample and two obtained from the nares. As blastomycosis is acquired by inhalation of spores, it is possible that an infection in the nares could be due to direct inoculation instead of dissemination from the lung and therefore we chose not to classify these cases as disseminated infections.

Another interesting finding of this study was that significantly more veterinary isolates (45 of 62, 73%), than human isolates (99 of 234, 42%) were genetic Group 2 organisms. In a recently conducted study, we found that genetic Group 2 infections are more often associated with disseminated disease in humans [14]. We were unable to directly assess association between genetic group and location of infection in veterinary cases in this study, because of the small number of Group 1 isolates. Despite this challenge, we observed a greater number of cases of disseminated disease in our veterinary cohort, with the predominant genetic group being Group 2 in these isolates. This is consistent with the associations established in the aforementioned study. If these results can be definitively verified across host species, it may indicate a molecular basis for differences in clinical phenotype observed between the two genetic groups.

Despite the patterns and correlations observed in this study, several limitations should be addressed in future research. The statistical power of this study is limited by the relatively small number of veterinary isolates and retrospective nature. Although culture is the primary diagnostic method for human blastomycosis, this may not be true of canine blastomycosis. This is supported by the fact that only 57 cultures of *B. dermatitidis* were isolated from canine sources by Marshfield Labs from 1999–2011. Certainly this collection of cultured isolates does not represent the clinical or genetic spectrum of canine blastomycosis, as veterinarians may be more likely to order culture under specific clinical circumstances such as more severe disease, obvious dissemination, value of pet, and a myriad of other factors. By partnering with veterinarians to obtain a less biased collection of isolates, future research can further explore the genetic diversity of veterinary blastomycosis. Although all clinical veterinary *B. dermatitidis* cultures used in this study were isolated and diagnosed in the same laboratory, over the same period of time and were obtained from the same geographic range as the human isolates to which they were compared, it must be recognized that this collection of isolates represents a limited geographic range of this global species. Despite the small number of veterinary isolates evaluated, this manuscript represents the largest genotyping study of canine *B. dermatitidis* isolates.

CONCLUSIONS

Genetic typing and comparison of veterinary and human isolates revealed that they were intermixed within previously identified genetic groups with no distinct resolution of veterinary isolates discrete from human isolates. Allelic diversity measures trended higher in veterinary samples, with a higher number of total and private alleles. Three haplotypes were identified as shared between human and canine isolates confirming that *B. dermatitidis* isolates capable of infecting both humans and canines occur in nature. Veterinary isolates of *B. dermatitidis* contributed a substantial amount of diversity to the overall population genetic structure demonstrating the importance of including veterinary isolates in genetic studies of evolution and virulence in this organism.

Another finding of this study was that significantly more veterinary isolates (73%), than human isolates (42%) were genetic Group 2 organisms and cultured isolates of *B. dermatitidis* from veterinary patients were more likely to be obtained from cases of disseminated disease than humans. This finding suggests that previous associations of genetic group of *B. dermatitidis* with specific clinical phenotypes in humans [14] may also hold true in veterinary infections.

The continued study of *B. dermatitidis* across its range of hosts and geography is key to a better understanding of the life cycle, evolution, and virulence, of this clinically important pathogen.

ACKNOWLEDGEMENTS

We especially thank Pamela Kivela for instructing us on how to retrieve clinical information associated with each veterinary case. We would also like to thank Jordan Dieckman for laboratory assistance and Dennis J. Baumgardner, MD (Aurora UW Medical Group) for providing a canine clinical isolate included in this study and for his thoughtful comments on this manuscript. The use of trade names in this manuscript does not imply endorsement by the U.S. Government. This work was supported by the Marshfield Clinic Research Foundation.

REFERENCES

1. Furcolow ML, Chick EW, Busey JF, Menges RW: Prevalence and incidence studies of human and canine blastomycosis. 1. Cases in the United States, 1885–1968. Am Rev Respir Dis. 1970, 102: 60-67.
2. Archer JR, Trainer DO, Schell RF: Epidemiologic study of canine blastomycosis in Wisconsin. J Am Vet Med Assoc. 1987, 190: 1292-1295.
3. Rudmann DG, Coolman BR, Perez C, Glickman LT: Evaluation of risk factors for blastomycosis in dogs: 857 cases (1980–1990). J Am Vet Med Assoc. 1992, 201: 1754-1759.
4. Arceneaux KA, Taboada J, Hosgood G: Blastomycosis in dogs: 115 cases (1980–1995). J Am Vet Med Assoc. 1998, 213: 658-664.
5. Minnesota Department of Health, Blastomycosis Statistics and Maps. http://www.health.state.mn.us/divs/idepc/diseases/blastomycosis/statistics.html.
6. Menges RW, Doto IL, Weeks RJ: Epidemiologic studies of blastomycosis in Arkansas. Arch Environ Health. 1969, 18: 956-971.
7. Baumgardner DJ, Paretsky DP, Yopp AC: The epidemiology of blastomycosis in dogs: north central Wisconsin, USA. J Med Vet Mycol. 1995, 33: 171-176. 10.1080/02681219580000361.
8. Brömel C, Sykes JE: Epidemiology, diagnosis, and treatment of blastomycosis in dogs and cats. Clin Tech Small Anim Pract. 2005, 20: 233-9. 10.1053/j.ctsap.2005.07.004.
9. Gilor C, Graves TK, Barger AM, O'Dell-Anderson K: Clinical aspects of natural infection with Blastomyces dermatitidis in cats: 8 cases (1991–2005). J Am Vet Med Assoc. 2006, 229: 96-9. 10.2460/javma.229.1.96.
10. Yates-Siilata KE, Sander DM, Keath EJ: Genetic diversity in clinical isolates of the dimorphic fungus *Blastomyces dermatitidis* detected by a PCR-based random amplified polymorphic DNA assay. J Clin Microbiol. 1995, 33: 2171-2175.
11. McCullough MJ, DiSalvo AF, Clemons KV, Park P, Stevens DA: Molecular epidemiology of *Blastomyces dermatitidis*. Clin Infect Dis. 2000, 30: 328-335. 10.1086/313649.
12. Meece JK, Anderson JL, Klein BS, Sullivan TD, Foley SL, Baumgardner DJ, Brummitt CF, Reed KD: Genetic Diversity in *Blastomyces dermatitidis*: Implications for PCR Detection in Clinical and Environmental Samples. Med Mycol. 2010, 48: 285-290. 10.3109/13693780903103952.

13. Meece JK, Anderson JL, Fisher MC, Henk DA, Sloss BL, Reed KD: Population Genetic Structure of Clinical and Environmental Isolates of *Blastomyces dermatitidis* based on 27 Polymorphic Microsatellite Markers. Appl Environ Microbiol. 2011, 17: 5123-5131.
14. Meece JK, Anderson JL, Gruszka S, Sloss BL, Sullivan B, Reed KD: Variation in clinical phenotype of human infection among genetic groups of *Blastomyces dermatitidis*. J Infect Disease. 2013, 207: 814-822. 10.1093/infdis/jis756.
15. Baumgardner DJ, Paretsky DP: Identification of *Blastomyces dermatitidis* in the stool of a dog with acute pulmonary blastomycosis. J Med Vet Mycol. 1997, 35: 419-421. 10.1080/02681219780001521.
16. Leberg PL: Estimating allele richness: Effects of sample size and bottlenecks. Mol Ecol. 2002, 11: 2445-2449.
17. Kalinowski ST: Counting alleles with rarefaction: Private alleles and hierarchical sampling designs. Conserv Genet. 2004, 5: 539-543.
18. Kalinowski ST: HP-Rare: a computer program for performing rarefaction on measures of allelic diversity. Mol Ecol Notes. 2005, 5: 187-189. 10.1111/j.1471-8286.2004.00845.x.
19. Peakall R, Smouse PE: GENALEX 6: genetic analysis in Excel. Population genetic software for teaching and research. Mol Ecol Notes. 2006, 6: 288-295. 10.1111/j.1471-8286.2005.01155.x.
20. Pritchard JK, Stephens M, Donnelly P: Inference of population structure using multilocus genotype data. Genetics. 2000, 155: 945-959.
21. Evanno G, Regnaut S, Goudet J: Detecting the number of clusters of individuals using the software STRUCTURE: a simulation study. Mol Ecol. 2005, 14: 2611-2620. 10.1111/j.1365-294X.2005.02553.x.
22. Earl DA: vonHoldt BM: STRUCTURE HARVESTER: a website and program for visualizing STRUCTURE output and implementing the Evanno method. Conserv Genet Resour. 2012, 2: 359-361.
23. Koufopanou V, Burt A, Taylor JW: Concordance of gene genealogies reveals reproduction isolation in the pathogenic fungus *Coccidioides immitis*. Proc Natl Acad Sci. 1997, 94: 5478-5482. 10.1073/pnas.94.10.5478.
24. Fisher MC, Hanage WP, de Hoog S, Johnson E, Smith MD, White NJ, Vanittanakom N: Low Effection Dispersal of Asexual Genotypes in Heterogeneous Landscapes by the Endemic Pathogen Penicillium marneffei. PLoS Pathog. 2005, 2: e20.

25. Wenjun L, Sullivan TD, Walton E, Averette AF, Sakthikumar S, Cuomo CA, Lein BS, Heitman J: Identification of the Mating-Type *(MAT)* Locus That Controls Sexual Reproduction of *Blastomyces dermatitids*. Eukaryot Cell. 2013, 12: 109-117. 10.1128/EC.00249-12.
26. Reed KD, Meece JK, Archer JR, Peterson AT: Ecologic niche modeling of Blastomyces dermatitidis in Wisconsin. PLoS One. 2008, 4: e2034.
27. Klein BS, Vergeront JM, DiSalvo AF, Kaufman L, Davis JP: Two outbreaks of blastomycosis along rivers in Wisconsin. Isolation of Blastomyces dermatitidis from riverbank soil and evidence of its transmission along waterways. Am Rev Respir Dis. 1987, 6: 1333-1338.
28. Denton JF, DiSalvo AF: Additional isolations of *Blastomyces dermatitidis* from natural sites. Am J Trop Med Hyg. 1979, 28: 697-700.
29. Baumgardner DJ, Paretsky DP, Baeseman ZJ, Schreiber A: Effects of season and weather on blastomycosis in dogs: Northern Wisconsin, USA. Med Mycol. 2011, 49: 49-55. 10.3109/13693786.2010.488658.

CHAPTER 21

Sarcoptes Mite Epidemiology and treatment in African Buffalo (Syncerus Caffer) Calves Captured for Translocation from the Kafue Game Management Area to Game Ranches

Hetron M Munang'andu[1,3], Victor M Siamudaala[2], Wigganson Matandiko[2], Musso Munyeme[3], Mwelwa Chembensofu[3] and Enala Mwase[3]

[1]Department of Basic Sciences and Aquatic Medicine, Section of Aquatic Medicine and Nutrition, Norwegian School of Veterinary Sciences, Ullevålsveien 72, NO-0033 Oslo, Norway,

[2]Zambia Wildlife Authority, Chilanga, Zambia

[3]School of Veterinary Medicine, University of Zambia, , Lusaka, Zambia

Citation: Munang'andu, H.M., Siamudaala, V.M., Matandiko, W. et al. Sarcoptes mite epidemiology and treatment in African buffalo (Syncerus caffer) calves captured for translocation from the Kafue game management area to game ranches. BMC Vet Res 6, 29 (2010). https://doi.org/10.1186/1746-6148-6-29

ABSTRACT

Background

In Zambia, translocation of wildlife from National Parks to private owned game ranches demands that only animals free of infectious diseases that could adversely affect the expansion of the wildlife industry should be translocated to game ranches. *Sarcoptes* mange (*Sarcoptes scarbiei*) has been involved in the reduction of wildlife populations in some species.

Results

Sarcoptes mange (*Sarcoptes scarbiei*) was detected and eradicated from two herds of African buffalo (*Syncerus caffer*) calves captured in the Kafue GMA in July 2004 and August 2005. The overall prevalence was estimated at 89.5% (77/86). Sex had no influence on the occurrence and severity of the disease. Of the 86 calves used in the study, 72.1% had good body condition scores, 20.9% were fair and 7.0% were poor. Of the 77 infected calves, 53.2% were mildly infected, 28.6% were moderately and 18.2% were severely infected. Body condition score was correlated to the severity of the infection ($r = 0.72$, $p < 0.000$, $n = 86$) at capture. Eradication of *Sarcoptes* mites from the entire herd using ivermetcin was dependant on the severity of the infection. The overall ability of ivermectin to clear the infection after the first treatment was estimated at 81.8% ($n = 77$). It increased to 94.8% and 100% after the second and third treatments respectively.

Conclusion

This is the first report on the epidemiology and treatment of *Sarcoptes* mange in African buffaloes in Zambia. This study improves our understanding about *Sarcoptes scabiei* epidemiology and treatment which will have further applications for the safe animal translocation.

BACKGROUND

Sarcoptic mange is a highly contagious disease affecting humans [1, 2], domestic animals [3] and wildlife [4]. The disease is caused by *Sarcoptes scabiei* mites that burrow in the epidermis of the skin where they form tunnels and deposit their eggs and digestive secretions causing intense skin irritations. The mites complete their lifecycle within a few weeks giving rise to high densities reaching up to 5000 mites/cm^2 in some species [5].

This leads to alopecia and hyperkeratosis of the skin [6]. The disease has been associated with high morbidity and mortality leading to population reductions in some species [7–10]. The efficacy of ivermectin in the treatment of *Sarcoptes* mange in different wildlife species has been carried out with variable responses [11–14]. Although the disease affects various wildlife species [15, 15], there are few reports describing its epidemiology in African game. In Zambia, it has been reported in humans and domestic animals with limited reports on wildlife [16].

The expansion of the game ranching industry in Zambia as an *ex-situ* conservation strategy has led to the translocation of several wildlife species from state owned National Parks (NPs) and game management areas (GMAs) to private owned game ranches in recent years. However, infectious diseases such as foot and mouth disease, *Sarcoptes* mange and bovine tuberculosis are a major constraint to the translocation of wildlife [17] because regulations demand that only animals not infected with such infectious diseases should be translocated to game ranches. This paper discusses the clinical observations made and the treatment regime used to eradicate sarcoptic mange from the entire herd of buffalo calves captured in the Kafue GMA prior to translocation.

METHODS

Animal data

The Study Was Carried Out At Nanzhila (26°03'E, 15°44'S) located on the western end of the Kafue flats GMA which is a wetland area designated as Ramsar Site No. 530 [18]. The area supports a transhumance grazing system for the traditional livestock farmers from the surrounding upland areas by providing grazing pastures for cattle in the dry season. GMAs are considered to be buffer zones being located between NPs and communal lands and are also referred to as livestock/wildlife interface areas because the co-existence of domestic animals and wildlife is permitted in these areas [19]. In August 2004 a total of 48 African buffalo calves were captured and used in the study while 38 were used in July 2005 with permission of the Zambia wildlife Authority (ZAWA) who are the custodians of wildlife in Zambia (Project authorization ref: 006/68-07-05). Age was estimated at four to six months for all calves using tooth development [20]. The calves were kept in quarantine facilities for 90 days and were sampled at 30 days intervals. At every sampling, calves were immobilized using M99 (etorphine hydrochloride, Novartis Ltd., Animal Health, South Africa) and reversed using M5050 revivon (dreprenorphine, Novartis SA Ltd., Animal Health, South Africa).

Sarcoptes diagnosis and treatment

The severity of the disease was classified into three categories as mild (+), moderate (++) and severe (+++) as described by Pence *et al* [21]. The mild group involved calves with focal lesions covering small patches of the body surface. Most of the lesions observed in this group were in the initial phase and they included erythema, pruritis, encrustations, suppuration and alopecia. The most affected body parts were the forehead, shoulders, hump and hock joints. The moderate group involved animals with actively progressing lesions covering large areas extending to almost half of the body surface. There were large surfaces of erythema, pruritis, alopecia, scabs, parakeratosis and fissures on some of the animals. Suffice to mention that some of the lesions observed in this group were similar to those observed in the mild group only that in the moderate group lesions were progressive and diffuse covering large areas. Severe groups were animals whose lesions covered more than half of the body surface. Alopecia, parakeratosis, thick and wrinkled skin surfaces were more pronounced in this group than in the mild and moderate groups. Body condition score was recorded as good (++), fair (+) or poor (-) depending on the degree of emaciation and skin pliability. Highly emaciated animals having little muscle with no sign of fat deposits and the skin falling slowly or staying in a fold after pulling a portion on the neck were scored poor (-). Animals having moderate deposition of muscle with the skin of the neck folding moderately faster than those in poor body conditions were scored as fair (+) while animals with excess fat and muscle with the skin quickly falling to its original position after the pliability test were scored as good (++). Skin scrapings were collected at every sampling. All claves were treated with 200 $\mu g\ kg^{-1}$ ivermectin (Novartis SA Ltd., Animal Health) at each sampling time point irrespective of whether they had mites detected or not. They were translocated to game ranches within three days after being declared free of mites after the fourth screening test. At the School of Veterinary Medicine at the University of Zambia, skin scrapings were examined after treatment with 20% potassium hydroxide [22]. Identification of mites was carried out as described elsewhere [23]. Sex, age, body condition score, severity score of the clinical condition and laboratory results for each calf were recorded in Exel® sheets at each sampling time point. Data was analyzed using STATA SE/10 for windows

(Stata Corp. College Station, TX, USA, http://www.stata.com). Chi-square was used to test the measures of association between sex and occurrence of the disease while the Pearson's correlation coefficient test was used to measure the correlation between severity of the disease and body condition score.

RESULTS

Sarcoptes mite epidemiology

Sex ratio (male:female) was estimated at 1:12. Our observations indicate that sex had no influence on the occurrence ($\chi^2 = 0.093$, $p > 0.70$) and severity ($\chi^2 = 0.70$, $p > 0.70$) of the disease. Overall prevalence for the two herds was estimated at 89.5% (77/86) at capture (Table 1). Of the 86 calves used in the study, 72.1% had good body condition score, 20.9% were fair and 6.9% were poor at capture (Table 2). Of the 77 infected calves, 53.2% were mildly infected, 28.6% were moderately and 18.2% were severely infected at capture. Body condition score was positively correlated to the severity of the infection ($r = 0.72$, $p < 0.000$, $n = 86$) at capture.

Table 1. Treatment regime and detection of mites at different sampling time points.

				Number of calves detected with presence of *Sarcoptes scabiei*mites at;			
Year	Number of calves	Severity of skin infection	Severity score	1st Sampling and treatment (Day 1)	2nd Sampling and treatment (Day 30)	3rd sampling and treatment (Day 60)	4th sampling (Day 90)
2004	48(45)*	Mild	+	25	2	0	0
		Moderate	++	12	4	0	0
		Severe	+++	8	6	2	0
2005	38(32)*	Mild	+	16	0	0	0
		Moderate	++	10	4	0	0
		Severe	+++	6	4	2	0
Totals	86(77)*			77	14	4	0

*Number of calves infected with Sarcoptes mange at capture (first sampling).

Table 2. Body condition score of the calves at different sampling time points.

				Body score condition at different sampling time points			
Year	***n***	**Cat-egory**	**Score**	**Day 1**	**Day 30**	**Day 60**	**Day 90**
				Number of calves	**Number of calves**	**Number of calves**	**Number of calves**
2004	48	Good	++	35	47	48	48
		Fair	+	11	1	-	-
		Poor	-	2	-	-	-
2005	38	Good	++	27	36	38	38
		Fair	+	7	2	-	-
		Poor	-	4	-	-	-
Totals	86			86	86	86	86

Sarcoptes mite treatment

Decrease in the number of infected calves after the first treatment was related to the severity of the infection observed at capture (Table 1). Of the 41 mildly infected calves, only two had *Sarcoptes* mites after the first treatment (4.98%) while in the moderately affected group 45.45% (10/22) needed a second treatment for the mites to be eliminated. In the severely affected group, there was a gradual decline with 28.57% (4/14) not having mites after the first treatment, 42.86% (10/14) after the second treatment and another 28.57% (4/14) after the third treatment. The overall drug ability to clear the infection was estimated at 81.8% ($n = 77$) after the first treatment. It increased to 94.8% after the second treatment and reached 100% after the third treatment. The calves were translocated to game ranches within three days after the fourth treatment after they were declared free of *Sarcoptes* mange.

DISCUSSION

In Zambia, sarcoptic mange has been reported in sitatunga (*Tragelaphus spekii*), hartebeest (*Alcephelus lichtensteini*) and silver-backed-jackal (*Canis mesomelas*) in wildlife [16]. In domestic animals the disease has been reported in cattle, goats and pigs [16]. It has not been reported from the African buffalo and no comprehensive study has been carried out describing the clinical findings in wildlife in Zambia. In line with observations made by

Pence and Ueckermann [24], our findings indicate that clinical conditions vary between animals depending on the severity of the infection. Tikaram and Ruprah [25]observed that the incidence of sarcoptic mange was highest in young animals below one year. They noted that prevalence decreased with advancement of age and that sex had no influence on the incidence of the disease [25]. The high prevalence 89.5%(77/86) obtained in the present study could be attributed to the fact that the animals involved were young ranging between four and six months of age at capture. We also observed that sex had no influence on the occurrence of the disease in the calves.

Fain *et al* [26] postulated that *Sarcoptes* mange originated as a human parasite that spread to domestic animals and subsequently transmitted to various wildlife species by contact with their domestic counterparts. It is not known how *Sarcoptes scabiei* was introduced in buffalo herds in the Kafue GMA although the disease has been present in domestic animals in the area [16]. Studies have shown that the transmission of mange was density dependant being more prevalent during times of higher environmental stress such as drought [27, 28]. In the present study, the high prevalence of *Sarcoptes* mange (89.5%, $n = 86$) reported in the calves could have been exacerbated by stress factors caused by the scarcity of water and grazing pastures during the dry season in the months of May to October on the Kafue flats. Livestock owners in the upland areas surrounding the Kafue flats practice a transhumance grazing system by moving their animals into the flats during the dry season leading to the sharing of grazing pastures and water between wildlife and livestock. This impacts a high population density on the flats. It is likely that *Sarcoptes* mange could have been transmitted from livestock to wildlife during such contacts. However, to fully understand the factors responsible for the epidemiology and transmission of the disease between species in the Kafue flats it is imperative that wildlife and livestock surveys be synchronized involving molecular studies of the gene flow of this parasite between different host species and geographical localities [29]. Perez *et al* [30] observed that *Sarcoptes* mange cases were higher in the cold season reducing in the hot months in summer when conditions were unfavorable for the survival of the mites. In Zambia, the cold season is in the months of June to August and this could have contributed to the increase in the number of calves infected given the timing of the present study that calves were captured in the cold months of August 2005 and July 2005. Treating the calves for the subsequent three months means that the treatment regime extended into the hot months of September to November which could have impacted negatively on the survival of the mites. It is likely that climatic factors had confounding effects on the treatment regime used in the current study. However, there is a need to carry out more detailed studies in order to determine the effect of climate on the epidemiology of sarcoptic mange on the Kafue basin.

The efficacy of ivermectin in the treatment of *Sarcoptes* mange in domestic animal has been studied by different scientists with some studies showing high responses after the first treatment [31, 32] although other studies have shown a gradual decrease in the number of mites after multiple doses of ivermetcin administered at regular intervals [33]. In Wildlife, there have been variable responses on the treatment of *Sarcoptes* mange using ivermectin in different animal species. Leon-Vizcaino *et al* [34] managed to eliminate the *Sarcoptes* mites from Spanish ibex (*Capra pyrenaica*) in the early stages of the infection, although double injections with high doses of ivermectin did not yield success in chronically infected animals. However, Skerratt [35] reported a complete resolution of clinical signs in wombats (*Wombatus ursinus*) after three injections of ivermectin administered 10 days apart although mites were not completely eliminated until the wambots were given a second regime of treatment. Yeruham *et al* [36] reported of complete recovery of severely infected wild ruminants kept in zoological gardens after mutlitple treament using ivermectin administered orally at regular intervals. These observations indicate that *Sarcoptes* mites can be eliminated from wild animals kept in captivity using ivermectin. Besides, ivermectin has been shown to be an effective drug for treating various endoparasites and parasitic skin diseases [37]. In the present study, we observed a gradual decrease in the number of infected calves from the first to the third treatment (Table 1). Our observations indicate that eradication of *Sarcoptes scabiei* mites using ivermectin was dependant on the severity of the infection (Table 1) similar to observations made by Skerratt [35].

CONCLUSION

Sarcoptes mites were eradicated from African buffalo calves captured from the Kafue game management area after the third treatment indicating that it is possible to eradicate some infectious diseases from infected wildlife populations prior to translocation by using the screening and treatment method. Hence, the risk of introducing *Sarcoptes* mange to naïve wildlife populations on the game ranches was avoided. Suffice to mention that such disease eradication methods are expensive as they require frequent immobilization of animals. Prolonged keeping of wildlife in captivity imposes a lot of stress on the animals. Therefore, we recommend that highly potent drugs aimed at reducing the need for repeat treatments should be explored for use in wildlife medicine. In addition, disease control strategies involving less strenuous handling procedures should be used with a view of promoting the success of *ex-situ* conservation.

ACKNOWLEDGEMENTS

The authors would like to thank Chaminuka game reserve as well as Eureka and Tim Ashworth game ranches for financing the Project. The authors also thank the Zambia wildlife Authority (ZAWA) for providing the study animals and their permission to carry out the project in the Kafue Basin game management area.

REFERENCES

1. Skerratt LF, Beveridge I: Human scabies of wombat origin. Aust Vet J. 1999, 77: 607-10.1111/j.1751-0813.1999.tb13202.x.
2. Terry BC, Kanjah F, Sahr F, Kortequee S, Dukulay I, Gbakima AA: *Sarcoptes scabiei* infestation among children in a displacement camp in Sierra Leone. Public Health. 2001, 115: 208-11. 10.1016/S0033-3506(01)00445-0.
3. Soll MD, Carmichael IH, Swan GE, Scherer H: Control of cattle mange in southern Africa using ivermectin. Trop Anim Health Prod. 1987, 19: 93-102. 10.1007/BF02297327.
4. Pence DB, Ueckermann E: Sarcoptic mange in wildlife. Rev Sci Tech. 2002, 21: 385-98.
5. Bornstein S, Mörner T, Samuel WM: *Sarcoptes scabiei* and sarcoptic mange. Parasitic Diseases of Wild Mammals. Edited by: Samuel WM, Pybus MJ, Kocan AA. Manson Publishing, London, UK, 2001, 107-19. full_text.
6. Newman TJ, Baker PJ, Harris S: Nutritional condition and survival of red foxes with sarcoptic mange. Can. J. Zool. 2002, 80: 154-61. 10.1139/z01-216.
7. Lindström ER: Pattern and spread and effects of sarcoptic mange among red fox populations in Sweden. Global Trends in Wildlife Management. Edited by: Bobek B, Perzanowski K, Regelin WL. Swiat Press, Krakow-Warszawa, Poland, 1991:591-5.
8. Morner T: Sarcoptic mange in Swedish wildlife. Rev Sci Tech. 1992, 11: 1115-21.
9. Soulsbury CD, Iossa G, Baker PJ, Cole NC, Funk SM, Harris S: The impact of sarcoptic mange *Sarcoptes scabiei* on the British fox *Vulpes vulpes* population. Mammal Rev. 2007, 37: 278-96.
10. Gonzalez-Candela M, Leon-Vizcaino L, Cubero-Pablo MJ: Population effects of sarcoptic mange in Barbary sheep (*Ammotragus lervia*) from Sierra Espuna Regional Park, Spain. J Wildl Dis. 2004, 40: 456-65.
11. Leon-Vizcaino L, Cubero MJ, Gonzalez-Capitel E, Simón MA, Pérez L, Rocío Ruiz deYbáñez M, Ortíz JM, González Candela M, Alonso F: Experimental ivermectin treatment of sarcoptic mange and establishment of a mange-free population of Spanish ibex. J Wildl Dis. 2001, 37: 775-85.

12. Skerratt LF: Clinical response of captive common wombats (*Vombatus ursinus*) infected with *Sarcoptes scabiei var. wombati*. J Wildl Dis. 2003, 39: 179-92.
13. Twomey DF, Birch ES, Schock A: Outbreak of sarcoptic mange in alpacas (*Vicugna pacos*) and control with repeated subcutaneous ivermectin injections. Vet Parasitol. 2009, 159: 186-91. 10.1016/j.vetpar.2008.10.023.
14. Yeruham I, Rosen S, Hadani A, Nyska A: Sarcoptic mange in wild ruminants in zoological gardens in Israel. J Wildl Dis. 1996, 32: 57-61.
15. Pence DB, Ueckermann E: Sarcoptic mange in wildlife. Rev Sci Tech. 2002, 21: 385-98.
16. Mwase ET, Baker AS: An annotated checklist of mites (Arachnida: Acari) of Zambia. Zootaxa. 2006, 1106: 1-24.
17. Munang'andu HM, Siamudaala VM, Nambota A, Bwalya JM, Munyeme M, Mweene AS, Takada A, Kida H: Disease constraints for utilization of the African buffalo (*Syncerus caffer*) on game ranches in Zambia. Jpn J Vet Res. 2006, 54: 3-13.
18. RIS: The Annotated Ramsar List: Zambia. The Ramsar Convention on Wetlands. Ramsar Convention Secretariat, Rue Mauverney 28, CH-1196 Gland, Switzerland. 1. [http://www.ramsar.org/cda/en/ramsar-pubs-annolist-annotated-ramsar-15789/main/ramsar/130-168%5E9_4000_0__]
19. Balakrishna M, Ndhlovu D: Wildlife utilization and local people; A case study in Upper Lupande game management area, Zambia. Environ cons. 1992, 19: 135-44. 10.1017/S0376892900030617.
20. Grimsdell JJR: Age determination of the African buffalo, *Syncerus caffer sparrman*. E Afr Wildlife J. 1973, 11: 54.
21. Pence DB, Windberg LA, Pence BC, Sprowls R: The epizootiology and pathology of sarcoptic mange in coyotes, *Canis latrans*, from south Texas. J Parasitol. 1983, 69: 1100-15. 10.2307/3280873.
22. Alasaad S, Rossi L, Soriguer RC, Rambozzi L, Soglia D, Pérez JM, Zhu XQ: *Sarcoptes* mite from collection to DNA extraction: the lost realm of the neglected parasite. Parasitol Res. 2009, 104: 723-32. 10.1007/s00436-009-1333-0.
23. Bowman DD: Georgis' Parasitology for Veterinarians. 6th edition Saunders. W.B. Saunders Co., Philadelphia, USA, 1995.

24. Pence DB, Ueckermann E: Sarcoptic mange in wildlife. Rev Sci Tech. 2002, 21: 385-98.
25. Tikaram SM, Ruprah NS: Incidence of sarcoptic mange in buffaloes in India. Trop Anim Health Prod. 1968, 18: 86-90. 10.1007/BF02359718.
26. Fain A: Epidemiological problems of scabies. Int J Dermatol. 1978, 17: 20-30. 10.1111/j.1365-4362.1978.tb06040.x.
27. Skerratt LF, Martin RW, Handasyde KA: Sarcoptic mange in wombats. Aust Vet J. 1998, 76: 408-10. 10.1111/j.1751-0813.1998.tb12389.x.
28. Walton SF, McBroom J, Mathews JD, Kemp DJ, Currie BJ: Crusted scabies: A molecular analysis of *Sarcoptes scabiei* variety hominis populations from patients with repeated infestations. Clin Infect Dis. 1999, 29: 1226-30. 10.1086/313466.
29. Rasero R, Rossi L, Maione S, Sacchi P, Rambozzi L, Sartore S, Soriguer R, Spalenza V, Alasaad S: Host taxon-derived *Sarcoptes* mites in European wildlife animals, revealed by microsetellite markers. Biol cons. 2010, 143: 1269-77. 10.1016/j.biocon.2010.03.001.
30. Perez JM, RuizMartinez I, Granados JE, Soriguer C, Fandos P: The dynmamic of sarcoptic mange in ibex of Sierra Nevada in Spain - Influence of climatic factors. J Wild Res. 1997, 2: 86-9.
31. Gill BS, Singh J, Gill BS, Singh A, Khehra SS, Rai A, Hussain O: Efficacy of ivermectin against mange and gastrointestinal nematodes of buffalo (*Bubalus bubalis*). Vet Parasitol. 1989, 31: 141-7. 10.1016/0304-4017(89)90029-0.
32. Rossi L, Rambozzi L, Ferroglio E: Diagnosi e monitoraggio della rogna sarcoptica nelsuino. Pfizer Animal health Proceedings. 1998, 18: 3-15.
33. Mitra J, Bardhan G, Basu A: Efficacy of oral ivermectin against sarcoptic mange in buffalo calves. Indian Vet J. 2005, 82: 666-7.
34. Leon-Vizcaino L, Cubero MJ, Gonzalez-Capitel E, Simón MA, Pérez L, Rocío Ruiz deYbáñez M, Ortíz JM, González Candela M, Alonso F: Experimental ivermectin treatment of sarcoptic mange and establishment of a mange-free population of Spanish ibex. J Wildl Dis. 2001, 37: 775-85.
35. Skerratt LF: Clinical response of captive common wombats (*Vombatus ursinus*) infected with *Sarcoptes scabiei var. wombati*. J Wildl Dis. 2003, 39: 179-92.
36. Yeruham I, Rosen S, Hadani A, Nyska A: Sarcoptic mange in wild ruminants in zoological gardens in Israel. J Wildl Dis. 1996, 32: 57-61.

37. Heukelbach J, Winter B, Wilcke T, Muehlen M, Albrecht S, de Oliveira FA, Kerr-Pontes LR, Liesenfeld O, Feldmeier H: Selective mass treatment with ivermectin to control intestinal helminthiases and parasitic skin diseases in a severely affected population. Bull World Health Organ. 2004, 82: 563-71.

INDEX

N

O

P

R

S